Advanced Analytics for Industry 4.0

The evolution of modern technology has affected all the industry dimensions. Mother industries play a critical role in providing the precursor materials for other industries, and a small improvement in these can make a big change in others. This book covers the analytics revolution in Industry 4.0 for the mother industries, such as mining, oil and gas, and steel. It focuses on the use of advanced analytics and artificial intelligence to improve business decisions aimed at increasing the quality and quantity of mother industries' products. It helps to design and implement their digital transformation strategies in these industries.

Key Features:

- Provides a concise overview of state of the art for mother industries' executives and managers.
- Highlights and describes critical opportunity areas for industry operations optimization.
- Explains how to implement advanced data analytics through case studies and examples.
- Provides approaches and methods to improve data-driven decision-making.
- Brings experience and learning in digital transformation from adjacent sectors.

This book is aimed at researchers, professionals, and graduate students in data science, manufacturing, automation, and computer engineering.

Advanced Analytics for Industry 4.0

Traditional Industries

Ali Soofastaei

CRC Press
Taylor & Francis Group
Boca Raton London New York

CRC Press is an imprint of the
Taylor & Francis Group, an **informa** business

Designed cover image: Shutterstock

First edition published 2025
by CRC Press
2385 NW Executive Center Drive, Suite 320, Boca Raton FL 33431

and by CRC Press
4 Park Square, Milton Park, Abingdon, Oxon, OX14 4RN

CRC Press is an imprint of Taylor & Francis Group, LLC

ISBN: 978-1-032-03344-0 (hbk)
ISBN: 978-1-032-03346-4 (pbk)
ISBN: 978-1-003-18682-3 (ebk)

DOI: 10.1201/9781003186823

Typeset in Times
by Apex CoVantage, LLC

Contents

Preface

The Fourth Industrial Revolution, or Industry 4.0, is not just a period of change; it is a transformative era in human history. The integration of digital technologies with traditional industrial practices is reshaping the global economic landscape. As industries evolve, they face unprecedented challenges and opportunities to innovate and thrive in a highly competitive environment. This book, *Advanced Analytics for Industry 4.0*, is a comprehensive guide that urgently addresses the need to understand and navigate this dynamic era.

Throughout this book, we delve into the profound impact of advanced analytics and digital technologies across a spectrum of traditional sectors, including mining, oil and gas, manufacturing, food production, construction, logistics, chemical engineering, agriculture, and insurance. By exploring these industries, we not only highlight the theoretical potential but also the practical applications of these cutting-edge technologies, such as driving efficiency and productivity and fostering sustainability and innovation. These technologies are actively shaping today's business landscape.

This book is intended for industry professionals, policymakers, academics, and anyone interested in the digital transformation journey. It aims to provide a holistic understanding of the challenges and opportunities presented by Industry 4.0, emphasizing the critical role of collaboration and the need for a strategic approach. We stress the importance of not just adopting these technologies, but doing so in a strategic and thoughtful manner. Here, planning and foresight are not just important; they are key to harnessing the full potential of these technologies.

I extend my gratitude to the numerous experts and practitioners who have contributed their insights and experiences to this book. Their collective knowledge has enriched the content and provided valuable perspectives on the practical applications of advanced analytics in traditional industries, making this book a relevant and valuable resource for professionals in these fields.

As we stand on the cusp of this revolutionary era, I hope this book will serve as a valuable resource, inspiring readers to embrace change, foster innovation, and drive sustainable growth in their respective fields.

Dr. Ali Soofastaei
CEO, Innovative AI
Australia 2024

Acknowledgments

Writing this book, *Advanced Analytics for Industry 4.0*, has been an extraordinary journey that would not have been possible without the support, guidance, and contributions of many individuals and organizations.

First and foremost, I would like to express my deepest gratitude to the experts and practitioners who generously shared their insights, experiences, and knowledge. Their contributions have enriched the content of this book and provided valuable perspectives on the practical applications of advanced analytics in traditional industries.

I am thankful to my colleagues and collaborators at various institutions and organizations. Their continuous support, encouragement, and feedback have been instrumental in shaping the ideas and concepts presented in this book. Special thanks to the academic and professional communities in mining, oil and gas, manufacturing, food production, construction, logistics, chemical engineering, agriculture, and insurance for their invaluable input and dedication to advancing Industry 4.0.

I sincerely appreciate my family and friends' unwavering support and understanding throughout this endeavor. Their encouragement and patience have been my anchor during the long hours of research and writing.

I am also grateful to the editorial team at CRC Press, Routledge Taylor & Francis Group for their meticulous attention to detail and commitment to excellence. Their professionalism and expertise have significantly enhanced the quality of this book.

Finally, I would like to acknowledge the inspiration and motivation provided by the rapidly evolving field of advanced analytics and digital technologies. The potential to drive innovation, efficiency, and sustainability in traditional industries is powerful, and I am honored to contribute to this exciting and transformative journey.

To all who have been a part of this process, directly or indirectly, thank you for your support and belief in this book's vision. We are paving the way for a future where advanced analytics and Industry 4.0 technologies lead to a more efficient, sustainable, and innovative world.

Dr. Ali Soofastaei
CEO, Innovative AI
Australia 2024

About the Author

Ali Soofastaei is the global projects leader at Vale Artificial Intelligence Centre. Vale is a multinational corporation engaged in metals and mining. It is one of the world's foremost producers of iron ore and the largest producer of nickel. Dr. Soofastaei leads innovative industrial projects in artificial intelligence (AI) applications to improve safety, productivity, and energy efficiency and reduce maintenance costs. He completed his Ph.D. at the University of Queensland in the field of AI applications in mining engineering, where he led a revolution in the use of deep learning and AI methods to increase energy efficiency, reduce operation and maintenance costs, and reduce greenhouse gas emissions in surface mines. As an assistant professor, he has provided undergraduate and postgraduate students with practical guidance in engineering and information technology. In the past 15 years, he has conducted various research studies in academic and industrial environments. He has acquired in-depth knowledge of energy efficiency opportunities and advanced analytics. He is an expert in using DL and AI methods in data analysis to develop predictive, optimization, and decision models of complex systems. Dr. Soofastaei has been involved in industrial research and development projects in several industries, including oil and gas (Royal Dutch Shell), steel (Danieli), and mining (BHP, Rio Tinto, Anglo American, and Vale). His extensive practical experience in the industry has equipped him to work with complex industrial problems in highly technical and multidisciplinary teams. As a research and development team member, Dr. Soofastaei has been actively involved in site inspections, business problem identification, and root cause analysis. He has experience in managing brainstorming sessions with operators, supervisors, managers, and original equipment manufacturers in the areas of automation (e.g., with electrical and computer systems engineers), maintenance (e.g., with mechanical engineers and maintenance supervisors), and production (e.g., process engineers, metallurgists). Dr. Soofastaei has more than ten years of academic experience as an assistant professor and a global research leader. His research and development projects have been published in international journals and keynote presentations. He has presented his practical achievements at conferences in the United States, Europe, Asia, and Australia.

Contributors

Carmen Carmona-Torres
Researcher
Institute of Regional Development, Scientific Documentation Centre
University of Granada
Granada, Spain

Guillermo Garcia-Garcia
Postdoctoral Research Fellow
Department of Chemical Engineering, Faculty of Sciences
University of Granada
Granada, Spain

Sandeep Jagtap
Senior Lecturer
Department of Mechanical Engineering Sciences, Faculty of Engineering
Lunds University
Lund, Sweden

Yang Luo
Assistant Professor
School of Intelligent Manufacturing Ecosystem
Xi'an Jiaotong-Liverpool University
Suzhou, Jiangsu, China

Federico Walas Mateo
CEO
Chaska Analytics
Buenos Aires, Argentina

Alper Ozpinar
Associate Professor
Boğaziçi University
Istanbul, Turkey

Ajaya Kumar Pani
Assistant Professor
Birla Institute of Technology and Science
Pilani, India

Carlos Parra-López
Tenured Researcher
Department of Agrifood System Economics
Institute of Agricultural and Fisheries Research and Training (IFAPA)
Granada, Spain

Anand S. Rao
Distinguished Service Professor of Applied Data Science and Artificial Intelligence
Carnegie Mellon University
Pittsburgh, Pennsylvania, United States

Andrés Redchuk
Researcher
URJC
Madrid, Spain

Ali Soofastaei
CEO
Innovative AI
Brisbane, Australia

Hana Trollman
Lecturer in Management
Department of Work, Employment, Management and Organisations, School of Business University of Leicester
Leicester, United Kingdom

Introduction

NAVIGATING THE FOURTH INDUSTRIAL REVOLUTION: THE ADVENT OF ADVANCED ANALYTICS IN TRADITIONAL INDUSTRIES

As we stand on the cusp of the Fourth Industrial Revolution, traditional industries find themselves at a pivotal juncture, faced with the dual challenge of adapting to a rapidly evolving technological landscape while preserving the core aspects of their operations that have stood the test of time. This inaugural chapter of our comprehensive exploration into advanced analytics for Industry 4.0 delves into the transformative impact of digital technologies across a spectrum of traditional sectors, including mining, oil and gas, manufacturing, food production, construction, logistics, chemical engineering, agriculture, and insurance.

We begin by laying the foundational knowledge of Industry 4.0, tracing its evolution from the steam-powered first revolution to today's digital era characterized by a fusion of technologies that blur the lines between the physical, digital, and biological spheres. The book then transitions into a detailed discussion on the pillars of advanced analytics—artificial intelligence, machine learning, big data, and the Internet of Things—and their critical role in driving the digital transformation of traditional industries.

Through a sector-by-sector analysis, we illuminate how these cutting-edge technologies are not merely disruptive forces but also catalysts for innovation, efficiency, and sustainability. From predictive maintenance and operational optimization in mining, oil, and gas sectors to precision agriculture and food safety in the agricultural and food industries, this book provides a bird's-eye view of the myriad ways advanced analytics are redefining production, management, and service delivery paradigms, sparking intrigue and engagement.

Moreover, this book sets the stage for subsequent in-depth discussions, offering readers a holistic understanding of the digital age's challenges and opportunities. It serves as a guidepost for industry professionals, policymakers, and academics alike, emphasizing the importance of a collaborative approach in harnessing the potential of Industry 4.0 technologies, making the audience feel included and valued in the process.

In this era of rapid technological advancement, integrating advanced analytics into traditional industries is not just a competitive advantage but also a necessity for survival and growth. As we embark on this journey through the chapters of this book, we invite you to explore the profound changes brought about by Industry 4.0 and envision a future where innovation, efficiency, and sustainability are the cornerstones of industrial success.

1 Navigating the Fourth Industrial Revolution

The Advent of Advanced Analytics in Traditional Industries

Ali Soofastaei

1.1 INTRODUCTION TO INDUSTRY 4.0

1.1.1 Defining Industry 4.0

Industry 4.0, often called the Fourth Industrial Revolution, marks a significant shift in the global industrial landscape. It heralds an era where digital technologies integrate with traditional industrial practices to create more intelligent, efficient, interconnected systems. Unlike previous revolutions, Industry 4.0 emphasizes combining physical production and operations with intelligent digital technologies, machine learning (ML), and big data to create a more holistic and better-connected ecosystem for companies focusing on manufacturing and supply chain management [1].

1.1.1.1 Historical Evolution

The journey from steam engines to digital automation outlines the transformative path of industrial revolutions. The first revolution (Industry 1.0) began in the 18th century with the introduction of mechanization and steam power. The second revolution (Industry 2.0) introduced mass production and electrical energy in the early 20th century. The third revolution (Industry 3.0), emerging in the late 20th century, saw the advent of computers and automation in manufacturing processes. Today, Industry 4.0 builds on these advancements, integrating cyber-physical systems (CPS) and the Internet of Things (IoT) to make factories more innovative and efficient [2].

1.1.1.2 Core Components of Industry 4.0

Industry 4.0 stands on several key technologies that synergize to enhance manufacturing and industrial processes [3]:

- **CPS integrates computation, networking, and physical processes. Embedded computers and networks monitor and control the physical**

DOI: 10.1201/9781003186823-1

processes, and there are feedback loops where physical processes affect computations and vice versa.

- **IoT:** IoT connects devices, machines, and people, enabling a seamless data exchange. This connectivity allows for more responsive and adaptive manufacturing processes.
- **Big Data and Analytics:** The ability to process and analyze vast amounts of data in real time can lead to more informed decision-making and predictive maintenance.
- **Artificial Intelligence (AI) and ML:** AI and ML algorithms can learn from data, allowing automation improvements and more efficient processes.
- **Cloud and Edge Computing:** These technologies provide the infrastructure for data storage, processing, and analytics, facilitating scalability and more agile responses to changing market demands.
- **Augmented Reality (AR) and Virtual Reality (VR):** AR and VR technologies offer innovative ways to train, maintain, and visualize complex industrial processes.
- **Additive Manufacturing (Three-Dimensional [3D] Printing):** This allows for more flexible and cost-effective production of parts and components, even in small batch sizes.
- **Advanced Robotics and Automation:** Enhanced robotics contribute to more efficient, safe, and flexible manufacturing environments.
- **Blockchain and Cybersecurity:** These technologies ensure data integrity and security, which are critical in a connected industrial ecosystem.

1.1.1.3 The Pillars of Industry 4.0

The foundational pillars of Industry 4.0 facilitate its implementation [4]:

- **Interconnectivity:** Allowing machines, devices, sensors, and people to connect and communicate (IoT and Internet of People [IoP]).
- **Information Transparency:** The ability of information systems to create a virtual copy of the physical world by enriching digital plant models with sensor data.
- **Technical Assistance:** The ability of assistance systems to support humans by aggregating and visualizing information comprehensively to make informed decisions and solve urgent problems on short notice.
- **Decentralized Decisions:** The ability of CPS to make decisions independently and perform their tasks as autonomously as possible.

1.1.1.4 Industry 4.0 in Practice

In practical terms, Industry 4.0 leads to "smart factories," where CPS monitors physical processes, creates a virtual copy of the physical world, and makes decentralized decisions. CPS communicates and cooperates with humans in real time through the IoT. This integration significantly improves production processes' efficiency, productivity, and flexibility [5].

1.1.1.5 Challenges and Opportunities

The transition to Industry 4.0 has challenges, including the need for significant investment in new technology and worker training. However, its opportunities—such as increased efficiency, reduced costs, and improved products and services—make it a pivotal step forward for traditional industries looking to remain competitive in the digital age [6].

1.1.1.6 The Future Outlook of Industry 4.0

Looking toward the future, Industry 4.0 is set to become even more integrated into our daily lives and work. This could lead to Industry 5.0, which will focus on the collaboration between humans and intelligent systems. The ongoing development of these technologies promises to revolutionize further how we produce, manufacture, and deliver services globally [7].

1.1.2 The Evolution from Industry 1.0 to 4.0

The journey from the First to the Fourth Industrial Revolution is a testament to human ingenuity and the relentless pursuit of efficiency and innovation. This section explores the pivotal transformations that have shaped the industrial landscape over the centuries, leading to the era of Industry 4.0 [8].

1.1.2.1 Industry 1.0: The Dawn of Mechanization

The First Industrial Revolution began in the late 18th century, characterized by the transition from hand production methods to machines through steam power and water power. The introduction of the steam engine revolutionized transportation and manufacturing processes, marking the beginning of industrialization. This era saw the rise of factories and the mechanization of textiles, fundamentally changing the structure of society and laying the groundwork for future industrial advancements.

1.1.2.2 Industry 2.0: The Age of Mass Production

Emerging in the late 19th and early 20th centuries, the Second Industrial Revolution was fueled by the discovery of electricity and the development of assembly line techniques, which significantly boosted production capabilities and efficiency. This period was marked by the widespread adoption of electrical power, which led to more extensive distribution systems and enabled factories to increase output. Henry Ford's introduction of the assembly line, most famously in the production of automobiles, made mass production possible, drastically reducing costs and making products more accessible to the general public [9].

1.1.2.3 Industry 3.0: The Digital Revolution

The Third Industrial Revolution, or the Digital Revolution, began in the late 20th century, driven by the advent of electronics, telecommunications, and, most importantly, computers and the internet. This era introduced automation into manufacturing through electronics and computer technology, leading to more efficient production processes and the beginning of the information age. The digitization of manufacturing processes allowed for more precise control and flexibility in production, paving the way for more complex and advanced manufacturing techniques.

1.1.2.4 Industry 4.0: The Era of Smart Technologies

Building upon the digitalization in Industry 3.0, the Fourth Industrial Revolution, known as Industry 4.0, integrates advanced digital technologies with physical production and operations to create connected, autonomous systems. This era is characterized by intelligent and autonomous systems powered by data and ML. Industry 4.0 is marked by the convergence of technologies such as the IoT, AI, robotics, cloud computing, and advanced data analytics, leading to unprecedented efficiency, flexibility, and customization in manufacturing. The hallmark of Industry 4.0 is the creation of "smart factories," where CPS monitors the physical processes, creates a virtual copy of the physical world, and makes decentralized decisions.

The evolution from Industry 1.0 to 4.0 represents the cumulative progress of humanity's industrial endeavors, with each revolution building on the achievements of its predecessors. This progression from steam-powered mechanization to intelligent, interconnected systems highlights the rapid pace of technological advancement and underscores industries' adaptability and resilience in the face of change. As we navigate the complexities of Industry 4.0, it is essential to acknowledge this rich history of innovation that has paved the way for the current digital transformation era and beyond [10].

1.2 THE PILLARS OF ADVANCED ANALYTICS

1.2.1 Understanding Advanced Analytics

Advanced analytics represents the frontier of data analysis. It utilizes sophisticated techniques and tools to extract valuable information from data, predict future trends, and provide actionable insights. This section delves into the essence of advanced analytics, its critical methodologies, and its transformative impact on industries in the era of Industry 4.0 [11].

1.2.2 Defining Advanced Analytics

At its core, advanced analytics goes beyond traditional data analysis and business intelligence techniques. It employs complex algorithms, ML, predictive modeling, and other advanced statistical methods to analyze and interpret vast datasets. This analytical depth enables organizations to gain deeper insights, foresee future scenarios, and make more informed decisions [12].

1.2.3 Critical Components of Advanced Analytics

- **Predictive Analytics:** This technique utilizes historical data to predict future outcomes. Predictive models can forecast trends and behaviors by analyzing trends, patterns, and relationships within data, empowering businesses to anticipate events and strategize accordingly.

- **ML** is a subset of AI that allows systems to learn and improve from experience without being explicitly programmed automatically. ML algorithms can uncover hidden insights without human intervention, adapt to new data, and evolve in accuracy and complexity over time.
- **Data Mining:** Involves exploring and analyzing large datasets to discover meaningful patterns, relationships, and anomalies. Data mining techniques include clustering, classification, regression, and association rules.
- **Big Data Analytics refers to examining large and varied data sets—or big data**—to uncover hidden patterns, unknown correlations, market trends, customer preferences, and other useful business information.
- **Text Analytics and Natural Language Processing (NLP):** These processes analyze text data and extract meaningful information. NLP allows machines to understand and interpret human language, facilitating sentiment analysis, topic detection, and text classification tasks.
- **Prescriptive Analytics:** Goes beyond predicting future outcomes by recommending actions to achieve desired results. It uses optimization and simulation algorithms to advise on possible outcomes and answer the question, "What should we do?"

1.2.4 Applications in Industry 4.0

In the context of Industry 4.0, advanced analytics plays a pivotal role in enabling smart manufacturing and operations. By harnessing the power of advanced analytics, businesses can optimize production processes, enhance quality control, reduce downtime, and improve supply chain efficiency. Some practical applications include [13]:

- **Predictive Maintenance:** ML algorithms predict equipment failures before they occur, reducing downtime and maintenance costs.
- **Supply Chain Optimization:** Analyzes patterns and trends to improve supply chain efficiency, from inventory management to distribution logistics.
- **Quality Control:** This department employs statistical models and ML to monitor and improve product quality, identifying real-time defects and non-conformities.
- **Customer Insights:** Analyzes customer data to understand preferences, behaviors, and trends, enabling personalized services and products.

1.2.5 Challenges and Considerations

While advanced analytics offers significant benefits, its implementation is challenging. These include ensuring data quality and integrity, managing data privacy and security, addressing the skills gap within organizations, and integrating advanced analytics into existing systems and processes [14].

Understanding advanced analytics is crucial for businesses seeking to thrive in the digital age, especially within the framework of Industry 4.0. By leveraging sophisticated analytical techniques, organizations can unlock deeper insights, drive innovation, and maintain a competitive edge in an increasingly data-driven world. As we continue to generate and collect data at an unprecedented scale, advanced analytics will only grow in importance, shaping the future of industries and economies globally [15].

1.3 DIGITAL TRANSFORMATION ACROSS SECTORS

Industry 4.0 has ushered in an era of digital transformation that transcends sectoral boundaries. This transformation influences every aspect of traditional industries, from manufacturing to agriculture. It is not only about adopting new technologies but also represents a fundamental shift in business models, operational processes, and organizational cultures [16].

1.3.1 The Universal Impact of Industry 4.0

Industry 4.0 has a universal impact, affecting industries in several profound ways:

- **Increased Efficiency and Productivity:** Automation, real-time data analytics, and CPS enhance operational efficiency, reducing human error and downtime while increasing productivity.
- **Enhanced Flexibility:** Digital technologies enable more agile production processes, allowing customization and rapid adaptation to changing market demands.
- **Improved Decision-Making:** Advanced analytics and IoT devices provide actionable insights from massive data sets, enabling more informed decision-making and strategic planning.
- **Customer-Centric Approaches:** The digital era facilitates a deeper understanding of customer needs and behaviors, leading to more personalized products and services.
- **Sustainability:** Digital transformation offers ways to optimize resource use, reduce waste, and minimize environmental impact, aligning with global sustainability goals.

These impacts are universal yet manifest differently across various sectors, reflecting each industry's unique challenges and opportunities.

1.3.2 Challenges and Opportunities in Traditional Industries

While digital transformation presents numerous opportunities, traditional industries face specific challenges in integrating Industry 4.0 technologies [17]:

1.3.2.1 Challenges

- **Cultural and Organizational Resistance:** Shifting from traditional to digital-first approaches requires significant cultural and organizational change, which can be met with resistance at various levels.
- **Skills Gap:** There is often a significant skills gap, with a need for training and development in digital competencies for both existing employees and new hires.

- **Infrastructure and Investment:** Upgrading existing infrastructure to support new technologies involves substantial investment, which can be a barrier, especially for small- and medium-sized enterprises.
- **Cybersecurity Risks:** Increased connectivity and reliance on digital systems raise concerns about data security and vulnerability to cyber-attacks.
- **Regulatory and Compliance Issues:** Navigating the complex landscape of digital regulations and ensuring compliance can be challenging for industries undergoing digital transformation.

1.3.2.2 Opportunities

- **Operational Excellence:** Industries can achieve new operational efficiency and productivity levels through automation, real-time monitoring, and predictive maintenance.
- **Market Expansion:** Digital platforms and e-commerce enable industries to reach new markets and customers, breaking geographical barriers.
- **Innovation and Competitiveness:** Embracing digital transformation fosters innovation, helping traditional industries develop new products, services, and business models to stay competitive.
- **Collaboration and Ecosystems:** Industry 4.0 encourages collaboration across industries and sectors, creating ecosystems that leverage collective strengths and innovations.
- **Resilience and Adaptability:** Digital capabilities enhance an industry's resilience to market fluctuations and external shocks, such as the COVID-19 pandemic.

Digital transformation under the umbrella of Industry 4.0 presents traditional industries with a dual aspect of challenges and opportunities. While the path may be fraught with obstacles, the potential rewards for efficiency, innovation, and competitiveness are substantial. Integrating new technologies with human ingenuity will be vital to unlocking future growth and sustainability as industries navigate this digital terrain. Embracing this transformation requires technological adoption and a strategic vision prioritizing flexibility, continuous learning, and a commitment to sustainability [18–20].

1.4 MINING IN THE AGE OF INFORMATION

The mining industry, traditionally known for its intensive labor and capital, has entered the Age of Information and is undergoing significant transformations under the influence of Industry 4.0. This digital revolution is reshaping the sector, focusing on predictive maintenance, operational efficiency, enhanced safety, and environmental sustainability.

1.4.1 Predictive Maintenance and Operational Efficiency

Predictive maintenance is a cornerstone of operational efficiency in modern mining. By leveraging data analytics, IoT devices, and ML, mining companies can predict equipment failures before they occur, minimizing downtime and reducing maintenance costs. Sensors installed on mining equipment collect real-time data on their condition and performance, feeding into advanced algorithms that analyze patterns and detect anomalies indicative of potential failures.

This proactive approach to maintenance extends the lifespan of expensive machinery and optimizes the scheduling of maintenance activities, ensuring that interventions are conducted without disrupting production. As a result, mining operations become more efficient, significantly reducing unplanned downtime and associated costs.

Moreover, operational efficiency in mining is further enhanced by optimizing various processes. For instance, using autonomous and remotely operated machinery guided by precise data analytics improves the accuracy and speed of drilling, blasting, and material handling operations. Integrating global positioning system and geospatial data allows for more accurate mapping and planning of mining activities, optimizing resource extraction and reducing waste.

1.4.2 Enhanced Safety and Environmental Sustainability

Integrating digital technologies in mining also brings paramount safety and environmental sustainability improvements. Advanced monitoring systems powered by IoT and AI play a crucial role in ensuring the safety of mining operations. These systems provide real-time insights into environmental conditions within mines, such as air quality, temperature, and hazardous gases, enabling timely interventions to prevent accidents and health issues among workers.

Drones equipped with cameras and sensors offer a safe and efficient means of conducting site surveys and inspections, especially in hazardous or inaccessible areas. This reduces personnel risk and provides detailed and accurate data for planning and decision-making.

From an environmental perspective, digital transformation in mining contributes to more sustainable practices. Precision mining techniques, supported by data analytics, ensure that resource extraction is carried out with minimal environmental impact, preserving the surrounding ecosystem. Digital tracking of materials from source to final product enhances traceability and accountability, promoting responsible sourcing and consumption.

Furthermore, using simulation models and digital twins in mine planning and water management helps minimize mining activities' environmental footprint. These technologies allow for the simulation of various scenarios, enabling the identification of the most sustainable approaches to resource extraction, waste management, and rehabilitation of mining sites.

The mining industry's journey into the Age of Information heralds a new era of efficiency, safety, and environmental stewardship. Through the strategic application of predictive maintenance, operational optimizations, and advanced monitoring systems, mining is shedding its traditional image and emerging as a sector that is more productive, safer, and more conscious of its environmental responsibilities. As digital technologies evolve, the potential for further advancements in mining remains vast, promising a future where mining is synonymous with innovation and sustainability.

1.5 REVOLUTIONIZING OIL AND GAS

The oil and gas industry, a cornerstone of the global economy, is undergoing a profound transformation driven by the digital revolution. This shift is characterized by integrating advanced technologies that optimize exploration and production processes, redefine supply chain management, and enable agile market adaptation.

1.5.1 Exploration and Production Optimization

In exploration, advanced analytics, high-performance computing, and geospatial data converge to enhance the accuracy and efficiency of identifying potential oil and gas reserves. Seismic imaging and data analysis technologies have become more sophisticated, allowing geologists to create more precise subsurface maps. This reduces the risks and costs of drilling exploratory wells, leading to more successful exploration endeavors.

Once reserves are identified, the focus shifts to optimizing production. Here, digital technologies such as IoT sensors and real-time data analytics play a pivotal role. Sensors placed on rigs and within wells provide continuous data on operational conditions, fluid properties, and equipment status. This influx of data, analyzed in real time, enables operators to make immediate adjustments to drilling operations, well-completion techniques, and production rates, maximizing output while minimizing environmental impact.

AI and ML algorithms are increasingly employed to predict equipment failures, optimize drilling paths, and enhance reservoir management. These technologies can analyze vast datasets from past and current operations to identify patterns and predict outcomes, leading to more informed decision-making and strategic planning.

1.5.2 Supply Chain Management and Market Adaptation

The oil and gas supply chain is complex, spanning diverse geographies and involving intricate product extraction, refining, and distribution logistics. Digital transformation is streamlining these processes, making them more transparent and efficient. Blockchain technology, for example, offers a secure and transparent way to track products from the wellhead to the consumer, enhancing traceability and reducing the risk of fraud.

Advanced analytics and IoT also improve inventory management, demand forecasting, and logistics planning. Companies can achieve a holistic view of their operations by integrating data from various sources. This enables them to anticipate supply chain disruptions, optimize inventory levels, and swiftly adapt to changing market conditions.

In terms of market adaptation, digital tools empower the oil and gas industry to respond more agilely to fluctuations in market demand and regulatory changes. Predictive analytics can forecast market trends, helping companies adjust their production and marketing strategies accordingly. In addition, digital platforms facilitate more direct engagement with customers and partners, fostering stronger relationships and enabling customized service offerings.

The digital revolution in the oil and gas industry is about adopting new technologies and transforming the industry's fabric. The sector is becoming more efficient, resilient, and sustainable by leveraging digital innovations for exploration and production optimization, supply chain management, and market adaptation. As the industry continues to navigate the challenges of energy transitions and environmental sustainability, the role of digital technologies in driving innovation and efficiency will only grow in importance, marking a new era of opportunity for the sector.

1.6 STEELING FOR THE FUTURE

The steel industry, fundamental to the global economy and infrastructure development, is poised for a new era of digitalization and sustainability. Embracing Industry 4.0, steel manufacturers leverage advanced technologies to drive process optimization, enhance energy efficiency, and significantly reduce waste, ensuring the sector's resilience and competitiveness in the 21st century.

1.6.1 Process Optimization in Steel Manufacturing

Integrating digital technologies, particularly IoT, AI, and advanced data analytics, is revolutionizing process optimization in steel manufacturing. These innovations facilitate real-time monitoring and control of steel production processes, from raw material handling to the final stages of rolling and finishing.

IoT sensors deployed across various stages of the steel production process collect vast amounts of temperature, pressure, chemical composition, and equipment status data. When analyzed using AI and ML algorithms, this data provides deep insights into the production process, enabling predictive maintenance, quality control, and operational efficiency.

AI-driven models can predict critical equipment's performance and potential failures, allowing for timely maintenance and reducing unplanned downtime. Furthermore, AI algorithms optimize production by adjusting operational parameters in real time, ensuring optimal resource use and consistent product quality. This level of precision and efficiency was previously unattainable with traditional methods.

1.6.2 Energy Efficiency and Waste Reduction

The steel industry is historically energy-intensive, with significant carbon emissions and waste generation. However, digital transformation is paving the way for substantial improvements in energy efficiency and waste reduction. Advanced analytics play a critical role in energy management, identifying patterns and inefficiencies in energy consumption and developing strategies to minimize waste and optimize energy use.

Energy optimization models powered by ML assess various energy inputs and outputs throughout the steel production process and offer recommendations for reducing energy consumption without compromising output quality. These models consider factors such as furnace temperatures, production rates, and material properties, providing a holistic approach to energy management.

Waste reduction is another critical area in which digital technologies are significantly impacting. The industry can reduce scrap rates and improve yield through enhanced process control and material tracking. Moreover, digital platforms facilitate the recycling and reuse of by-products, turning waste into valuable resources and contributing to a circular economy.

Digitalization also promotes using alternative, less carbon-intensive energy sources in steel production. For instance, integrating renewable energy sources, such as solar and wind, is made more feasible through innovative grid technologies and energy storage solutions, further enhancing the sector's sustainability profile.

The digital era presents an unprecedented opportunity for the steel industry to redefine its operational paradigms. By embracing process optimization, energy efficiency, and waste reduction through advanced technologies, the sector is improving its environmental footprint and ensuring its long-term viability and competitiveness. "Steeling for the Future" encapsulates this journey toward a more efficient, sustainable, and resilient steel industry underpinned by Industry 4.0's transformative power.

1.7 THE MANUFACTURING RENAISSANCE

The dawn of Industry 4.0 has ushered in a manufacturing renaissance characterized by unprecedented customization, agility, and efficiency. This transformative period is fueled by the convergence of digital technologies, such as printing, IoT, AI, and advanced robotics, redefining manufacturing paradigms to meet the dynamic demands of the modern market and consumer.

1.7.1 Customization and Agile Manufacturing

In an era of increasingly personalized consumer preferences, the ability to efficiently and at scale customize products is a significant competitive advantage. Digital manufacturing technologies, mainly additive manufacturing (3D printing), have emerged as critical enablers of mass customization. They allow manufacturers to produce complex and customized products without needing expensive molds or tooling, significantly reducing lead times and costs associated with product variations.

Agile manufacturing extends beyond customization, encompassing the flexibility and responsiveness of the entire production process. Advanced digital systems, integrated through IoT and powered by AI, enable manufacturers to rapidly adjust production schedules, methods, and even plant configurations in response to fluctuating market demands or supply chain disruptions. This agility ensures that manufacturers can maintain high levels of efficiency and productivity, even in the face of uncertainty.

1.7.2 Quality Control and Asset Management

Quality control is paramount in the manufacturing renaissance, emphasizing meeting and exceeding consumer expectations. Digital technologies have transformed traditional quality control methods, introducing real-time monitoring and predictive analytics to detect anomalies and prevent defects before they occur. Machine vision systems and sensors continuously inspect products during the manufacturing process. At the same time, AI algorithms analyze the data to identify patterns or deviations from quality standards, ensuring high product quality and consistency.

Asset management, an often-overlooked aspect of manufacturing, has also been revolutionized by digitalization. IoT sensors and AI-driven analytics provide comprehensive insights into the condition and performance of manufacturing assets. Predictive maintenance algorithms analyze this data to forecast potential equipment failures, scheduling maintenance activities proactively to avoid costly downtime and extend the lifespan of valuable assets. Moreover, digital twins—virtual replicas of physical assets—allow manufacturers to simulate and optimize asset performance, further enhancing operational efficiency.

The manufacturing renaissance heralded by Industry 4.0 is a shift toward more customized, agile, and quality-focused production processes. As manufacturers embrace these digital technologies, they can meet consumers' evolving demands and achieve new operational excellence. This renaissance is not just about technological advancement; it represents a fundamental change in how products are designed, produced, and delivered, paving the way for a future where manufacturing is more responsive, sustainable, and aligned with the needs of a digital age.

1.8 FEEDING THE FUTURE: AGRICULTURE AND FOOD INDUSTRY

The agriculture and food industry are at a pivotal moment where the demands of a growing global population and the imperative for sustainability converge. Digital technologies, central to Industry 4.0's ethos, are pivotal in transforming these challenges into opportunities, heralding a new era of precision agriculture, sustainable practices, enhanced food safety, and supply chain traceability.

1.8.1 Precision Agriculture and Sustainable Practices

Precision agriculture epitomizes the application of digital technologies in farming. It combines data analytics, IoT, satellite imagery, and AI to cultivate crops more efficiently and sustainably. This approach enables farmers to monitor and manage their fields with unprecedented detail. Sensors in the soil and drones flying overhead provide real-time information on crop health, soil conditions, and microclimates.

By harnessing this wealth of data, farmers can make informed decisions about where, when, and how much to water, fertilize, or apply pest control, minimizing waste and environmental impact. AI-driven predictive models further enhance these capabilities, forecasting weather patterns, pest invasions, and crop diseases, allowing for preemptive measures that safeguard yield and quality.

Sustainable practices extend beyond the field, encompassing water conservation, energy efficiency, and reducing carbon footprints through more competent resource management. Digital tools facilitate these practices, optimizing irrigation systems, managing renewable energy sources, and employing automated equipment that reduces manual labor and increases efficiency.

1.8.2 Food Safety and Supply Chain Traceability

Safety and traceability are paramount in the food industry. The journey from farm to fork is complex, involving numerous stakeholders and processes that can impact food quality and safety. Digital transformation within the food supply chain introduces a previously unattainable level of transparency and traceability.

Blockchain technology, for instance, offers a secure and immutable ledger to record every transaction and movement of produce through the supply chain. This capability allows all parties, from farmers to retailers and consumers, to trace food products' origin, processing, and handling, ensuring their authenticity and safety.

Moreover, IoT devices and radio-frequency identification tags enable real-time monitoring of food products during transit, recording temperature, humidity, and other conditions critical to maintaining quality and safety. In the event of a food safety concern, these digital tools can rapidly identify and isolate affected products, significantly reducing the scope and scale of food recalls.

Data analytics also play a crucial role in predicting food safety. They analyze historical data to identify potential risk factors and prevent contamination before it occurs. AI algorithms can monitor social media and other digital platforms for early warning signs of foodborne illness outbreaks, enabling swift action to protect public health.

The digital transformation in the agriculture and food industry, driven by precision agriculture and enhanced food safety and traceability measures, is not just about technological adoption. It represents a fundamental shift toward more sustainable, efficient, and consumer-responsive practices. As the sector continues to evolve under the influence of Industry 4.0, the promise of feeding the future in a way that respects both people and the planet becomes increasingly attainable. This marks a significant step forward in pursuing global food security and sustainability.

1.9 BUILDING SMARTER: CONSTRUCTION INDUSTRY INNOVATIONS

The construction industry is on the cusp of a technological revolution, with digital innovations paving the way for more innovative, efficient, and sustainable building practices. Advanced construction management methodologies and adopting building information modeling (BIM) are at the heart of this transformation, complemented by the integration of innovative materials and a commitment to sustainable construction practices.

1.9.1 Construction Management and BIM

Adopting BIM has significantly accelerated the digital transformation of construction management. BIM goes beyond traditional two-dimensional drafting, offering a 3D digital representation of spaces' physical and functional characteristics. It enables architects, engineers, and construction professionals to collaboratively design, visualize, simulate, and manage buildings and infrastructure projects with unprecedented detail and coordination.

BIM facilitates improved decision-making throughout the construction lifecycle, from the earliest concept stages to design, construction, operation, and maintenance. It enhances the efficiency and accuracy of the design and construction processes. It significantly reduces costs and project timelines by identifying potential issues and clashes early in the design phase, thus preventing costly corrections during construction.

Moreover, integrated with BIM and powered by AI and ML, construction management software optimizes resource allocation, scheduling, and logistics. These digital tools provide real-time updates and analytics, improving stakeholder communication and collaboration and ensuring projects are completed on time and within budget.

1.9.2 Innovative Materials and Sustainable Practices

The rise of intelligent materials is redefining the possibilities within the construction industry, contributing to the development of more adaptive, resilient, and sustainable buildings. These materials, ranging from self-healing concrete to phase-changing materials and photovoltaic glass, are engineered to respond to environmental changes and improve energy efficiency, durability, and the overall lifecycle of construction projects.

Self-healing concrete, for example, contains bacteria that produce limestone to fill cracks that develop over time, significantly extending the material's lifespan and reducing maintenance costs. Phase-changing materials incorporated into building fabrics can absorb, store, and release heat, maintaining comfortable indoor temperatures and reducing reliance on heating and cooling systems.

Using renewable materials and green construction practices further evidences the industry's commitment to sustainability. The industry increasingly prioritizes materials with lower carbon footprints, such as bamboo, recycled steel, and reclaimed wood. In addition, green construction practices, including optimizing site selection, water and energy efficiency, and waste reduction, are being adopted to minimize environmental impact.

The construction industry's embrace of digital innovations, intelligent materials, and sustainable practices heralds a new era of efficiency, resilience, and environmental stewardship. Through the synergistic use of BIM, advanced construction management tools, and innovative materials, the sector is poised to address some of the most pressing challenges of our time, including urbanization, climate change, and resource scarcity. As these technologies and methodologies evolve, the construction industry is set to redefine the landscapes of cities and communities worldwide, building more intelligent and greener for future generations.

1.10 ON THE MOVE: TRANSFORMING TRANSPORT LOGISTICS

The transportation and logistics sector is undergoing a profound transformation, propelled by the advent of Industry 4.0 technologies. This transformation is characterized by integrating autonomous vehicles into fleets and enhancing supply chain visibility and efficiency, setting new benchmarks for speed, safety, and reliability in logistics operations.

1.10.1 Autonomous Vehicles and Fleet Management

The rise of autonomous vehicles is revolutionizing fleet management in the transport logistics industry. These self-driving vehicles, equipped with advanced sensors, cameras, and AI algorithms, can navigate without human intervention, offering many benefits, including increased safety, efficiency, and cost savings.

In fleet management, autonomous technology enables more precise control over vehicle operations, reducing the likelihood of human error-related accidents and optimizing fuel consumption and maintenance schedules. Autonomous trucks and drones for cargo transport, particularly in repetitive and predictable routes, enhance operational efficiency, allowing 24/7 operations without human driver needs and limitations.

Moreover, integrating IoT devices and telematics in fleet management provides real-time vehicle location, condition, and performance data. This continuous stream of data, coupled with predictive analytics, enables logistic companies to anticipate maintenance needs, avoid potential breakdowns, and optimize routes, further improving fleet efficiency and reducing downtime.

1.10.2 Supply Chain Visibility and Efficiency

Enhanced supply chain visibility is a cornerstone of modern transport logistics, underpinned by digital technologies such as IoT, blockchain, and cloud computing. These technologies enable unprecedented transparency and real-time tracking of goods across the supply chain, from the point of origin to the final destination.

IoT sensors affixed to cargo can monitor temperature, humidity, and location, ensuring that sensitive products, like perishables and pharmaceuticals, are maintained in optimal conditions throughout their journey. This real-time monitoring capability, combined with advanced data analytics, allows for proactive supply chain management, enabling companies to respond swiftly to any disruptions or changes in demand.

Blockchain technology further enhances supply chain efficiency by providing a secure and immutable ledger for recording transactions and tracking assets as they move through the supply chain. This increases stakeholder trust and streamlines customs clearance and compliance processes, reducing delays and improving the overall speed of logistics operations.

Cloud-based supply chain management platforms offer a centralized hub for data and analytics, facilitating collaboration and information sharing among stakeholders. These platforms enable more effective planning, forecasting, and inventory management, reducing waste and ensuring that products are delivered to the right place at the right time.

The transformation of transport logistics through the integration of autonomous vehicles, enhanced fleet management, and improved supply chain visibility and efficiency is driving a new era of innovation in the sector. These advancements promise to optimize logistics operations and create more resilient, responsive, and sustainable supply chains. As these technologies continue to evolve and mature, the potential for further innovation in transport logistics remains vast, paving the way for even greater efficiency and effectiveness in the movement of goods around the globe.

1.11 CHEMICAL INDUSTRY: A CATALYST FOR CHANGE

The chemical industry, a critical component of the global manufacturing ecosystem, is undergoing a transformative phase propelled by the imperatives of digitalization, safety, environmental compliance, and sustainability. This transformation is driven by advancements in process optimization and safety enhancements alongside a growing commitment to sustainable production practices.

1.11.1 Process Optimization and Safety Enhancements

Digital technologies like the IoT, AI, and advanced analytics are pivotal in process optimization. These technologies enable chemical manufacturers to achieve unprecedented efficiency, precision, and control over their production processes. IoT sensors deployed throughout chemical plants gather real-time data on equipment performance, process conditions, and product quality, facilitating a granular understanding of operational dynamics.

This wealth of data is analyzed by leveraging AI and ML to identify patterns, predict potential disruptions, and optimize process parameters. Such predictive capabilities are crucial for enhancing efficiency and preempting process deviations that could lead to safety incidents. Moreover, digital twin technology, which creates virtual replicas of physical assets, allows for the simulation and testing of process changes in a risk-free environment, further contributing to process innovation and safety.

Safety enhancements in the chemical industry are increasingly data-driven, focusing on process safety and occupational health. Advanced monitoring systems and wearable technology ensure the well-being of personnel by providing real-time alerts on hazardous conditions. In contrast, automated emergency response systems enhance preparedness and mitigate the impact of incidents.

1.11.2 Environmental Compliance and Sustainable Production

The chemical industry is at the forefront of addressing environmental challenges, driven by stringent regulatory standards and a corporate ethos aligned with sustainability. Digital technologies are instrumental in achieving these goals, enabling more efficient resource use, reducing emissions, and facilitating compliance with environmental regulations.

One key area where digitalization significantly impacts energy and resource efficiency is process optimization tools, which streamline production and minimize energy consumption and waste generation, contributing to a smaller environmental footprint. Furthermore, real-time monitoring and analytics support the effective treatment and management of waste and emissions, ensuring compliance with environmental standards.

Sustainable production in the chemical industry also encompasses the development and use of greener materials and processes. Research and digital innovation drive the prevalence of bio-based chemicals and catalysis technologies that require less energy and produce fewer byproducts. In addition, adopting circular economy principles, facilitated by digital platforms that enable material tracking and supply chain transparency, promotes the industry's reuse and recycling of materials.

The chemical industry's journey toward digital transformation, enhanced safety, and environmental sustainability is emblematic of its role as a catalyst for change in the manufacturing world. By embracing digital innovations for process optimization, safety enhancements, and sustainable production, the industry is improving its operational efficiency and compliance and contributing to a more sustainable and resilient global economy. As these digital and sustainable practices evolve, the chemical industry will lead by example, demonstrating how technological advancements and environmental stewardship can go hand in hand.

1.12 CULTIVATING GROWTH: THE NEW AGE OF AGRICULTURE

The agricultural sector is entering a new age characterized by integrating intelligent farming techniques and a heightened focus on sustainability and resource management. This shift is driven by the need to feed a growing global population, address climate change, and ensure the efficient use of natural resources. In this new age, technology is pivotal in optimizing crop production, mitigating climate impacts, and managing resources more effectively.

1.12.1 Clever Farming Techniques and Crop Optimization

Smart farming, at the heart of modern agricultural practices, leverages digital technologies like IoT, AI, drones, and satellite imagery to enhance crop yield, optimize inputs, and reduce environmental impact. These technologies enable precision agriculture, where farmers can monitor and manage their fields with unprecedented detail.

IoT sensors in the soil and mounted on equipment provide real-time data on soil moisture, nutrients, and temperature, allowing for precise irrigation, fertilization, and pest control. Drones and satellites offer aerial imagery, giving insights into crop health, growth patterns, and areas requiring attention. AI and data analytics tools process this information, enabling predictive insights that guide planting decisions, optimize crop rotations, and enhance yield forecasts.

These innovative farming techniques improve productivity and crop quality, minimize waste, and reduce the reliance on chemical inputs, contributing to more sustainable agricultural practices.

1.12.2 Climate Impact Mitigation and Resource Management

Agriculture is both a victim of and a significant contributor to climate change, making climate impact mitigation and resource management critical components of the new age of agriculture. Innovative farming technologies are crucial in these areas, enabling farmers to adapt to changing climate conditions and manage resources more efficiently.

Water management is a prime example of how technology can help mitigate climate impacts. Precision irrigation systems, informed by soil moisture data from IoT sensors, ensure that water is delivered in the right amounts at the correct times, minimizing waste and conserving water resources. Similarly, AI-driven models can predict weather patterns and climate impacts, helping farmers to adapt their practices accordingly and build resilience against extreme weather events.

Sustainable resource management extends beyond water to include soil health, biodiversity, and the efficient use of energy. Technologies such as cover cropping, no-till farming, and crop diversification, supported by digital tools, enhance soil health and carbon sequestration, contributing to climate change mitigation. Renewable energy sources, such as solar-powered irrigation systems, reduce farming operations' carbon footprints.

Furthermore, digital platforms facilitate sharing knowledge and best practices among farmers, researchers, and agricultural professionals, fostering a collaborative approach to addressing the challenges of climate change and resource management in agriculture.

The new age of agriculture, marked by intelligent farming techniques and a focus on sustainability, represents a transformative period for the sector. By harnessing the power of technology, farmers can optimize crop production, mitigate the impacts of climate change, and manage resources more efficiently, ensuring food security and environmental sustainability for future generations. As these practices evolve, the agricultural sector is set to become more resilient, productive, and sustainable, contributing to the well-being of people and the planet.

1.13 CONCLUSION: EMBRACING ADVANCED ANALYTICS FOR SUSTAINABLE GROWTH

As we journey through the transformative landscape of Industry 4.0, it becomes evident that advanced analytics is a beacon for traditional industries navigating the complexities of the modern era. This concluding section reflects on the path forward for these sectors and the imperative of building a future that is not only resilient and innovative but also sustainable and inclusive.

1.13.1 The Path Forward for Traditional Industries

Traditional industries, from manufacturing and agriculture to construction and mining, are at a crossroads. The rapid pace of technological change, increasing environmental concerns, and the shifting dynamics of global markets present both challenges and opportunities. The path forward is paved with digitalization, where advanced analytics, IoT, AI, and other Industry 4.0 technologies become the tools of choice for transforming operational processes, enhancing efficiency, and driving innovation.

Embracing these technologies enables traditional industries to transcend legacy constraints, optimize resource use, reduce waste, and significantly improve productivity. Moreover, advanced analytics' predictive capabilities empower these sectors to anticipate market trends, adapt to changing consumer demands, and mitigate risks associated with supply chain disruptions, environmental regulations, and other external factors.

However, this journey is not solely about technological adoption. It also involves a cultural shift toward data-driven decision-making, continuous learning, and agility.

Traditional industries must foster an environment that encourages innovation, values digital skills, and embraces change as a constant.

1.13.2 Building a Resilient and Innovative Future

The future of traditional industries hinges on their ability to integrate resilience and innovation into their core operations. Resilience refers to industries' capacity to withstand and adapt to external shocks, such as economic downturns, environmental disasters, or pandemics. Digital technologies, particularly advanced analytics, are crucial in building this resilience. They offer real-time insights and scenario-planning tools that enable businesses to navigate uncertainties confidently.

On the other hand, innovation is the engine of growth and competitiveness. By leveraging the vast possibilities of advanced analytics and other digital tools, traditional industries can develop new products, services, and business models that meet society's and the environment's evolving needs. This innovation extends beyond product development, encompassing sustainable practices that minimize environmental impact and contribute to communities' well-being.

Moreover, the commitment to sustainability, underpinned by advanced analytics, opens up new avenues for growth. Sustainable practices, driven by data and insights, reduce operational costs and resonate with increasingly eco-conscious consumers and stakeholders, enhancing brand value and market positioning.

The journey toward embracing advanced analytics for sustainable growth is both a challenge and an opportunity for traditional industries. By harnessing the power of digital technologies, these sectors can redefine their operations, enhance their resilience, and unlock new paths to innovation. The future envisioned is one where traditional industry survives and thrives, contributing to a more sustainable, efficient, and equitable world. The call to action is clear: to embrace change, invest in digital capabilities, and commit to sustainability, paving the way for a resilient and innovative future.

REFERENCES

[1] Bartodziej, C.J. and C.J. Bartodziej, *The concept Industry 4.0*. 2017: Springer.
[2] Dalenogare, L.S., et al., The expected contribution of Industry 4.0 technologies for industrial performance. *International Journal of Production Economics*, 2018. 204: p. 383–394.
[3] Davis, R., B.-O. Sessions, and A.R. Check, *Industry 4.0. Digitalization for productivity and growth*, European Parliament, members' research service, 2015.
[4] Frank, A.G., L.S. Dalenogare, and N.F. Ayala, Industry 4.0 technologies: implementation patterns in manufacturing companies. *International Journal of Production Economics*, 2019. 210: p. 15–26.
[5] Soofastaei, A., *Data analytics applied to the mining industry*. 2020: CRC Press.
[6] Ghobakhloo, M., Industry 4.0, digitization, and opportunities for sustainability. *Journal of Cleaner Production*, 2020. 252: p. 119869.
[7] Soofastaei, A., The application of artificial intelligence to reduce greenhouse gas emissions in the mining industry. *Green Technologies to Improve the Environment on Earth*, 2018. 25: p. 234–245.
[8] Gubán, M. and G. Kovács, Industry 4.0 conception. *Acta Technica Corviniensis-Bulletin of Engineering*, 2017. 10(1).

[9] Soofastaei, A., et al., Energy-efficient loading and hauling operations. In *Energy efficiency in the minerals industry: best practices and research directions*, 2018: p. 121–146. Springer.
[10] Soofastaei, A., Introductory chapter: virtual assistants. In *Virtual assistant*, Vol. 1. 2021: IntechOpen.
[11] Javaid, M., et al., Industry 4.0 technologies and their applications in fighting COVID-19 pandemic. *Diabetes & Metabolic Syndrome: Clinical Research & Reviews*, 2020. 14(4): p. 419–422.
[12] Lasi, H., et al., Industry 4.0. *Business & Information Systems Engineering*, 2014. 6: p. 239–242.
[13] Lele, A. and A. Lele, Industry 4.0. *Disruptive Technologies for the Militaries and Security*, 2019: p. 205–215.
[14] Soofastaei, A., *Numerical simulation: advanced techniques for science and engineering*. 2023: BoD–Books on Demand.
[15] Rojko, A., Industry 4.0 concept: background and overview. *International Journal of Interactive Mobile Technologies*, 2017. 11(5).
[16] Santos, R.C. and J.L. Martinho, An Industry 4.0 maturity model proposal. *Journal of Manufacturing Technology Management*, 2020. 31(5): p. 1023–1043.
[17] Saucedo-Martínez, J.A., et al., Industry 4.0 framework for management and operations: a review. *Journal of Ambient Intelligence and Humanized Computing*, 2018. 9: p. 789–801.
[18] Sony, M. and S. Naik, Key ingredients for evaluating Industry 4.0 readiness for organizations: a literature review. *Benchmarking: An International Journal*, 2020. 27(7): p. 2213–2232.
[19] Xu, L.D., E.L. Xu, and L. Li, Industry 4.0: state of the art and future trends. *International Journal of Production Research*, 2018. 56(8): p. 2941–2962.
[20] Zheng, T., et al., The applications of Industry 4.0 technologies in manufacturing context: a systematic literature review. *International Journal of Production Research*, 2021. 59(6): p. 1922–1954.

2 Transforming Mining Operations

Harnessing Advanced Analytics for Optimal Decision-Making

Ali Soofastaei

2.1 INTRODUCTION

The mining industry is pivotal in an era marked by unprecedented global demand for natural resources. With populations expanding, urbanization accelerating, and technological advancements driving the hunger for raw materials, the imperative for mining operations to optimize their processes has never been more pressing. This chapter delves into the evolving landscape of mining, where the convergence of burgeoning demand and technological innovation propels the industry toward a new paradigm: Mining 4.0 [1].

At the heart of this transformation lies the application of advanced analytics techniques, a burgeoning field that promises to revolutionize how mining enterprises approach decision-making. Like many other industries, the mining sector has realized the potential of harnessing vast troves of data to inform strategic and operational decisions. By leveraging sophisticated analytical tools and methodologies, mining companies stand poised to extract actionable insights from the deluge of information generated by their operations.

However, despite the promise of advanced analytics, the mining industry grapples with many challenges in its quest for optimization. Traditional decision-making processes, often characterized by siloed data streams and disjointed workflows, hinder the seamless integration of analytics into operational workflows. From plant history to maintenance logs and mine planning data to logistical insights, the wealth of information available to mining operations must be mainly used, relegated to disparate databases and systems [2].

Moreover, the complexity of mining data analysis poses a formidable barrier to effective decision-making. Extracting meaningful insights from heterogeneous datasets requires advanced analytical capabilities and specialized domain knowledge—expertise often in short supply within the industry. As a result, despite the proliferation of data collection technologies, mining companies cannot fully capitalize on the wealth of information at their disposal.

DOI: 10.1201/9781003186823-2

Yet, amid these challenges lie opportunities for transformation. By embracing advanced analytics, mining enterprises can close the loop between data analysis and actionable decision-making, ushering in a new era of efficiency and optimization. Through seamless analytics integration into business processes, decision-makers—from senior management to frontline operators—can access timely, data-driven insights to drive strategic initiatives and operational improvements.

In the following sections, we will delve deeper into the application of advanced analytics across the mining value chain. We will explore how analytics reshapes the mining industry from asset management to supply chain optimization, from predictive maintenance to automated decision-making. Through real-world case studies and practical insights, we will uncover the transformative potential of advanced analytics and chart a course toward a future where data-driven decision-making reigns supreme in the mining sector [3].

Furthermore, applying advanced analytics in mining is not just about improving efficiency; it fundamentally redefining how mining operations are managed and executed. By harnessing the power of analytics, mining companies can unlock new levels of productivity, enhance safety protocols, and minimize environmental impacts—all while sustaining the burgeoning demand for natural resources.

One of the primary drivers behind adopting advanced analytics in mining is the need to optimize decision-making processes across the entire value chain. Traditionally, decisions in mining operations have been made in isolation, with each department or function optimizing its processes without considering the broader implications for the entire operation. This siloed approach often leads to suboptimal outcomes, with decisions that may be locally optimized but fail to maximize the mine's overall value.

Advanced analytics has the potential to break down these silos and facilitate a more holistic approach to decision-making. By integrating data from across the value chain—from exploration and development to extraction, processing, and transportation—mining companies can gain a comprehensive understanding of their operations and identify opportunities for optimization that would have been impossible to discern using traditional methods [4].

Moreover, advanced analytics enables mining companies to move beyond reactive decision-making toward a more proactive, predictive approach. By analyzing historical data and identifying patterns and trends, analytics tools can anticipate future events and recommend preemptive actions to mitigate risks or capitalize on opportunities. This shift from reactive to proactive decision-making is particularly valuable in the mining industry, where unplanned downtime, equipment failures, and supply chain disruptions can have significant financial and operational consequences.

In addition to optimizing decision-making processes, advanced analytics also holds the key to unlocking new sources of value within mining operations. For example, predictive maintenance algorithms can help mining companies identify potential equipment failures before they occur, allowing them to schedule maintenance proactively and minimize downtime. Similarly, advanced supply chain analytics can help companies optimize logistics operations, reducing transportation costs and improving delivery times.

However, realizing the full potential of advanced analytics in mining requires more than just technological investment; it also requires a cultural shift within mining organizations. Decision-makers at all levels must be willing to embrace data-driven decision-making and trust in the insights generated by analytics tools. This may require upskilling existing staff, hiring new talent with expertise in data analytics, and fostering a culture of collaboration and innovation across the organization [5].

In conclusion, adopting advanced analytics represents a paradigm shift for the mining industry. It offers the potential to optimize decision-making processes, unlock new sources of value, and drive sustainable growth. By embracing analytics, mining companies can position themselves for success in an increasingly competitive and complex global market while contributing to a more sustainable and responsible future for the industry.

2.2 CURRENT STATE OF DECISION-MAKING IN MINING

In the traditional decision-making paradigm within the mining industry, processes have often been categorized, with decisions made in isolation within specific departments or functions. While this approach effectively addresses immediate concerns, it has led to a fragmented view of operations and missed opportunities for optimization across the entire value chain.

2.2.1 Analysis of Decision-Making Processes

Decision-making in mining typically involves many stakeholders, each with their priorities and objectives. At the strategic level, senior management makes decisions regarding long-term investments, resource allocation, and overall operational strategy. Meanwhile, at the operational level, decisions are made on a day-to-day basis by frontline supervisors and operators, focusing on issues such as production targets, equipment utilization, and safety protocols.

However, despite the involvement of multiple stakeholders, decision-making processes in mining often need more integration and coordination to maximize efficiency and value. Data silos are expected, with information stored in disparate systems and databases that are not easily accessible or shareable across departments. As a result, decision-makers may need more comprehensive insights to make informed choices, leading to suboptimal outcomes [6].

2.2.2 Limitations of Current Decision-Making Processes

The current state of decision-making in mining is characterized by several critical limitations, which hinder the industry's ability to maximize its potential:

1. **Lack of Data Integration:** One of the primary challenges facing mining decision-makers is the need to integrate various data sources. Information related to plant history, maintenance logs, mine planning, logistics, and engineering data is often stored in separate databases, making it difficult to correlate and analyze holistically. As a result, decision-makers may need

a comprehensive view of operations, leading to inefficiencies and missed opportunities for optimization.

2. **Requirement for Specialist Skills:** Another limitation of current decision-making processes is the requirement for specialist skills and knowledge to analyze and interpret data effectively. Mining operations generate vast amounts of data, ranging from geological surveys to equipment telemetry, which must be processed and analyzed to extract meaningful insights. However, the expertise needed to perform this analysis is often scarce within the industry, leading to delays and inefficient decision-making.
3. **Reactive Decision-Making:** In many cases, decision-making in mining remains reactive rather than proactive. Issues such as equipment failures, supply chain disruptions, and safety incidents are often addressed as they arise rather than anticipated and mitigated proactively. This reactive approach can lead to increased downtime, higher costs, and reduced productivity, undermining mining operations' overall competitiveness.

Fragmented processes, data silos, and a reactive approach to problem-solving characterize the current state of decision-making in mining. To overcome these limitations and unlock the full potential of the industry, mining companies must embrace advanced analytics techniques that enable integrated, data-driven decision-making across the entire value chain. By leveraging analytics to break down silos, empower decision-makers, and anticipate future challenges, mining companies can position themselves for success in an increasingly complex and competitive global market [7].

2.3 ADVANCED ANALYTICS: CLOSING THE DECISION-MAKING LOOP

In response to the limitations of traditional decision-making processes in the mining industry, advanced analytics emerges as a transformative solution, offering the potential to revolutionize the way decisions are made at all organizational levels. By harnessing the power of data and analytics, mining companies can overcome the shortcomings of current decision-making processes and drive efficiency, productivity, and value across the entire value chain.

2.3.1 Addressing Shortcomings of Current Decision-Making Processes

Advanced analytics offers several critical advantages over traditional decision-making processes, enabling mining companies to address longstanding challenges and unlock new opportunities for optimization [8]:

1. **Data Integration:** One of the primary benefits of advanced analytics is its ability to integrate data from disparate sources and provide a holistic view of operations. By aggregating and analyzing data from plant history, maintenance logs, mine planning, logistics, and engineering, analytics platforms can provide decision-makers with comprehensive insights into the factors influencing performance and productivity. This integrated approach enables

decision-makers to identify correlations, trends, and patterns that may have been overlooked in siloed data environments, facilitating more informed and strategic decision-making.

2. **Predictive Capabilities:** Another critical advantage of advanced analytics is its predictive capabilities, which enable mining companies to anticipate future events and trends and proactively respond to emerging challenges. By analyzing historical data and identifying patterns and anomalies, analytics platforms can forecast equipment failures, supply chain disruptions, and other potential risks, allowing decision-makers to take preemptive action to mitigate their impact. This proactive approach reduces downtime and operational costs and improves safety and reliability, enhancing the overall competitiveness of mining operations.
3. **Empowering Decision-Makers:** Advanced analytics empowers decision-makers at all organizational levels by providing timely, relevant, and actionable insights. From senior management to frontline supervisors and operators, decision-makers can access analytics platforms to monitor performance, track key metrics, and make real-time data-driven decisions. By democratizing access to data and analytics, mining companies can foster a culture of informed decision-making and collaboration, enabling employees to work together toward common goals and objectives.

2.3.2 Integration into Business Processes

To realize the full potential of advanced analytics, mining companies must integrate analytics into their business processes and workflows, embedding data-driven decision-making into the organization's fabric. This requires more than just technological investment; it requires a cultural shift toward data-driven decision-making and a commitment to continuous improvement and innovation.

At the strategic level, advanced analytics can inform long-term planning and resource allocation, enabling senior management to identify opportunities for growth and investment and optimize the overall operational strategy. By analyzing market trends, customer preferences, and competitor behavior, analytics platforms can provide strategic insights that guide decision-making and drive competitive advantage.

At the operational level, advanced analytics can support day-to-day decision-making by providing frontline supervisors and operators with real-time insights into equipment performance, production targets, and safety protocols. By monitoring key metrics and key performance indicators (KPIs), analytics platforms enable decision-makers to identify trends, anomalies, and opportunities for optimization, facilitating more efficient and effective operations [9].

Moreover, by integrating analytics into business processes, mining companies can establish feedback loops that enable continuous improvement and optimization. By analyzing the outcomes of previous decisions and actions, decision-makers can identify areas for improvement and refine their strategies and tactics, accordingly driving incremental gains in performance and productivity over time.

Advanced analytics allows mining companies to close the decision-making loop, enabling integrated, data-driven decision-making at all organizational levels. By addressing the shortcomings of current decision-making processes and empowering decision-makers with timely, relevant, and actionable insights, advanced analytics can drive efficiency, productivity, and value across the entire value chain, positioning mining companies for success in an increasingly competitive and complex global market.

2.4 OPPORTUNITIES FOR ADVANCED ANALYTICS IN MINING

Adopting advanced analytics presents many opportunities for mining enterprises to unlock value, enhance efficiency, and drive innovation across various operations. By leveraging advanced analytics tools and techniques, mining companies can gain deeper insights into their operations, optimize processes, and make more informed decisions. This section explores critical areas within the mining enterprise where advanced analytics can provide significant value and the potential of advanced analytics to enhance the effectiveness of remote operations centers (ROCs) [10].

2.4.1 Asset Management

Effective asset management is critical for mining companies to maximize the lifespan and performance of their equipment while minimizing downtime and maintenance costs. Advanced analytics can play a pivotal role in asset management by enabling predictive maintenance strategies. By analyzing equipment telemetry data, historical maintenance records, and environmental factors, analytics platforms can identify patterns indicative of potential equipment failures. This allows mining companies to schedule maintenance proactively, reducing unplanned downtime and optimizing asset utilization.

2.4.2 Reconciliation

Reconciliation is crucial to mining operations, ensuring that production targets are met while maintaining accurate inventory records. However, reconciling production data with geological models and mine plans can take time and effort. Advanced analytics can streamline reconciliation efforts by automating data collection, analysis, and reporting. By integrating data from various sources and applying advanced algorithms, analytics platforms can identify discrepancies and anomalies, enabling mining companies to reconcile their production data more efficiently and accurately [11].

2.4.3 Planning

Effective planning is essential for optimizing production, minimizing costs, and maximizing resource utilization in mining operations. Advanced analytics can enhance planning processes by giving decision-makers actionable insights into ore grades, equipment performance, and market demand. By simulating various scenarios and analyzing the potential impacts of different decisions, analytics platforms can help mining companies develop more robust and resilient plans that adapt to changing conditions and uncertainties.

2.4.4 Supply Chain Management

Optimizing the supply chain is critical for ensuring the timely delivery of raw materials, equipment, and supplies to mining operations while minimizing costs and mitigating risks. Advanced analytics can improve supply chain management by providing real-time visibility into inventory levels, transportation routes, and supplier performance. By analyzing historical data and external factors such as weather patterns and geopolitical events, analytics platforms can help mining companies identify potential bottlenecks, optimize logistics routes, and enhance overall supply chain resilience.

2.4.5 Enhancing ROCs

ROCs enable mining companies to monitor and manage their operations from centralized locations. Advanced analytics can enhance the effectiveness of ROCs by providing decision-makers with real-time insights into KPIs and operational metrics. By integrating data from various sources, including sensors, drones, and satellite imagery, analytics platforms can enable ROC operators to detect anomalies, identify trends, and make data-driven decisions to optimize operations and mitigate risks [12].

Advanced analytics presents significant opportunities for mining companies to optimize operations, improve decision-making, and drive innovation. By leveraging advanced analytics tools and techniques, mining companies can enhance asset management, streamline reconciliation processes, optimize planning efforts, and improve supply chain management. In addition, advanced analytics can enhance the effectiveness of ROCs by providing decision-makers with real-time insights into operational performance and enabling data-driven decision-making. By embracing advanced analytics, mining companies can position themselves for success in an increasingly competitive and dynamic industry landscape.

2.5 PREDICTIVE TOOLS AND AUTOMATION

Anticipating and responding to potential challenges is paramount in mining operations' dynamic and high-stakes environment. Predictive tools powered by advanced analytics offer a transformative solution, enabling mining companies to proactively identify risks, optimize processes, and enhance overall efficiency. This section explores how predictive tools can suggest or automate courses of action in mining operations. It discusses the integration of predictive analytics into decision-making processes to enhance efficiency and effectiveness [13].

2.5.1 Predictive Tools in Mining Operations

Predictive tools utilize historical data, ML algorithms, and statistical models to forecast future events or trends within mining operations. These tools range from simple predictive maintenance algorithms to complex optimization models considering multiple variables and scenarios. By analyzing patterns, anomalies, and correlations in historical data, predictive tools can identify potential risks, opportunities, and optimization strategies, enabling mining companies to make more informed decisions.

One common application of predictive tools in mining operations is predictive maintenance. By analyzing equipment telemetry data, maintenance logs, and environmental factors, predictive maintenance algorithms can identify patterns indicative of potential equipment failures. This allows mining companies to schedule maintenance proactively, reducing downtime, minimizing costs, and optimizing asset utilization. Predictive maintenance can also extend the lifespan of equipment and improve safety by addressing issues before they escalate into serious problems.

Another application of predictive tools in mining operations is production forecasting. By analyzing historical production data, geological models, and market trends, predictive models can accurately forecast future production volumes and grades. This enables mining companies to anticipate changes in demand, optimize production schedules, and allocate resources more effectively. Predictive production forecasting can help mining companies identify potential bottlenecks, optimize workflows, and improve operational efficiency [14].

2.5.2 Integration into Decision-Making Processes

Integrating predictive analytics into decision-making processes is essential for realizing the full potential of predictive tools in mining operations. Predictive models must seamlessly integrate into existing workflows and decision-making frameworks to ensure that their insights are actionable and impactful. This requires collaboration between data scientists, domain experts, and decision-makers to develop models that address specific business challenges and deliver tangible results.

One approach to integrating predictive analytics into decision-making processes is to embed predictive models into operational systems and workflows. By incorporating predictive models into existing data management systems, mining companies can automate the generation of insights and recommendations based on real-time data. For example, predictive maintenance alerts can be automatically generated when equipment telemetry data indicates a potential failure, enabling maintenance crews to take preemptive action before an issue escalates [15].

Another approach to integrating predictive analytics into decision-making processes is to empower decision-makers with user-friendly analytics tools and dashboards. By providing decision-makers with access to intuitive analytics platforms, mining companies can enable them to explore data, visualize trends, and generate insights without relying on specialized expertise. This democratization of analytics empowers decision-makers at all levels of the organization to make data-driven decisions and take proactive action to optimize operations.

Predictive tools powered by advanced analytics offer mining companies a powerful means of anticipating and responding to potential challenges within their operations. Mining companies can optimize asset management, streamline production processes, and enhance efficiency by leveraging predictive maintenance algorithms, production-forecasting models, and other predictive tools. However, successfully integrating predictive analytics into decision-making processes requires collaboration, communication, and a commitment to data-driven organizational decision-making. By embracing predictive analytics, mining companies can position themselves for success in an increasingly competitive and dynamic industry landscape.

2.6 APPLICATION OF ADVANCED ANALYTICS ACROSS THE MINE VALUE CHAIN

The mine value chain encompasses interconnected stages critical to mining operations' success and efficiency. From exploration and extraction to processing and transportation, advanced analytics offers opportunities for optimization, innovation, and value creation at every step. This section provides a detailed exploration of how advanced analytics can be applied to different stages of the mine value chain. It presents case studies or examples demonstrating the practical application of advanced analytics in each stage.

2.6.1 Exploration

Exploration is the initial stage of the mine value chain, where mining companies identify and evaluate potential mineral deposits. Advanced analytics can enhance the effectiveness of exploration efforts by analyzing geological data, remote sensing imagery, and historical exploration data to identify prospective areas for further investigation. For example, predictive modeling algorithms can analyze geological datasets to identify geological anomalies indicative of potential mineralization, enabling mining companies to prioritize exploration efforts and allocate resources more effectively [16].

Case Study:

A mining company operating in a remote region leveraged advanced analytics to optimize its exploration strategy. By analyzing geological datasets and satellite imagery, the company identified promising areas for further investigation, reducing the time and resources required for exploration activities. As a result, the company discovered new mineral deposits and expanded its resource base, ultimately enhancing its competitiveness and profitability.

2.6.2 Extraction

Extraction removes valuable minerals from the earth's crust and transports them to the surface for further processing. Advanced analytics can optimize extraction processes by analyzing equipment telemetry data, geological models, and operational metrics to identify opportunities for efficiency improvements and cost savings. For example, predictive maintenance algorithms can proactively analyze equipment performance data to anticipate potential failures and schedule maintenance, reducing downtime and minimizing production disruptions.

Case Study:

A mining company operating a fleet of heavy equipment implemented predictive maintenance algorithms to optimize its extraction operations. By analyzing equipment telemetry data and historical maintenance records, the company proactively identified patterns indicative of potential equipment failures and scheduled maintenance. As a result, the company reduced unplanned downtime, improved equipment reliability, and increased overall production efficiency.

2.6.3 Processing

Processing is the stage of the mine value chain where mined ore is crushed, ground, and refined to extract valuable minerals. Advanced analytics can optimize processing operations by analyzing process data, metallurgical test results, and historical production data to identify improvement and optimization opportunities. For example, predictive modeling algorithms can analyze process data to optimize operating parameters such as temperature, pressure, and pH, maximizing recovery rates and minimizing processing costs [17].

Case Study:

A mining company operating a mineral processing plant implemented advanced analytics to optimize its processing operations. The company identified opportunities for improvement and optimization by analyzing process data and historical production records. Adjusting operating parameters based on predictive modeling recommendations increased recovery rates, reduced processing costs, and improved overall operational efficiency.

2.6.4 Transportation

Transportation is the final stage of the mine value chain, where processed ore is transported from the mine site to processing facilities or end-users. Advanced analytics can optimize transportation operations by analyzing logistics data, transportation routes, and vehicle performance metrics to identify opportunities for cost savings and efficiency improvements. For example, predictive modeling Algorithms can analyze transportation data to optimize routing, scheduling, and vehicle utilization, minimizing transportation costs and reducing delivery times [18].

Case Study:

A mining company operating a fleet of transport vehicles implemented advanced analytics to optimize its transportation operations. By analyzing logistics data and vehicle performance metrics, the company identified opportunities for route optimization and vehicle scheduling. By adjusting routing and scheduling based on predictive modeling recommendations, the company minimized transportation costs, reduced delivery times, and improved overall transportation efficiency.

Advanced analytics offers mining companies a powerful tool for optimizing operations, improving efficiency, and enhancing profitability across the entire mine value chain. By leveraging advanced analytics techniques and technologies, mining companies can unlock new opportunities for innovation, value creation, and competitive advantage, ultimately driving success in an increasingly dynamic and competitive industry landscape.

2.7 CHALLENGES AND CONSIDERATIONS

Implementing advanced analytics in mining operations holds immense potential for driving efficiency, optimizing processes, and unlocking value. However, this transformative journey has its challenges and considerations. In this section, we analyze the potential obstacles and complexities that mining companies may encounter when implementing advanced analytics and discuss strategies for overcoming these barriers.

2.7.1 Data Integration Challenges

One primary challenge in implementing advanced analytics in mining operations is data integration. Mining operations generate vast amounts of data from various sources, including sensors, equipment telemetry, geological surveys, and operational records. Integrating and harmonizing these disparate datasets to create a unified data infrastructure can be complex and time-consuming. Moreover, legacy systems and siloed data environments may further exacerbate data integration challenges, hindering the seamless flow of information across the organization.

Mining companies must adopt a comprehensive approach encompassing data governance, management, and architecture to address data integration challenges. Establishing clear data governance policies and standards ensures data quality, consistency, and reliability, laying the foundation for effective data integration. Investing in modern data management technologies and platforms enables mining companies to aggregate, store, and analyze diverse datasets in a centralized and scalable manner. In addition, implementing robust data architecture principles, such as data lakes and warehouses, facilitates the integration and interoperability of disparate datasets, enabling seamless data flow across the organization.

2.7.2 Skill Requirements and Talent Shortages

Another significant challenge in implementing advanced analytics in mining operations requires specialized skills and expertise. Data science, machine learning, and analytics are highly specialized fields that demand a unique blend of technical knowledge, domain expertise, and analytical prowess. However, the mining industry often needs talent shortages in these areas, with a limited pool of data scientists, statisticians, and analytics professionals available to meet the growing demand for advanced analytics capabilities [19].

Mining companies must invest in talent development and workforce training initiatives to overcome skill requirements and talent shortages. Developing internal training programs and partnerships with educational institutions can help upskill existing employees and equip them with the necessary skills to leverage advanced analytics effectively. In addition, mining companies can leverage external partnerships and collaborations with analytics firms, consulting agencies, and technology providers to access specialized expertise and fill skill gaps. By cultivating a continuous learning and innovation culture, mining companies can build a talented and diverse workforce capable of accelerating advanced analytics initiatives.

2.7.3 Cultural and Organizational Challenges

Implementing advanced analytics in mining operations often requires a cultural shift and organizational change. Traditional decision-making processes and hierarchical structures may refrain from adopting data-driven decision-making approaches, posing challenges to successfully implementing advanced analytics initiatives. Moreover, resistance to change, fear of technology, and skepticism toward analytics-driven insights may impede progress and hinder the realization of advanced analytics' full potential.

Mining companies must foster a culture of data-driven decision-making and innovation to address cultural and organizational challenges. Leadership buy-in and commitment are critical for driving cultural change and promoting the adoption of advanced analytics initiatives. By demonstrating the value of advanced analytics through pilot projects, proofs-of-concept, and success stories, mining companies can build trust and confidence in analytics-driven insights among decision-makers and stakeholders. In addition, fostering cross-functional collaboration and communication ensures alignment and cooperation across departments and functions, facilitating the integration of advanced analytics into business processes and workflows.

2.7.4 Infrastructure and Technology Constraints

Infrastructure and technology constraints pose additional challenges to implementing advanced analytics in mining operations. Legacy systems, outdated technologies, and inadequate IT infrastructure may need more scalability, flexibility, and computational power to support advanced analytics initiatives effectively. Moreover, data privacy, security, and regulatory compliance concerns may further complicate the adoption of cloud-based analytics platforms and technologies.

To overcome infrastructure and technology constraints, mining companies must invest in modernizing their IT infrastructure and embracing cloud-based analytics platforms. Adopting scalable and flexible cloud infrastructure enables mining companies to access computational resources on-demand, accommodate growing data volumes, and support advanced analytics workloads efficiently. In addition, implementing robust cybersecurity measures and compliance protocols ensures the protection of sensitive data and mitigates risks associated with data breaches and regulatory violations. By embracing digital transformation and investing in cutting-edge technologies, mining companies can overcome infrastructure and technology constraints and unlock the full potential of advanced analytics.

Implementing advanced analytics in mining operations presents numerous challenges and considerations, ranging from data integration and skill requirements to cultural and organizational barriers. However, by addressing these challenges proactively and adopting a strategic approach to implementation, mining companies can overcome obstacles, drive innovation, and unlock value across the entire value chain. Embracing advanced analytics as a strategic imperative enables mining companies to gain actionable insights, optimize processes, and enhance competitiveness in an increasingly dynamic and competitive industry landscape.

2.8 FUTURE DIRECTIONS AND CONCLUSION

As we look toward the mining industry's future, advanced analytics will become increasingly pivotal in driving innovation, efficiency, and sustainability. In this final section, we explore the future directions of advanced analytics in the mining industry and summarize the key insights and conclusions drawn from this chapter.

2.8.1 Outlook on the Future of Advanced Analytics in the Mining Industry

The future of advanced analytics in the mining industry holds immense promise, with significant opportunities for continued innovation and transformation. As technological advancements accelerate and data becomes increasingly abundant, the adoption and integration of advanced analytics will become ubiquitous across all facets of mining operations. Here are some key trends and developments shaping the future of advanced analytics in the mining industry [20]:

1. **Artificial Intelligence (AI) and Machine Learning (ML):** Integrating AI and ML technologies will revolutionize decision-making processes in mining operations. Advanced ML algorithms will enable mining companies to extract insights from complex datasets, identify patterns and trends, and optimize operations in real time.
2. **Internet of Things (IoT) and Sensor Technologies:** The proliferation of IoT devices and sensor technologies will fuel the generation of real-time data streams from mining operations' equipment, vehicles, and infrastructure. Integrating IoT data with advanced analytics platforms will enable predictive maintenance, asset optimization, and safety monitoring.
3. **Edge Computing:** The rise of edge computing technologies will enable mining companies to process and analyze data closer to the source, reducing latency and enabling real-time decision-making at the network's edge. Edge analytics capabilities will empower frontline operators to respond swiftly to changing conditions and optimize operations.
4. **Cloud-Based Analytics:** Adopting cloud-based analytics platforms will enable mining companies to leverage scalable computational resources, advanced analytics tools, and data storage capabilities cost-effectively and flexibly. These solutions will also facilitate industry collaboration, data sharing, and innovation.

5. **Predictive Simulation:** Integrating predictive simulation techniques with advanced analytics will enable mining companies to simulate various scenarios, assess the impact of different decisions, and optimize operations in virtual environments. Predictive simulation capabilities will also support strategic planning, risk management, and scenario analysis.
6. **Autonomous Operations:** The advancement of autonomous technologies, such as autonomous vehicles, drones, and robotics, will transform mining operations into highly efficient and autonomous ecosystems. Integrating autonomous technologies with advanced analytics will enable self-optimizing and self-adaptive mining operations that maximize productivity and safety.
7. **Sustainability and Environmental Analytics:** The focus on sustainability and environmental stewardship will drive the adoption of advanced analytics for ecological monitoring, compliance reporting, and resource optimization. Environmental analytics capabilities will enable mining companies to minimize their ecological footprint, reduce emissions, and enhance sustainability performance.

In summary, the future of advanced analytics in the mining industry is characterized by innovation, integration, and transformation. By embracing advanced analytics technologies and methodologies, mining companies can unlock new opportunities for optimization, efficiency, and sustainability and position themselves for success in an increasingly competitive and dynamic industry landscape.

2.8.2 Conclusion: Key Insights and Conclusions

In conclusion, this chapter has provided a comprehensive overview of the role of advanced analytics in the mining industry, exploring its applications, challenges, and future directions. Key insights and conclusions drawn from this chapter include:

- Advanced analytics holds immense potential for optimizing decision-making processes, driving efficiency, and unlocking value across the entire mine value chain.
- Data integration, skill requirements, cultural barriers, and infrastructure constraints are among the key challenges and considerations in implementing advanced analytics in mining operations.
- Overcoming these challenges requires a strategic approach, including investment in talent development, technology infrastructure, and organizational change initiatives.
- The future of advanced analytics in the mining industry is characterized by innovation, integration, and transformation, with significant opportunities for continued growth and evolution.

As mining companies embark on their journey toward digital transformation and embrace advanced analytics as a strategic imperative, they will position themselves for success in an increasingly competitive and dynamic industry landscape. By leveraging advanced analytics to optimize operations, drive innovation, and enhance sustainability, mining companies can thrive in the digital age and unlock new growth and value-creation opportunities.

2.9 PRACTICAL AND COMPLETED ADVANCED ANALYTICS PROJECTS IN MINING 4.0

2.9.1 Case Study 1: Shipping Advanced Analytics Opportunities

2.9.1.1 Introduction

Pricing for spot-fixing vessels is highly variable and fluctuates daily. There is no proven way to determine which day or time to procure the vessel.

This project aims to develop a forecasting model to determine which period (current or future) is best for procuring spot vessels to reduce shipping costs.

As a potential solution, consolidate shipping and vessel data from multiple intelligence sources (brokers, third parties) and costing information and use ML to develop models that will give a directional indication of the best vessel-fixing period.

2.9.1.2 Short-Term Objectives (Quick Wins)

- Identify all effective parameters on time charting (TC).
- Complete sensitivity analyses and related hypothesis test.
- Find the correlation between identified parameters and TC.
- TC short-term forecasting ($/day)
  - Weekly
  - Monthly
  - Quarterly
- Control the market modeling
  - Approach → Number of fixed vessels (per day, per week, per month)
- Make a model to compare the received information from brokers and available heuristic data for ballasting vessels per region and loaded Cargoes.

2.9.1.3 Methodology

Table 2.1 shows the case study methodology.

2.9.1.4 Effective Parameters on TC

The effective parameters mentioned below have been identified as critical parameters.

2.9.1.5 Data Collection

Data from mining companies, brokers, and third parties for an extended period (2014–2018) have been collected. Overall, we have access to two different datasets:

1. **Fixture Historical:** This dataset details the company's fixed contract from 2014.
2. **Dashboard:** This dataset provides TC price and some relevant data from brokers, mostly from 2014 until now.

TABLE 2.1
Shipping Advanced Analytics Opportunities: Methodology

Objective	Suggested Approach
Identify all effective parameters on TC	Data analysis, technical meetings with business partners and shipping specialists
Sensitivity analyses and related hypothesis tests	Advanced statistical methods
Find the correlation between identified parameters and TC	Nonlinear regression models, machine learning
TC short-term forecasting	Artificial neural network (ANN)
Control the market	ANN and genetic algorithm (GA)
Compeering brokers' information with heuristically data	Statistical methods

TABLE 2.2
Shipping Advanced Analytics Opportunities: Effective Parameters on TC

1	Fixture date	9	Congestion in the port
2	CP date	10	GAP
3	Fixed stem	11	Number of fixed vessel
4	Estimate intake	12	Vessel speed
5	Rout	13	Vessel type
6	LAY-CAN	14	Loaded volume
7	Promptness	15	Market demand
8	DWT (size of the vessel)	16	Vessel availability

The AI center's research team developed a Python application that cleaned and merged all received datasets. The forecasting model's target value is TC.

2.9.1.6 Analysis of TC Fluctuations

Figure 2.1 presents the variation of TC from 2014 to the present. There has been a rising trend since 2016. The value of TC at the beginning of each year is lower than in the last months of the year. This fact is highlighted in Figure 2.2, where the variation of TC in each year is plotted on the same graph.

Neglecting TC values in 2014 (blue line), TC values rise from January till:

- August 2015
- November 2016
- September 2017 (although there is another peak in March)
- August 2018 (till now)

Hence, the peak of TC occurred between August and November.

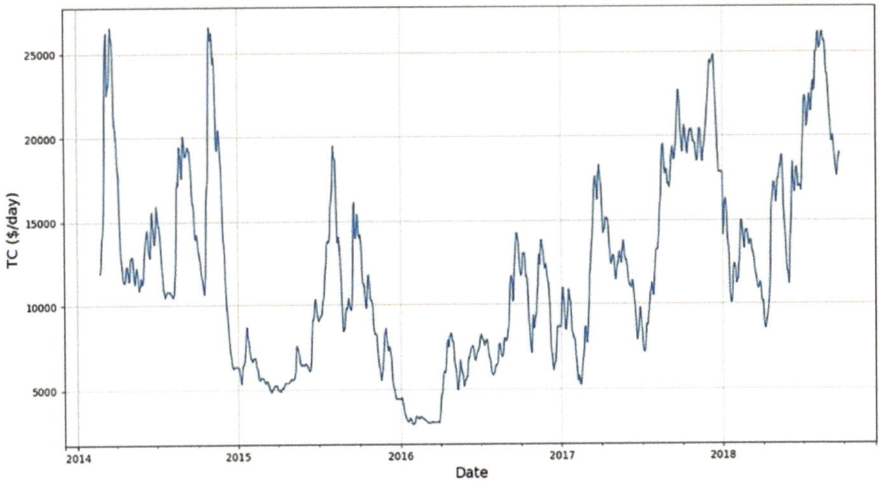

FIGURE 2.1 TC Fluctuations from 2014 to the present.

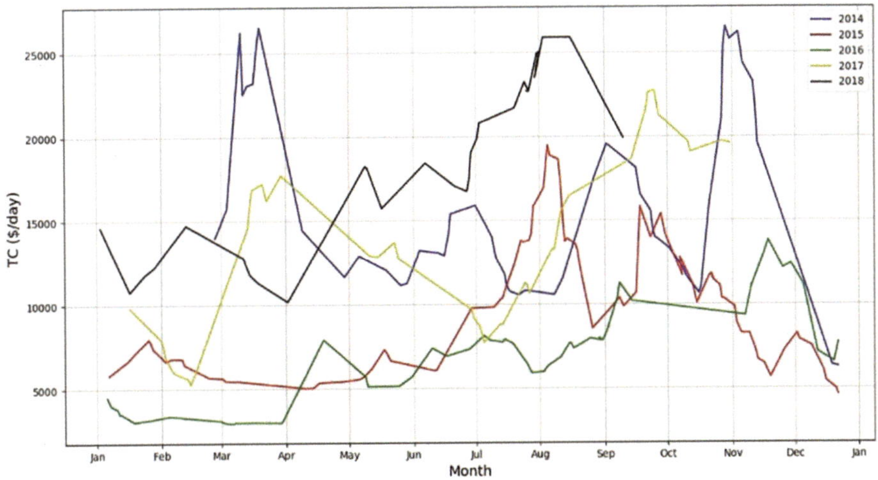

FIGURE 2.2 Yearly variation of TC.

A moving average approach was used to study the overall trend of the TC each year. The moving average calculates the signal (TC) average on a specified window (days). The window size of the moving average function has been set to:

- 5: to present the weekly variations of TC
- 20: to present the monthly variations of TC
- 60: to present the quarterly variations of TC

Figure 2.3 demonstrates the 5-day (A), 20-day (B), and 60-day (C) average of TC in each year, respectively. The yearly variation of the 60-day average of TC, presented

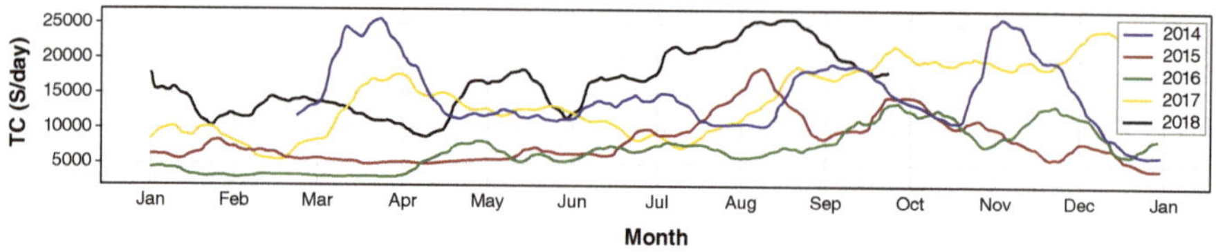

FIGURE 2.3A Yearly variation of the average of TC: 5-day average to present weekly TC fluctuations.

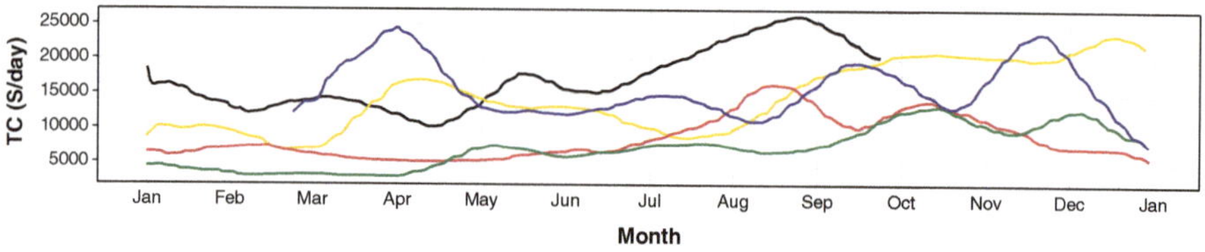

FIGURE 2.3B Yearly variation of the average of TC: 20-day average to present monthly TC fluctuations.

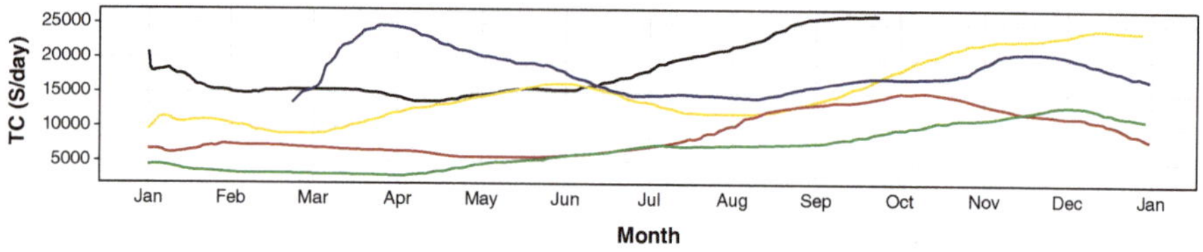

FIGURE 2.3C Yearly variation of the average of TC: 60-day average to present quarterly TC fluctuations.

in Figure 2.3C, supports our hypothesis that the TC value is higher from August to November, on average, each year. There is no significant pattern in the 5-day and 20-day average of TC.

The variation of the average of TC in each year is represented in Figure 2.4. The figure illustrates that the TC average value is increasing from 2016, progressively and the amount of increase from 2016 to 2017 is higher than that of 2017 to 2018. However, there must be a clear trend in the average TC per year.

2.9.1.7 Analysis of Parameters

This section examined the correlation matrix and correspondence of TC and other available parameters.

The studied parameters include:

- **TC:** TC price, which is extracted from the Dashboard dataset for fixture date
- **Fix Year:** Year of the fixture date
- **Fix Month:** Month of the fixture date

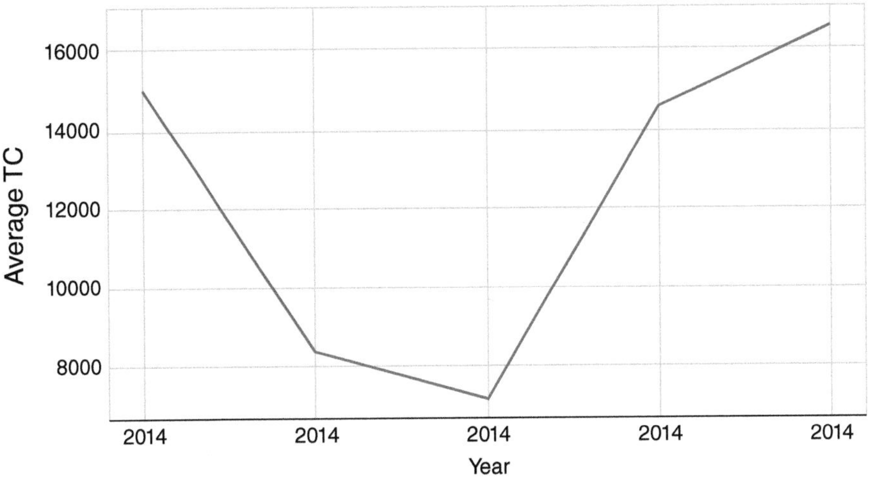

FIGURE 2.4 Variation of average TC in each year.

- **Fix Week:** Number of the week of the fixture date
- **Congestion in the port:** The value of congestion in Tubarao Port.
- **Promptness:** The difference between LAY and fixture date. LAY-CAN is the window through which the vessel can arrive. The contract will be canceled if the vessel arrives after that period (Latest Cancelling). In other words, LAY-CAN is the vessel's arrival range.
- **DWT:** Size of the vessel
- **Fixed Stem**
- **Fixed Vessels, Daily:** The number of fixed vessels in each fixture date, according to the Fixture Historical dataset.
- **Fixed Vessels, Weekly:** The number of fixed vessels in each week, according to the Fixture Historical dataset.
- **Fixed Vessels, Monthly:** The number of fixed vessels in each month, according to the Fixture Historical dataset.

The number of fixed vessels (daily, weekly, and monthly) is extracted from the Fixture Historical dataset according to the fixture dates.

Correlation matrix of the parameters is represented in Figure 2.5. It is essential to mention that correlation values span between −1 and +1. If the correlation value between the two parameters is close to 0 (yellow), there is no relationship between the two parameters. Otherwise, if the correlation value between the two parameters is close to +1 (dark green) or −1 (dark red), the two parameters are linearly correlated. A positive correlation value means that an increase in one parameter leads to a rise in the other. A negative correlation value means that an increase in one parameter leads to a decrease in the other.

Correlation values are coded by color in the correlation matrix. The correlation value of each color is shown on the color bar on the right side of the correlation matrix (see Figure 2.5).

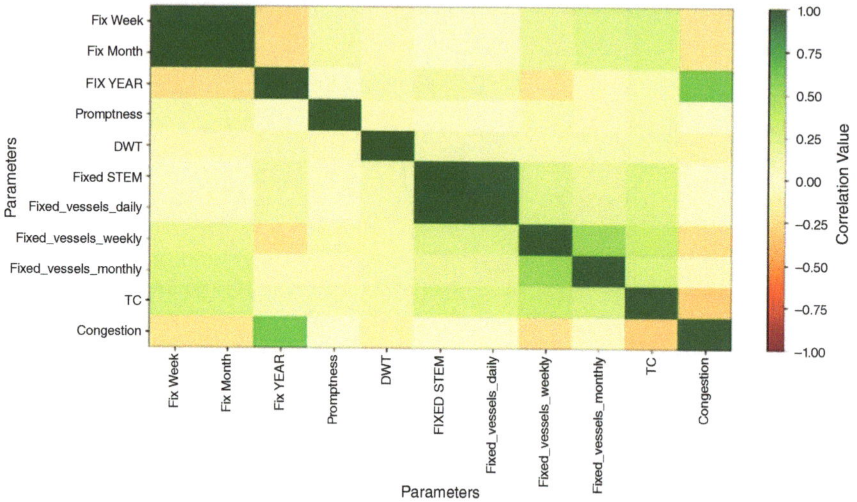

FIGURE 2.5 Correlation matrix of the parameters.

Figure 2.5 illustrates that there is a correlation between TC and the following parameters:

- Fix Week and Fix Month (correlation is about +0.50)
- Fixed Stem (correlation is about +0.50)
- Fixed Vessels: daily, weekly, and monthly. TC is more correlated with Fixed Vessels weekly (correlation is about +0.75)
- Congestion. There is a negative correlation between TC and congestion (correlation is about −0.5)

The relationship between the abovementioned parameters and TC will be explained as follows.

Also, there is a relatively high correlation between port congestion and the fixed year (about +0.75). So, the relationship between these parameters will also be investigated.

Note: The high correlation between fixed week and fixed month (correlation value is +1) is reasonable, as these parameters are linearly correlated. Also, the high correlation value between weekly and monthly fixed vessels (about +0.75) shows that these two parameters are very similar.

Figure 2.6 individually demonstrates the relationship between TC and Fixture Month in a Boxplot view for each year (2014–2018).

Figure 2.6 shows that the highest TC in each year happens from August to October. Also, the highest fluctuation in TC was in October 2014. Also, the fluctuation of TC in July, August, and September of 2014, 2015, and 2017 are significant.

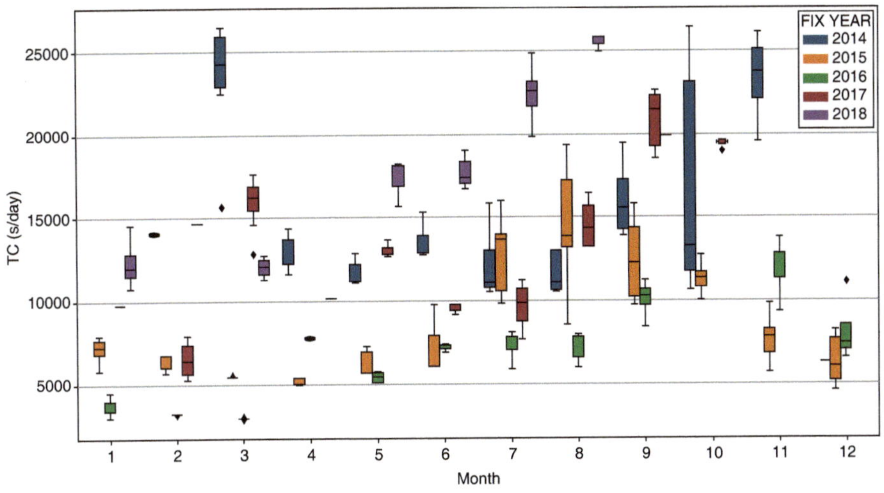

FIGURE 2.6 Variation of TC versus fixture month (2014–2018).

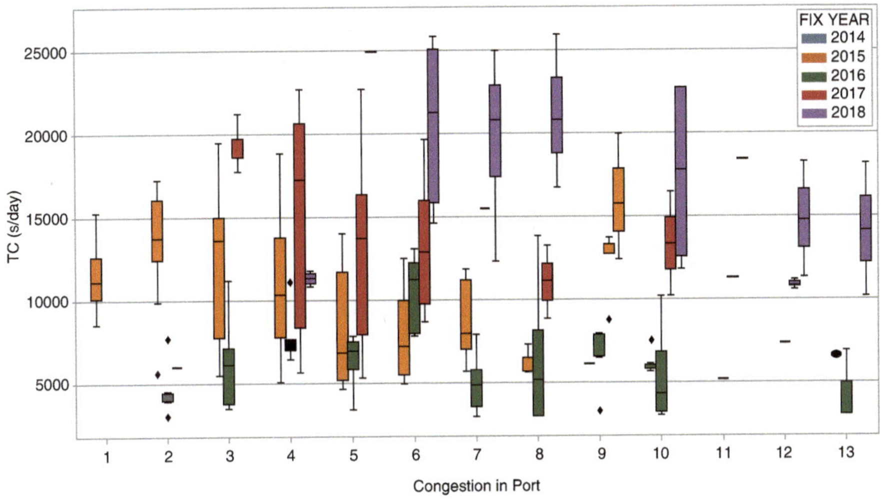

FIGURE 2.7 Variation of TC versus congestion in Tubarao Port (2015–2018).

Note: The TC versus fixture week variation has not been plotted because its behavior is similar to that of Figure 2.6.

TC versus congestion in Tubarao Port is illustrated in Figure 2.7. Nonlinear regression models cannot link TC and the abovementioned parameters. The correlation matrix highlighted this relationship. So, other models should be tested to find a correlation.

Note: The congestion data is not available for 2014.

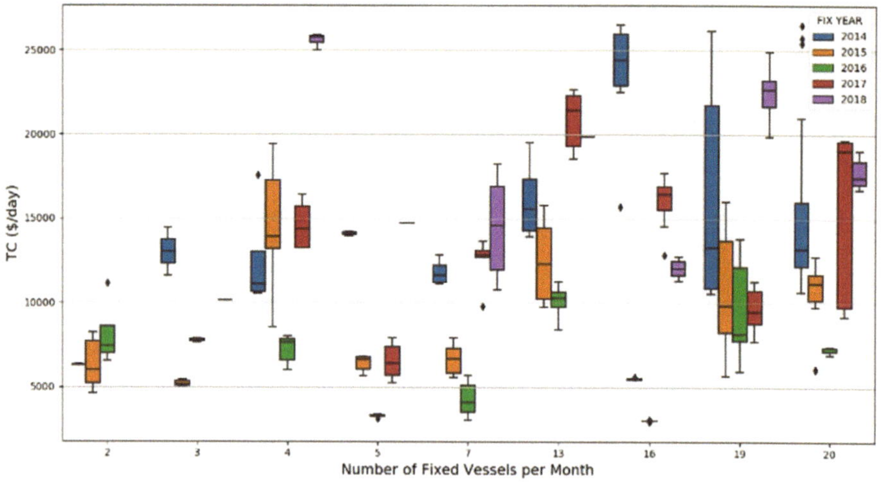

FIGURE 2.8 Boxplot view of TC versus number of fixed vessels per month (2014–2018).

Figures 2.8 to 2.13 show the monthly, weekly, and daily relationship between TC and Fixed Vessels. Nonlinear regression models can show a clear relationship between TC and the abovementioned parameters.

Figure 2.14 presents a TC versus Fixed Stem plot. Nonlinear regression models cannot correlate TC and Fixed Stem. The correlation matrix highlighted this relationship, so other models should be tested to find the related correlation.

2.9.1.8 Data Analysis

The target parameter in the forecasting model is TC, and the related datasets for this parameter have been available since January 2014. However, some other parameters' values have not been valid since that date. The TC value and related dates are parameters that must be considered in the model inherently. The number of months and weeks for every date will be used to develop the model. Therefore, Month, Week, and TC are extracted from the TC dataset.

Other parameters extracted from the Fixture historical dataset (DWT, Promptness, Fixed vessels per day, Fixed vessels per week, and Fixed vessels per month) will be considered when TC values exist.

Month and Week will be used as two parameters to simulate the effect of date in the model. Therefore, the Fixture date and CP data are omitted.

Estimate Intake is another parameter weakly correlated to TC and will not be considered for more investigation.

This report does not consider route parameters and vessel type because the Tubarao/Qin route and Capsize vessel were selected for investigation in this project phase.

The LAY-CAN parameter combines two dates that define the vessel's availability period at Tubarao Port. Therefore, these parameters are not considered in the model.

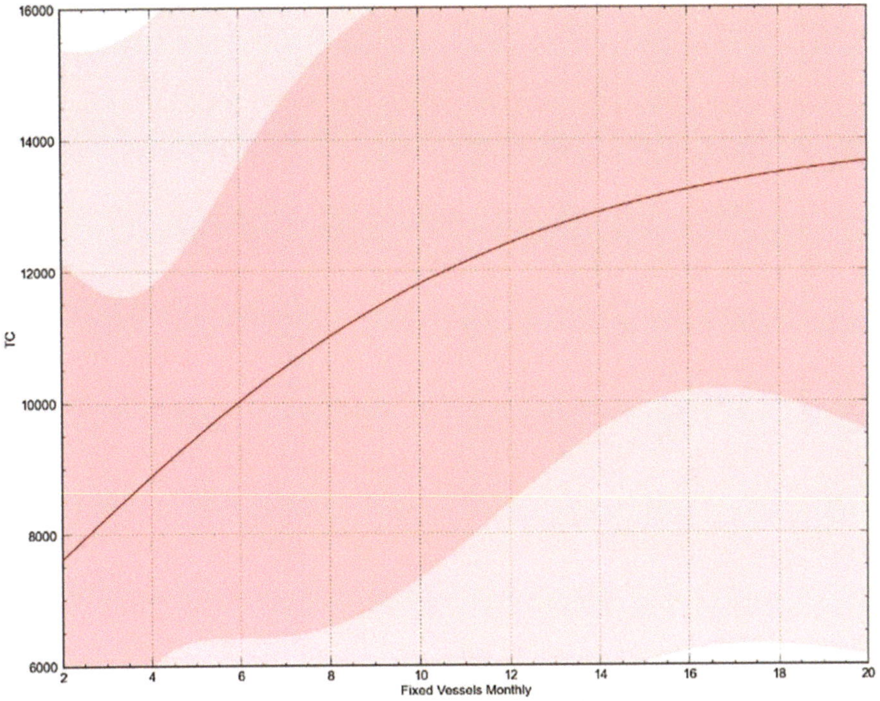

FIGURE 2.9 Identified correlation between TC and the number of fixed vessels per month (2014–2018).

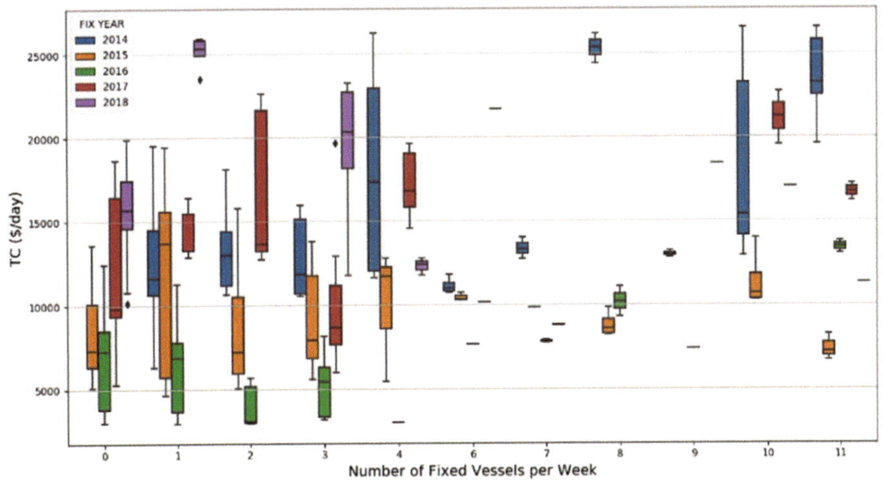

FIGURE 2.10 Boxplot view of TC versus number of fixed vessels per week (2014–2018).

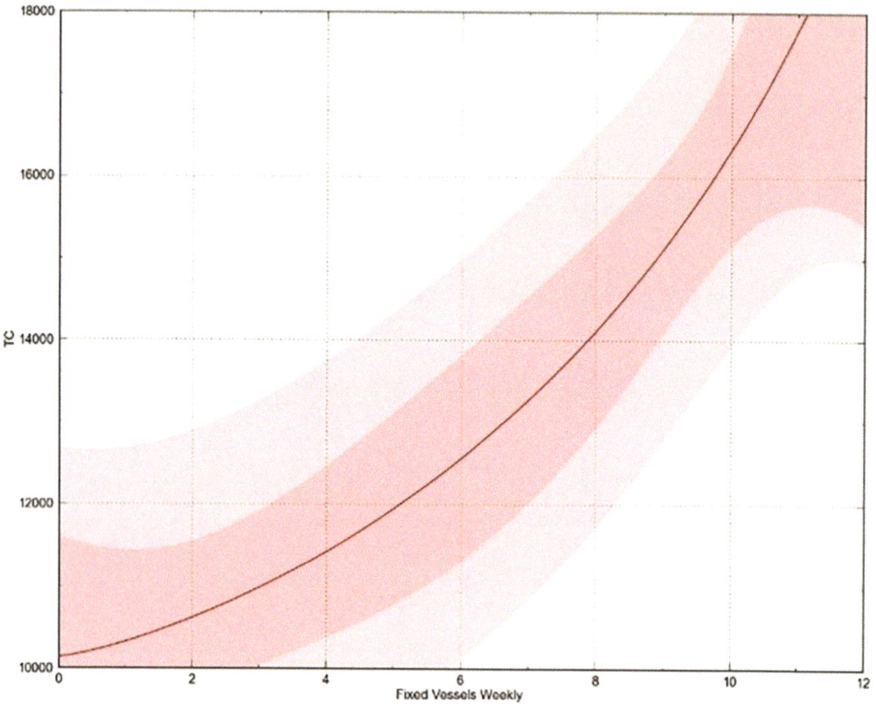

FIGURE 2.11 Identified correlation between TC and the number of fixed vessels per week (2014–2018).

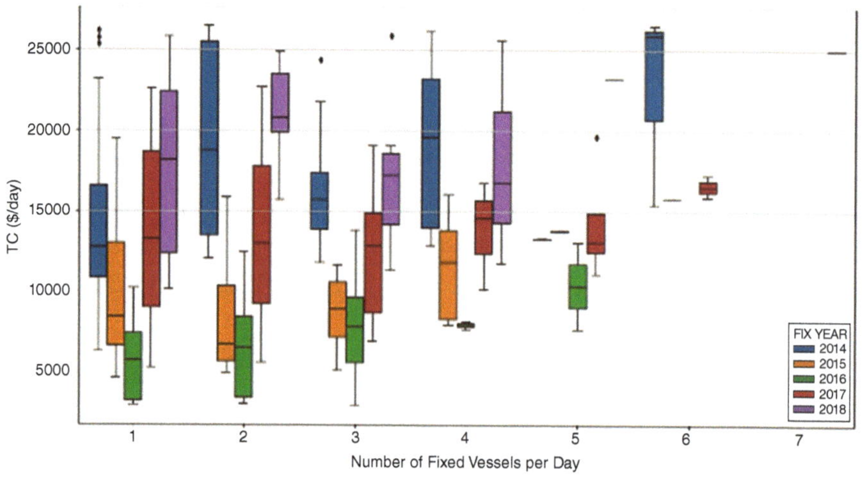

FIGURE 2.12 Boxplot view of TC versus number of fixed vessels per day (2014–2018).

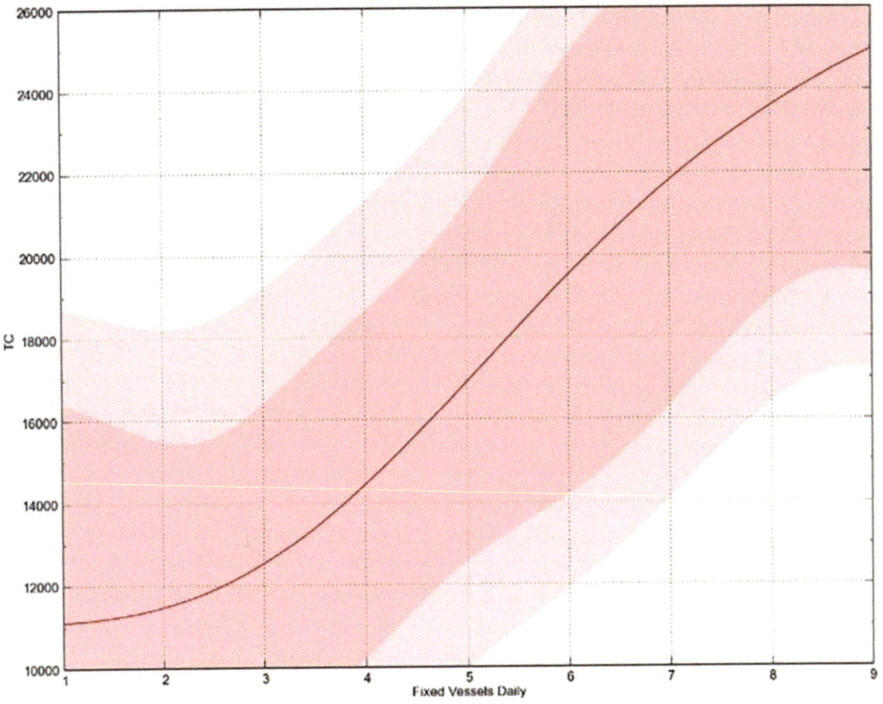

FIGURE 2.13 Identified correlation between TC and the number of fixed vessels per day (2014–2018).

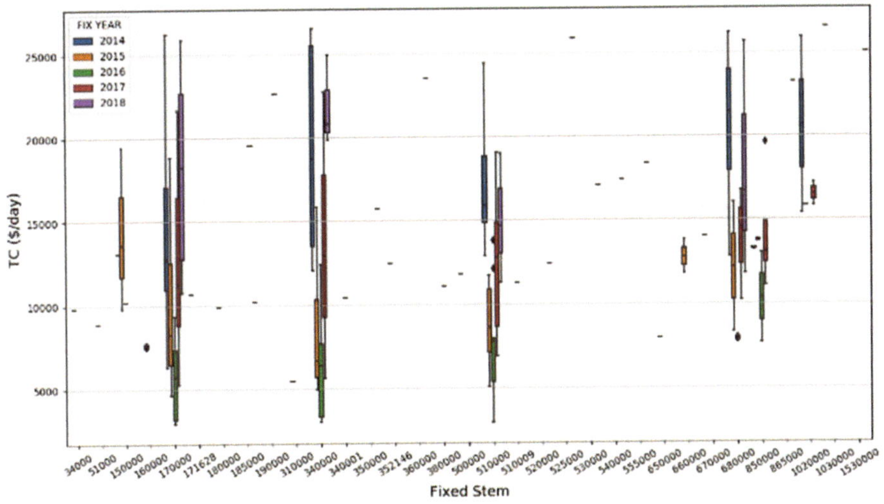

FIGURE 2.14 Boxplot view of TC versus Fixed Stem (2014–2018).

However, the vessel's availability period at Tubarao Port is slightly correlated with TC; more analysis is required to select it as an input parameter in the forecasting model. A matrix of the following parameters will be used as input parameters to develop the forecasting model:

- Month (*available from Jan 2014*)
- Week (*available from Jan 2014*)
- DWT (*available from Feb 2014*)
- Fix stem (*available from Feb 2014*)
- Promptness (*available from Jan 2014*)
- Fixed vessels per day (*available from Feb 2014*)
- Fixed vessels per week (*available from Feb 2014*)
- Fixed vessels per month (*available from Feb 2014*)
- Congestion in Tubarao Port (*available from Jan 2015*)
- Gap (*available from Jan 2017*)
- TC (*available from Jan 2014*)

There are 1196 available values for TC. To illustrate the data situation and parameter sufficiency, the number of empty values for each parameter has been calculated (Table 2.3).

Note: All data for Month, Week, and TC are available.

Regarding the number of missing data for each parameter, most parameters are useless to be utilized as inputs to the predictive model. There are a lot of missing values (75% for DWT, Fix Stem, Promptness, and Fixed vessels per day, and 63% for Gap) that make it impossible to use them because enough data for each parameter is needed to train the predictive model sufficiently. Other parameters, including TC, Month, Week, Fixed vessels per month, and Congestion, are valid data for input parameters to develop a predictive model.

Regarding previous correlation analysis, Congestion in Tubarao Port and Gap are critical parameters strongly correlated to TC.

TABLE 2.3
Input Parameters Availability

Parameter	Empty Values	Data Availability
DWT	903	25%
Fix Stem	903	25%
Promptness	903	25%
Fixed vessels per day	903	25%
Fixed vessels per week	431	64%
Fixed vessels per month	84	93%
Congestion	225	81%
Gap	757	37%

The forecasting model cannot utilize missing data, and a solution must be found to preprocess the data. To manage the empty values, there are three different approaches:

1. Calculate the average of the available values of the parameter and set all missing values to the average.
2. Remove the row of data, wherever there is missing data in a row.
3. Set the missing value to a defined value far away from that parameter's actual values.

The first approach may bias the forecasting model, as too much data is missing (especially DWT, fixed stem, promptness, fixed vessels per day, and gap). Moreover, the available data percentage needs to be increased to calculate the average value. In other words, it is not logical to calculate the average of a parameter when only 25% of its values exist and 75% are missed.

The second approach could not be utilized because of a large amount of missing data. Only 1196 rows of data are available for training, testing, and validation. Therefore, removing about 75% is not rational because much data is needed to utilize advanced machine-learning algorithms for forecasting.

The third approach was applied based on the abovementioned details, and a significant negative value was defined to replace the missing data. If a parameter's value is empty, a significant negative value (-99,999) is replaced because models cannot utilize empty values.

After that, a regression model was used to develop a forecasting model, which examined the effect of missing data and selected suitable input parameters.

Regression is the simplest model with the lowest cost. It is an essential and commonly used type of predictive analysis. Regression attempts to model the relationship between two variables by fitting a simple equation to observed data. One variable is considered an explanatory variable, and the other is a dependent variable.

The overall idea of regression is to examine two things:

1. Does a set of predictor variables make an excellent prediction of an outcome (dependent) variable?
2. Which variables, in particular, are significant predictors of the outcome variable?

These regression estimates explain the relationship between one dependent variable and one or more independent variables.

A regression model predicts the TC value five days later. In the first step, the parameters with no missing data or a few missing values are used as inputs to the model. These parameters include:

- Month
- Week
- Fixed Vessels per month
- Congestion
- TC

The data is split into train data (80%) and test data (20%) to utilize a framework to train and test the forecasting model. Some data is also needed to evaluate the model. This data should not be presented to the model during training or testing.

For the 5-day forecasting model, the number of rows of train data was 945, and the number of rows of test data was 237.

The coefficient of determination (R^2) has been used to evaluate the performance of the forecasting model.

The R^2 or score is defined as:

$$R^2 = 1 - \frac{\Sigma(y - \hat{y})^2}{\Sigma(y - \bar{y})^2} \tag{2.1}$$

where y is the actual value of TC on a specified date, $\hat{y}$ is the forecasted value of TC for that date, and $\bar{y}$ is the average value of TC values. The best possible score is 1.00, which can be negative (because the model can be arbitrarily worse). A constant model that always predicts the expected value of y, disregarding the input features, would get an R^2 score of 0.00.

After five runs of the model, the average value of the prediction's coefficient of determination (R^2) was 0.85.

In the second step, the model considered other parameters as inputs to the previous inputs, one by one. In other words, one of the different parameters (DWT, Fixed Stem, Promptness, Fixed vessels per day, Fixed vessels per week, and Gap) was added to the model as input. The average value of the forecasting model's R^2 with a new set of inputs, including Gap, increased to 0.87, while other parameters did not affect R^2 and did not change.

Therefore, as another input to the forecasting model, the gap along with the months, weeks, fixed vessels per month, congestion, and TC will be considered. However, 63% of the missing data is regarding the gap.

Furthermore, a different subset of DWT, Fixed Stem, Promptness, Fixed vessels per day, and Fixed vessels per week were fed to the model in combination with Month, Week, Fixed Vessels per month, Congestion, Gap, and TC. However, it is approved that the parameters with too much missing data cannot affect the forecasting model's performance. The forecasting model's R^2 did not change, and it was about 0.85.

In the third step, a similar approach to training the model was utilized to forecast the TC values for the next 20 days (four weeks). Month, Week, Fixed Vessels per month, Congestion, Gap, and TC are fed to the model as input. The average coefficient of determination (R^2) of the model is 0.50, which is much lower than that of the 5-day forecasting model.

The effect of the Gap parameter on the 20-day forecasting model was examined. By omitting this parameter and using Month, Week, Fixed Vessels per month, Congestion, and TC as the input parameters to the forecasting model, the R^2 decreased to 0.47. Hence, it is again approved that Gap is a significant parameter with a considerable effect on TC.

A different combination of parameters mentioned above with DWT was examined: Fixed Stem, Promptness, Fixed vessels per day, and Fixed vessels per week as input to the model. The results revealed that those parameters with too much missing data could not affect the performance of the 20-day forecasting model. The forecasting model's R^2 did not change and is about 0.47 for all subsets of input parameters.

Meanwhile, the simulations illustrated that the forecasting model needs to react to the parameters, and there is a lot of missing data.

Some parameters are extracted from the Fixture Historical dataset and have no value on many days because there is no fixed vessel. So, another approach was utilized. In the new approach, if the value of the following parameters at a date is missing, the 0.00 value will be replaced.

- Fix Stem
- Fixed vessels per day
- Fixed vessels per week
- Fixed vessels per month

The same approach was applied to train the 5-day forecasting model with a different subset of the parameters. But the outcome did not improve. The following parameters are used as input parameters:

- Month
- Week
- Fixed vessels per month
- Congestion
- Gap
- TC

The prediction's average coefficient of determination (R^2) is 0.88, a bit higher than the previous approach of conditioning the missing data (setting missed data to -99,999). Therefore, the missing data value in the Fixed vessels per month parameter was set to 0.00, while the other parameters were set to -99,999.

2.9.1.9 ML Algorithms

The performance of several ML algorithms has been investigated in this project to forecast the TC value.

2.9.1.9.1 Polynomial Regression

One typical pattern in ML is using linear models trained on nonlinear functions of the data. This approach maintains the generally fast performance of linear methods while allowing them to fit a more comprehensive range of data.

2.9.1.9.2 Support Vector Machine (SVM)

SVMs are supervised learning methods used for classification, regression, and detection of outliers. The model produced by support vector regression depends only on a subset of the training data because the cost function for building the model ignores any training data close to the model prediction. SVM with radial basis function kernel is used to develop the predictive model.

2.9.1.9.3 k-Nearest Neighbors (KNN)

The principle behind nearest neighbor methods is to find a predefined number of training samples closest to the new point and predict the label. The number of samples can be a user-defined constant (KNN learning) or vary based on the local density of points (radius-based neighbor learning). The distance can be any metric measure: standard Euclidean distance is the most common choice.

Neighbors-based regression can be used in cases where the data labels are continuous rather than discrete variables. The label assigned to a query point is computed based on the mean of its nearest neighbors' labels.

2.9.1.9.4 Decision Tree

Decision Trees are a nonparametric supervised learning method for classification and regression. The goal is to create a model that predicts the value of a target variable by learning simple decision rules inferred from the data features.

2.9.1.9.5 Bagging

Bagging methods form a class of algorithms that build several instances of a black-box estimator on random subsets of the original training set and then aggregate their predictions to create a final prediction. These methods reduce the variance of a base estimator (e.g., a decision tree) by introducing randomization into its construction procedure and then making an ensemble out of it. In many cases, bagging methods constitute a straightforward way to improve a single model without adapting the underlying base algorithm.

2.9.1.9.6 Random Forest

In Random Forests, each tree in the ensemble is built from a sample drawn with a replacement from the training set. Also, when splitting a node during the tree's construction, the chosen split is no longer the best among all features. Instead, the selected split is the best split among a random subset of the features. As a result of this randomness, the bias of the forest usually slightly increases (concerning the bias of a single non-random tree). Still, due to averaging, its variance decreases, usually more than compensating for the bias increase, yielding a better model overall.

2.9.1.9.7 Extremely Randomized Trees

Randomness goes one step further in how splits are computed in highly randomized trees. As in random forests, a random subset of candidate features is used; but instead of looking for the most discriminative thresholds, thresholds are drawn at random for each candidate feature, and the best of these randomly generated thresholds is picked as the splitting rule.

2.9.1.9.8 AdaBoost

The core principle of AdaBoost is to fit a sequence of weak learners (i.e., models that are only slightly better than random guessings, such as small decision trees) on repeatedly modified versions of the data. All predictions are combined through a weighted majority vote (or sum) to produce the final prediction. The data modifications at each boosting iteration involve applying weights w_1, w_2, . . ., and w_N to each training sample. Initially, those weights are all set to $w_i = 1/N$ so that the first step trains merely a weak learner on the original data. The sample weights are individually modified for each successive iteration, and the learning algorithm is reapplied to the reweighted data. Those training examples incorrectly predicted by the boosted model induced at the previous step increase their weights at a given step. In contrast, the weights are decreased for those that were predicted correctly. As iterations proceed, examples that are difficult to predict receive ever-increasing influence. Each subsequent weak learner is forced to concentrate on the examples missed by the previous ones in the sequence.

To sum up, an AdaBoost regressor is a meta-estimator that begins by fitting a regressor on the original dataset and then fits additional copies of the regressor on the same dataset but where the weights of instances are adjusted according to the error of the current prediction. As such, the following regressors focus more on complex cases.

2.9.1.9.9 *Gradient Tree Boosting*

Gradient Tree Boosting or Gradient Boosted Regression Trees (GBRT) is a generalization of boosting to arbitrary differentiable loss functions. GBRT is an accurate and effective off-the-shelf procedure that can be used for both regression and classification problems.

2.9.1.9.10 *Artificial Neural Network (ANN)*

ANNs, also known as neural networks (NNs), simulated neural networks, or "parallel distributed processing," represent methods the brain uses for learning. ANNs are a series of mathematical models that imitate a few known characteristics of natural nerve systems and sketch the analogies of adaptive natural learning. The critical component of a particular ANN paradigm could be the unusual structure of the data processing system. ANNs are utilized in various computer applications to solve complex problems. They are fault-tolerant and straightforward models that do not require information to identify the related factors and do not require a mathematical description of the phenomena involved in the process.

The central part of a neural network structure is a "node." Biological nodes generally sum the signals received from numerous sources differently and then perform a nonlinear action on the results to create the outputs. Neural networks typically have an input layer, one or more hidden layers, and an output layer. Each input is multiplied by its connected weight, and in the simplest state, these quantities and biases are combined; they then pass through the activation functions to create the output (see Equations 2, 3, 4). Figure 2.15 shows the data treatment in a node (it should be noted that the hidden layer nodes may use any differentiable activation function to generate their output).

$$E_k = \sum_{j=1}^{q} (w_{i,j,k} x_j + b_{i,k}) \quad k = 1, 2, ..., m \qquad (2.2)$$

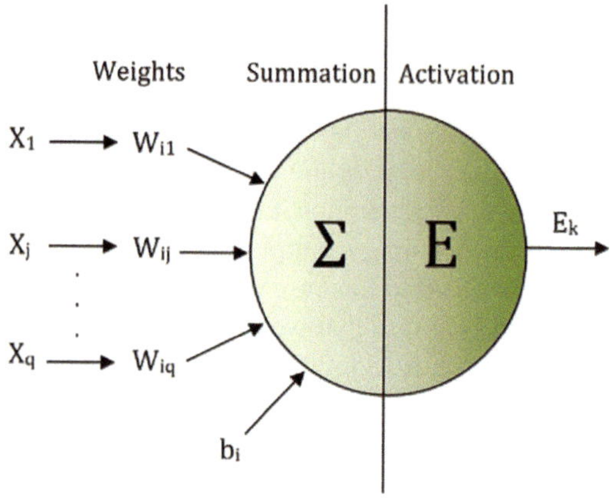

FIGURE 2.15 Data processing (treatment) in a neural network cell (node).

Here x is the normalized variable, w is the weight of that variable, i is the input, b is the bias, q is the number of input variables, and k and m are the counter and number of neural network nodes, respectively, in the hidden layer.

In general, the activation functions consist of both linear and nonlinear equations. The coefficients associated with the hidden layer are grouped into matrices $W_{i,j,k}$ and $b_{i,k}$. Equation 3 can be used as the activation function between the hidden and the output layers (in this equation, f is the transfer function).

$$F_k = f(E_k) \tag{2.3}$$

The output layer computes the weighted sum of the signals provided by the hidden layer. The associated coefficients are grouped into matrices $W_{o,k}$ and b_o. Using the matrix notation, the network output can be given by Equation 4.

$$Out = \left(\sum_{k=1}^{m} w_{o,k} F_k\right) + b_o \tag{2.4}$$

Network training is the most important part of neural network modeling and is carried out using controllable and uncontrollable training. The most common training algorithm is backpropagation. A training algorithm is a procedure that consists of adjusting a network's coefficients (weights and biases) to minimize the error function between the estimated network outputs and the actual outputs.

2.9.1.9.11 Stacking Regression

Stacking Regression combines the best forecasting models among the algorithms mentioned above. Stacking regression is an ensemble learning technique to combine multiple regression models via a meta-regressor. The individual regression models are trained based on the complete training set; then, the meta-regressor is fitted based

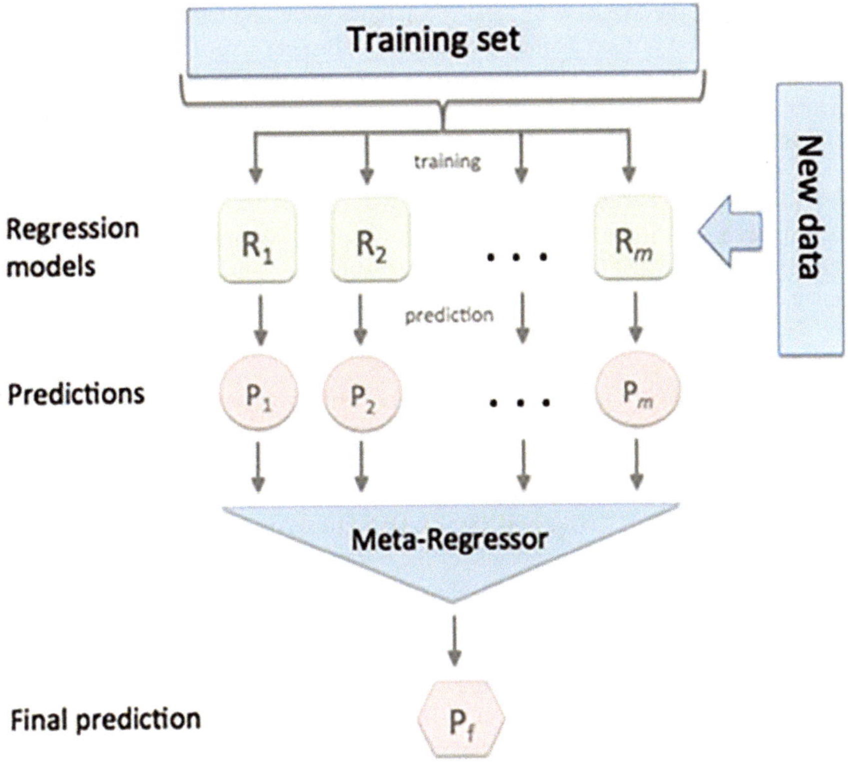

FIGURE 2.16 A schematic presentation of Stacking Regression.

on the outputs—meta-features—of the individual regression models in the ensemble. Figure 2.16 illustrates Stacking Regression schematically.

MSE, RMSE, R^2, and MAE are the statistical criteria utilized to evaluate the accuracy of the predictive model results according to the following equations:

$$MSE = \frac{1}{p}\sum_{r=1}^{p}(y_r - z_r)^2 \tag{2.5}$$

$$RMSE = (\frac{1}{p}\sum_{r=1}^{p}(y_r - z_r)^2)^{\frac{1}{2}} \tag{2.6}$$

$$R^2 = 1 - \frac{\sum_{r=1}^{p}(y_r - z_r)^2}{\sum_{r=1}^{p}(y_r - \overline{y})^2} \tag{2.7}$$

$$MAE = \frac{\sum_{i=1}^{n}(y_i - x_i)}{n} \tag{2.8}$$

In this project, the MAE, MSE, and R^2 methods were applied to examine the error and performance of forecasting models. Mean absolute error (MAE) is the average of the absolute difference between real TC and forecasted TC ($/day); mean squared error (MSE) is the average of the squared difference between real TC and forecasted TC; and R^2 is the coefficient of determination or score.

2.9.1.10 5-Day Forecasting Model

Data is split into training and testing data to train the predictive model. 80% of the data has been used for training, and the rest has been used for testing the model. Some data have also been used as evaluation data. The length of this data was twice the forecasting period. So, this data was seen by the model during training and testing.

The ML algorithms above have created predictive models to forecast the TC in the next five days. The accuracy of the developed forecasting models has been evaluated through MAE, MSE, and R^2. Table 2.4 represents the accuracy values for 5-day forecasting models. The values are an average of five times the model's run.

According to Table 2.4, results show that AdaBoost and Extremely Randomized Trees are the best predictive models for the 5-day forecasting of TC. These models' R^2 is about 0.96, and their mean absolute error is 647 $/day and 701 $/day, respectively. The prediction results of these models for the validation data and absolute TC values are illustrated in Figures 2.17 and 2.18, respectively.

According to Figure 2.17, the forecasted values of TC for the next five days are very close to the fundamental values of TC using the AdaBoost algorithm. Particularly, the forecasted TC graph is adjacent to the accurate TC graph in the first three days.

Two critical parameters should be set on AdaBoost. The Decision Tree is the base estimator from which the boosted ensemble is built. The maximum number of estimators at which boosting is terminated is 500.

The Extremely Randomized Tree algorithm is due to a closed graph of forecasted TC values to the actual TC graph, as seen in Figure 2.18. The maximum difference

TABLE 2.4
Forecasting Models' Performance (5-Day Forecasting)

Model	R^2	MSE	MAE ($/day)
Polynomial Regression	0.9386	2,043,600	1,051.25
SVM	0.9319	2,344,646	983.32
KNN	0.9373	2,224,373	939.57
Decision Tree	0.9312	2,363,504	957.23
Bagging	0.9326	2,188,747	861.44
Random Forest	0.9443	1,753,335	803.61
Extremely Randomized Trees	0.9597	1,294,027	701.37
AdaBoost	0.9628	1,195,681	647.00
Gradient Tree Boosting	0.9185	2,519,931	1,141.01
ANN	0.9103	2,704,312	1,306.05

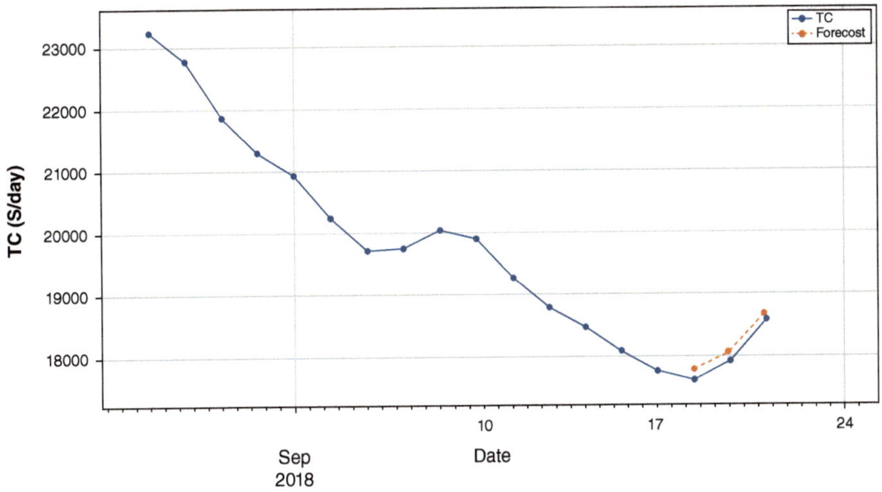

FIGURE 2.17 Variation of TC and 5-day forecasting model using AdaBoost.

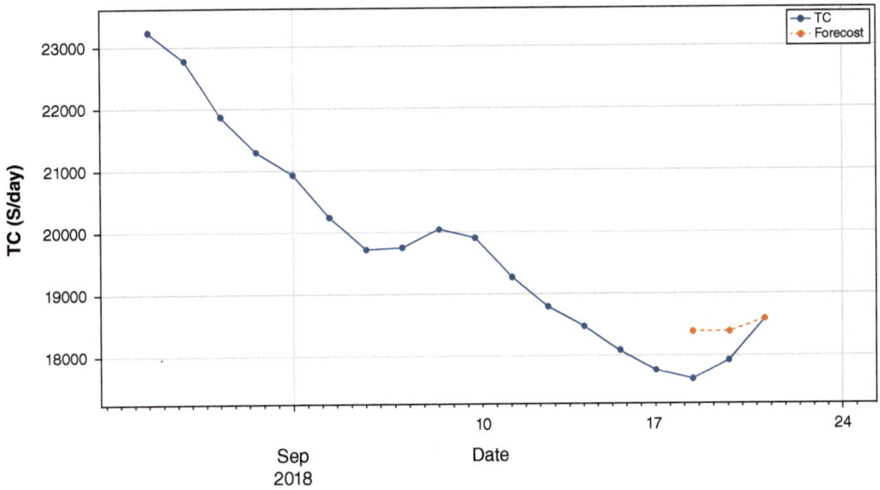

FIGURE 2.18 Variation of TC and 5-day forecasting model using Extremely Randomized Tree.

between forecasted TC and real TC is about 900 $/day, which is less than 5% of the actual value of TC on that day. So, it performs very well in short-term forecasting of TC (5 days).

The most significant parameter of this algorithm is the number of trees in the forest set to 500.

The Random Forest model is the third model regarding R^2 and MAE. Its score is 0.94, and its MAE is ~ 800 $/day. Figure 2.19 shows the short-term forecasted values of TC using a Random Forest.

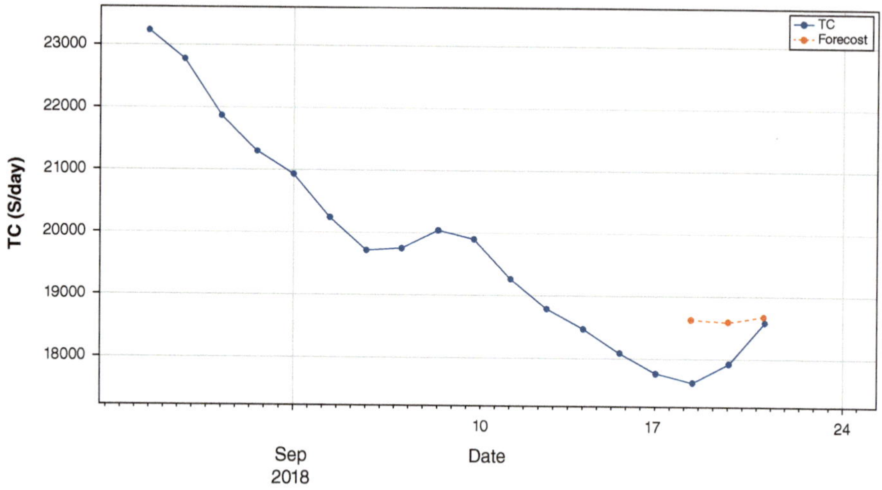

FIGURE 2.19 Variation of TC and 5-day forecasting model using Random Forest.

According to Figure 2.19, AdaBoost is better than Random Forest in the 5-day forecasting of TC. However, the forecasted TC of validation data shows a maximum difference of 1000 $/day compared to the real TC value.

Polynomial Regression, SVM, KNN, Decision Tree, and Bagging algorithms due to R^2 of ~0.93 are used to forecast TC for the next five days. While the mean absolute error of the Polynomial Regression, SVM, KNN, and Decision Tree is ~ 1000 $/day, the MAE of the bagging algorithm is ~ 850 $/day. Again, note that these values result from calculating the average of the criteria five times running the code.

Figure 2.20 represents the variation of TC and 5-day forecasted TC of the validation data. The forecasted TC values for the next three days are close to the actual TC values of those dates. However, real TC and forecasted TC diverge, and the difference between real TC and forecasted TC is about 1,700 $/day on the fifth day. This means the prediction error on the fifth day is ~9%.

The base estimator to fit on random subsets of the dataset is a decision tree in the Bagging algorithm. Also, the number of base estimators in the ensemble is set to 500.

Figure 2.21 shows the variation of TC and 5-day forecasted TC using the Decision Tree algorithm. The Decision Tree algorithm uses simple decision rules inferred from training data. So, the forecasting result could be more visually efficient, as is seen in Figure 2.21.

The graph of forecasted TC is a horizontal line that passes through the accurate TC graph most of the time. The forecasting results of the third and fourth days are superb. However, the algorithm's overall performance could be more satisfying. The maximum forecasting error is less than 10% on the fifth day of the forecasting period.

A graph of TC over the last month and forecasted TC values for the last five days (validation data) using the KNN algorithm is exhibited in Figure 2.22. The forecasted value graph is far from absolute TC values initially and converges to the

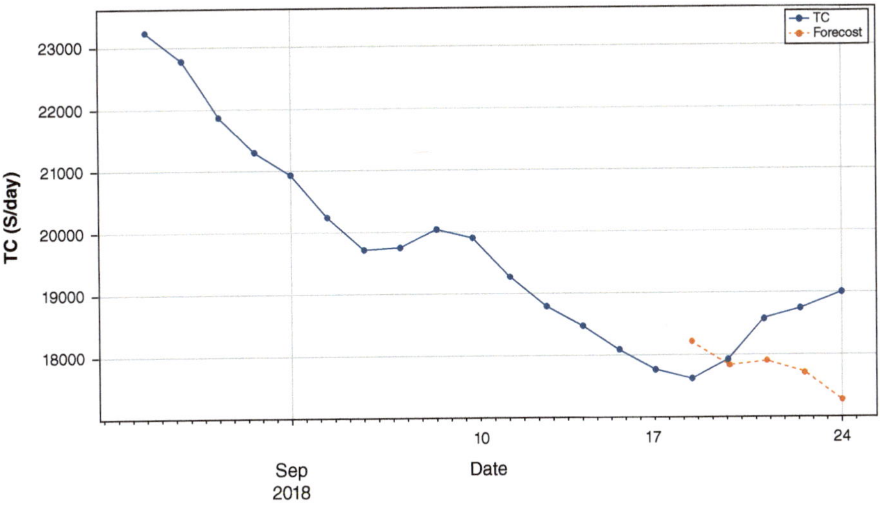

FIGURE 2.20 Variation of TC and 5-day forecasting model using Bagging.

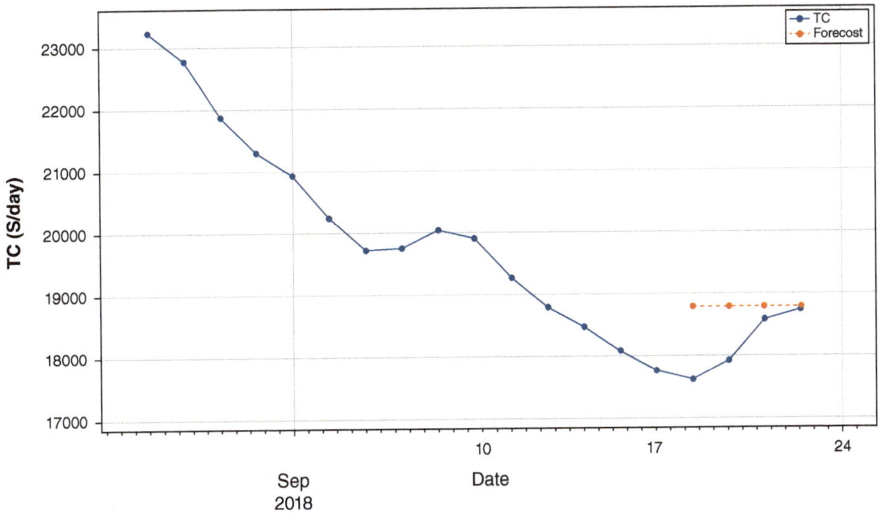

FIGURE 2.21 Variation of TC and 5-day forecasting model using Decision Tree.

actual TC values on the fifth day. The maximum prediction error is more than 20% on September 18, which is unacceptable.

Figure 2.23 illustrates the TC and forecasted TC variation of the validation data using SVM. The graph of forecasted TC is higher than actual TC values, and the maximum error of prediction is about 15%, which belongs to the first day of the prediction period (Sep. 18th).

Figure 2.24 represents the graph of TC last month and the forecasted TC graph utilizing Polynomial Regression. The degree of the polynomial features is set to four.

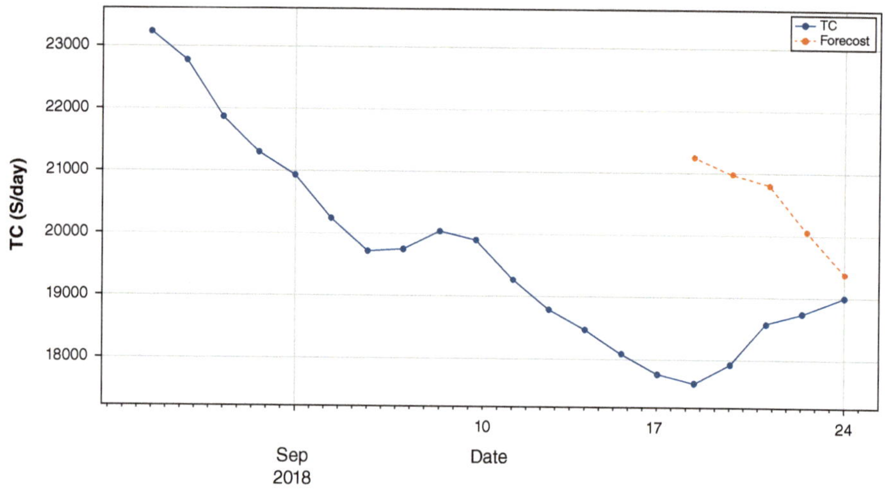

FIGURE 2.22 Variation of TC and 5-day forecasting model using KNN.

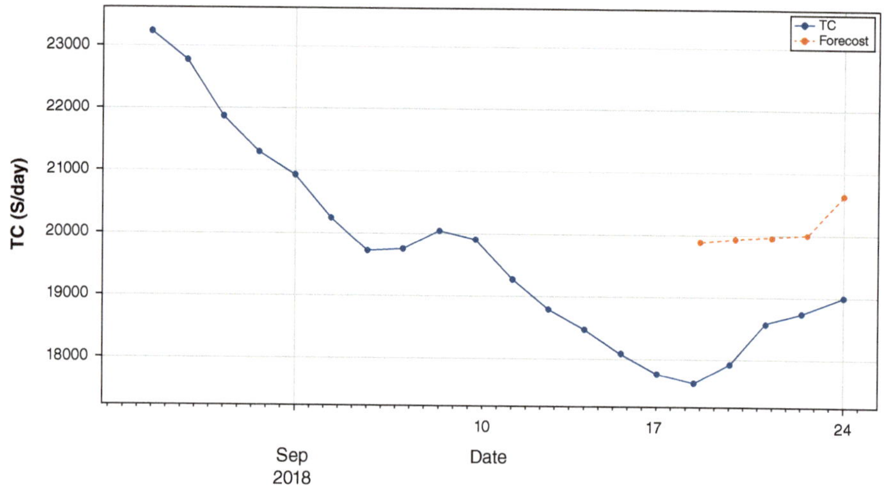

FIGURE 2.23 Variation of TC and 5-day forecasting model using SVM.

The forecasted TC values are very close to the average of actual TC values in the five days. The maximum error of the forecasting is about 3%, which occurred on September 18. The 5-day predicted TC graph is close to the actual TC graph.

Figure 2.25 shows the TC graph and forecasted values of TC using Gradient Tree Boosting. The maximum forecasting error is about 10% on September 24.

Figure 2.26 exhibited the variation of TC and forecasted TC graph of the validation data using ANN. The forecast graph is far from the original TC graph on the first days of the forecasting period and seems to converge to the actual TC graph on the fifth day. However, the maximum error is about 15% invalidation data.

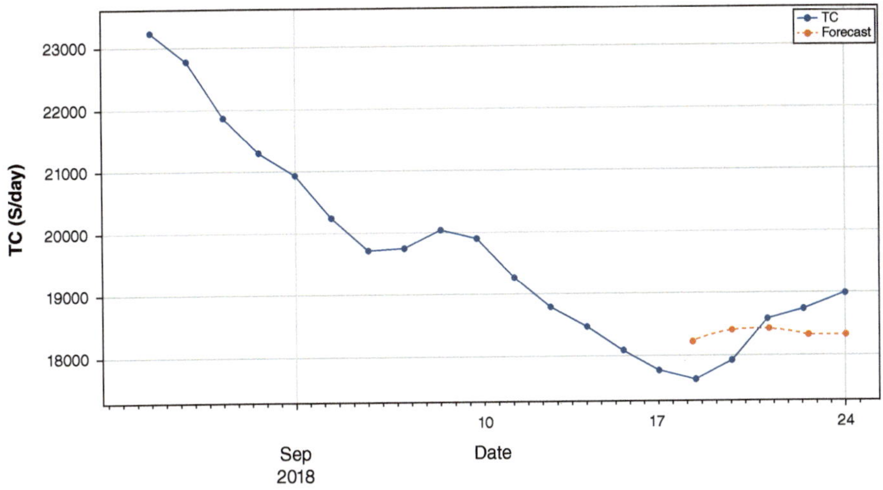

FIGURE 2.24 Variation of TC and 5-day forecasting model using Polynomial Regression.

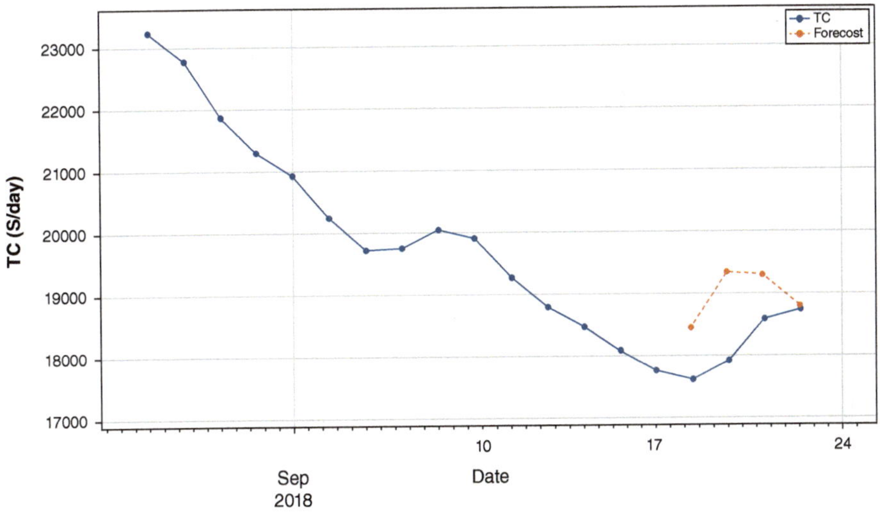

FIGURE 2.25 Variation of TC and 5-day forecasting model using Gradient Tree Boosting.

In the next step, Stacking Regression was utilized to combine the best forecasting algorithms for the 5-day prediction of TC and examine the outcome. According to Table 2.5, AdaBoost and Extremely Randomized Tree performed better than all other algorithms. Using Stacking Regression and combining AdaBoost and Extremely Randomized Trees for forecasting did not significantly improve the model's accuracy. However, it provided more stability and lower variance on the forecasted outcome. The result is presented in Table 2.5.

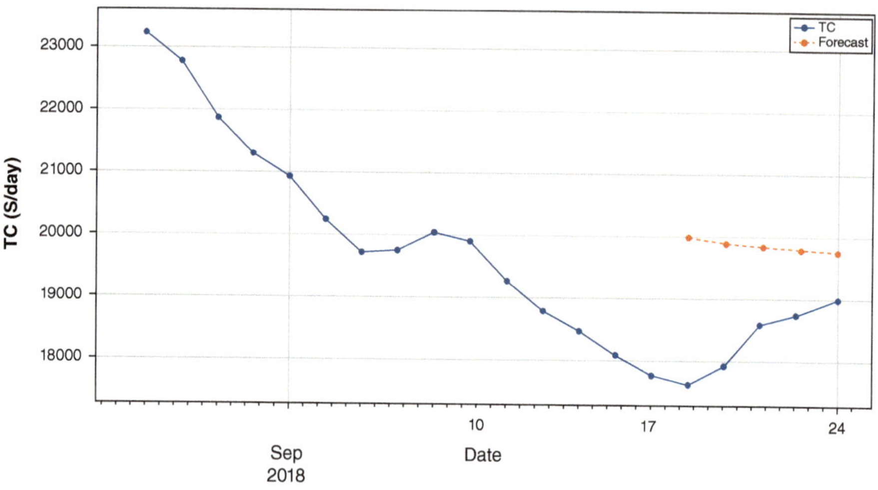

FIGURE 2.26 Variation of TC and 5-day forecasting model using ANN.

TABLE 2.5
Stacking Regression Model Accuracy for 5-Day Forecasting of TC (Combining AdaBoost and Extremely Randomized Trees)

Model	R^2	MSE	MAE ($/day)
Stacking Regression	0.9634	1,146,753	680.62

Figure 2.27 illustrates the variation of TC and the forecasted TC graph by Stacking Regression. The result could be more visually appealing, and the maximum error of the forecasted TC is about 7%.

2.9.1.11 20-Day Forecasting Model

The 20-day forecasting is equivalent to 4-week or 1-month forecasting of TC. The previous ML algorithms have been extended and utilized to forecast the TC in the next month (20 days). The accuracy of the developed forecasting models has been evaluated through MAE, MSE, and R^2. The results of the predictive models of the 20-day forecasting model are shown in Table 2.6. The values are an average of five times the run of the model.

Results show that AdaBoost and Extremely Randomized Trees are the best predictive models for 20-day forecasting of TC, according to Table 2.6. The R^2 in these models is about 0.96, and the mean absolute error is about 600 $/day. These models' prediction results for the validation data and actual TC values are illustrated in Figures 2.28 and 2.29, respectively.

For the AdaBoost model, the base estimator from which the boosted ensemble is built is set to Decision Tree. The maximum number of estimators at which boosting is terminated is also set to 500.

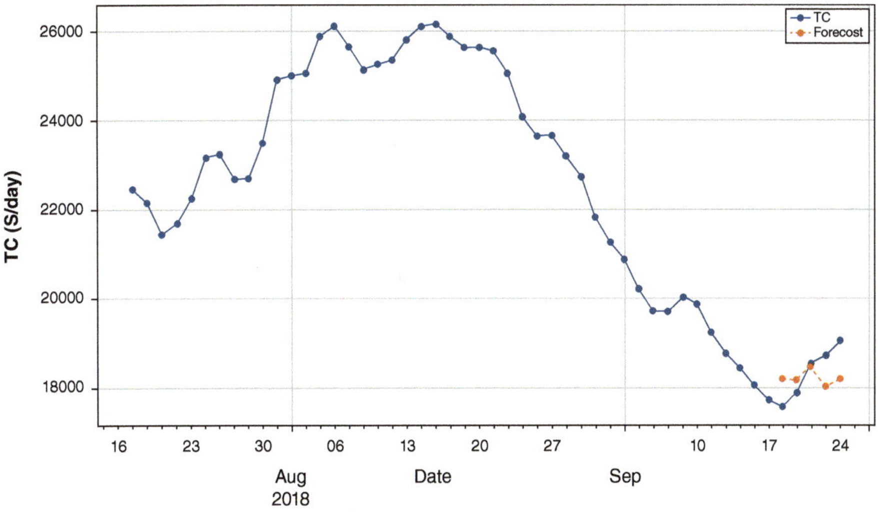

FIGURE 2.27 Variation of TC and 5-day forecasting model using Stacking Regression and combining AdaBoost and Extremely Randomized Tree.

TABLE 2.6
Forecasting Modes' Performance (20-Day Forecasting)

Model	R^2	MSE	MAE ($/day)
Polynomial Regression	0.7959	6,338,758	1,635.24
SVM	0.8685	3,986,626	1,373.38
KNN	0.9022	3,014,842	921.59
Decision Tree	0.9110	2,966,986	889.52
Bagging	0.9537	1,547,303	808.71
Random Forest	0.9570	1,321,459	763.99
Extremely Randomized Trees	0.9612	1,181,025	605.19
AdaBoost	0.9648	1,084,352	594.23
Gradient Tree Boosting	0.9093	2,434,558	1,146.77
ANN	0.8029	5,786,856	1,839.18

According to Figure 2.28, the AdaBoost model accurately predicts the TC in the middle of the 20 days. In this period, the TC is about 20,000 $/day (September 4–10th). However, the model could not accurately forecast the TC on other days. The maximum forecasting error is on September 24, about 20% of the real TC (3,600 $/day).

The Extremely Randomized Tree algorithm and an AdaBoost algorithm forecast the TC from September 4 to 10. These days, the real TC value is about 2,000 $/day. The maximum error in forecasting evaluation data is about 20% of the real TC (3,500 $/day) on September 18.

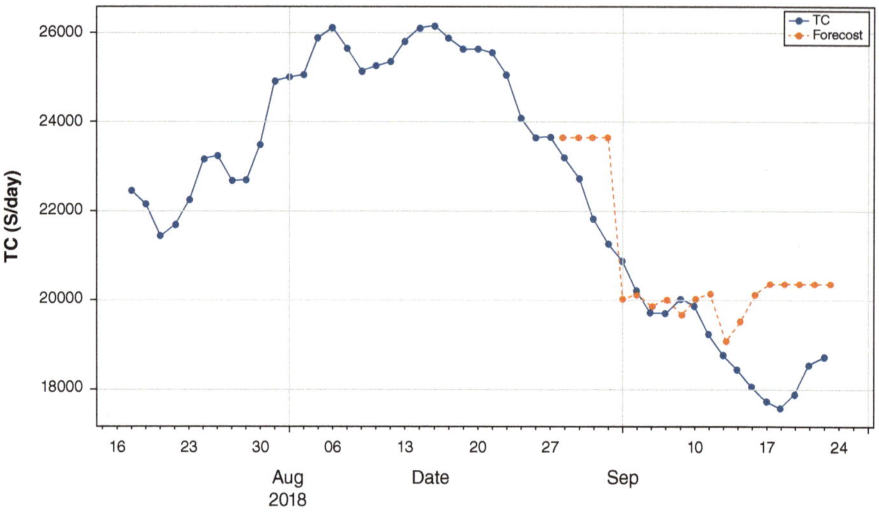

FIGURE 2.28 Variation of TC and 20-day forecasted TC using the AdaBoost algorithm.

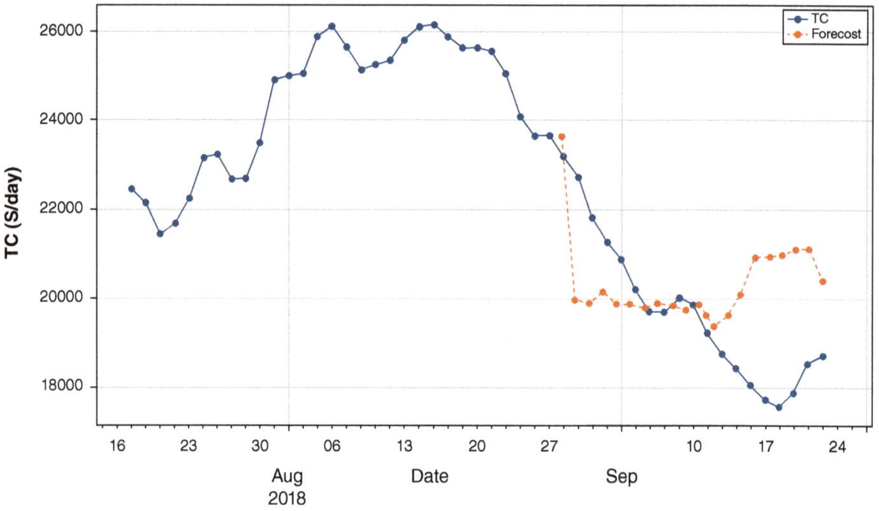

FIGURE 2.29 Variation of TC and 20-day forecasted TC using an Extremely Randomized Tree algorithm.

For this model, the maximum number of estimators at which boosting is terminated is 500.

According to Table 2.5, Bagging and Random Forests models show good forecasting performance regarding R^2 and MAE, and their score is more than 0.95. The mean absolute error is ~ 800 \$/day.

Figure 2.30 shows the graph of TC and 20-day forecasted TC utilizing the Random Forest algorithm. The forecasted TC is very close to real TC in the first six days

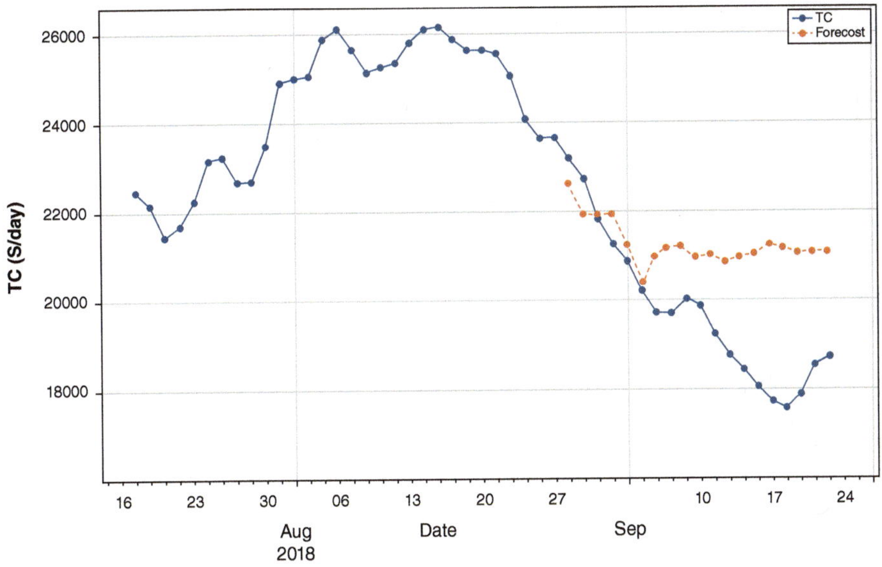

FIGURE 2.30 Variation of TC and 20-day forecasted TC using Random Forest algorithm.

of the 20 days. But the forecasting model's outcome is about 21,000 \$/day for the rest of the 20 days. The maximum forecasting error is about 3,750 \$/day (20%) on September 18.

Variation of TC and forecasted TC value by the Bagging algorithm is presented in Figure 2.31. The graph of the forecasted TC is close to the Real TC graph in the first 12 days of the 20 days of forecasting. The maximum error of the forecasting (in validation data) is about 17% (3250 \$/day) on September 24 (the last day of the 20-day forecasting period).

Gradient Tree Boosting, KNN, and Decision Tree have produced a score of ~0.90. Their mean absolute error is about 1,000 \$/day, back to Table 2.6.

Figure 2.32 presents the real TC and forecasted TC graphs by the Decision Tree algorithm. Although the forecasted TC graph passes the actual TC graph, the error is high. The maximum prediction error is about 4,000 \$/day (22%) on August 29.

The variation of TC and forecasted TC by the KNN algorithm is shown in Figure 2.33. The KNN algorithm outcome is a horizontal line close to the average value of real TC in a 20-day forecasting period. The maximum forecasting error is about 9,000 \$/day (48%) on September 24.

Figure 2.34 demonstrates the TC and forecasted TC variation by the Gradient Tree Boosting algorithm. The forecasted TC graph oscillates much more than the actual TC graph. On September 18, the maximum error of the predicted TC is 6,000 \$/day (35%).

According to Table 2.6, Polynomial Regression, ANN, and SVM presented the worst R^2: 0.79, 0.80, and 0.86, respectively. These predictive models' mean absolute error is too high. However, they performed much better than the linear regression model, whose R^2 is about 0.50.

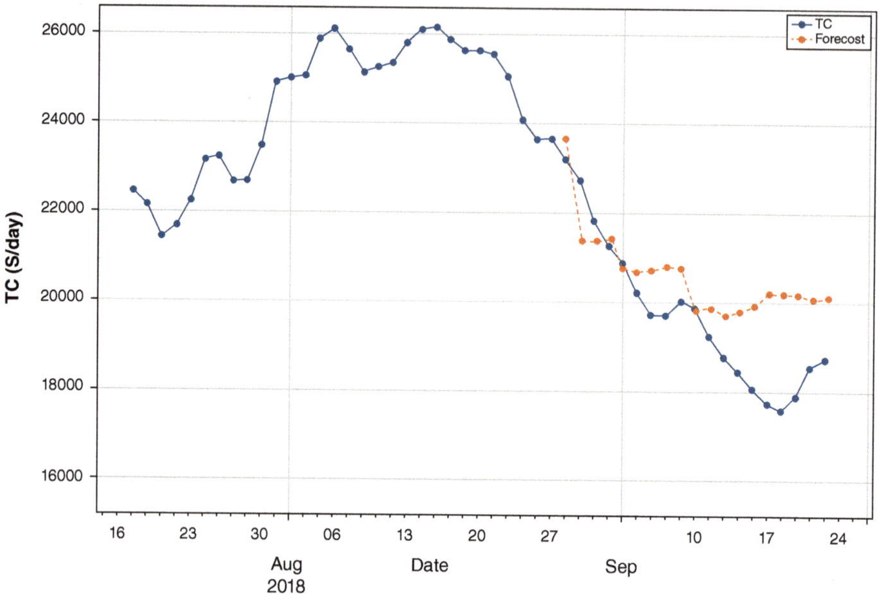

FIGURE 2.31 Variation of TC and 20-day forecasted TC using the Bagging algorithm.

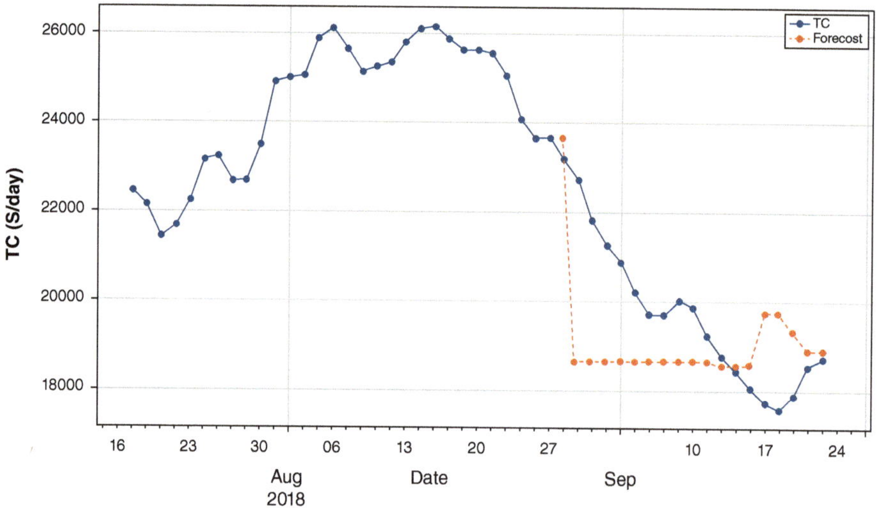

FIGURE 2.32 Variation of TC and 20-day forecasted TC using Decision Tree algorithm.

Figure 2.35 illustrates the variation of TC and forecasted TC of the validation data using the SVM algorithm. The graph of forecasted TC is mainly below the actual TC graph, and the maximum error of prediction is about 22% (5000 $/day), which belongs to the second day of the prediction period (August 29).

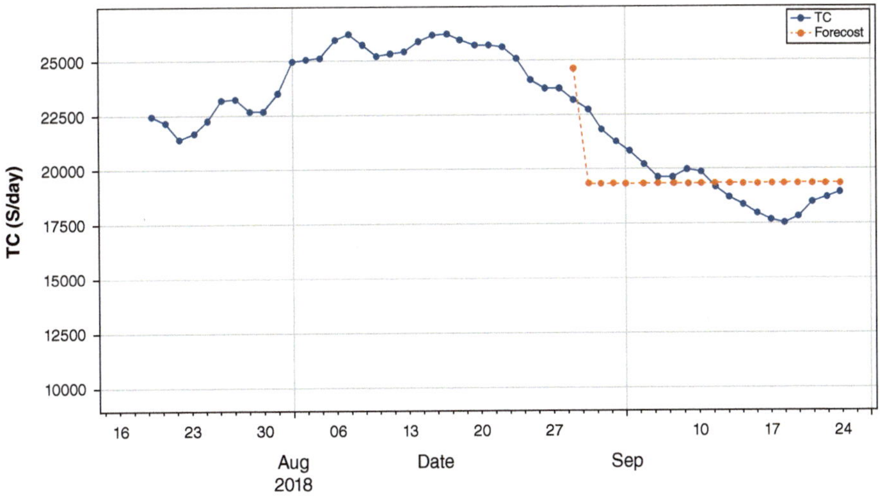

FIGURE 2.33 Variation of TC and 20-day forecasted TC using the KNN algorithm.

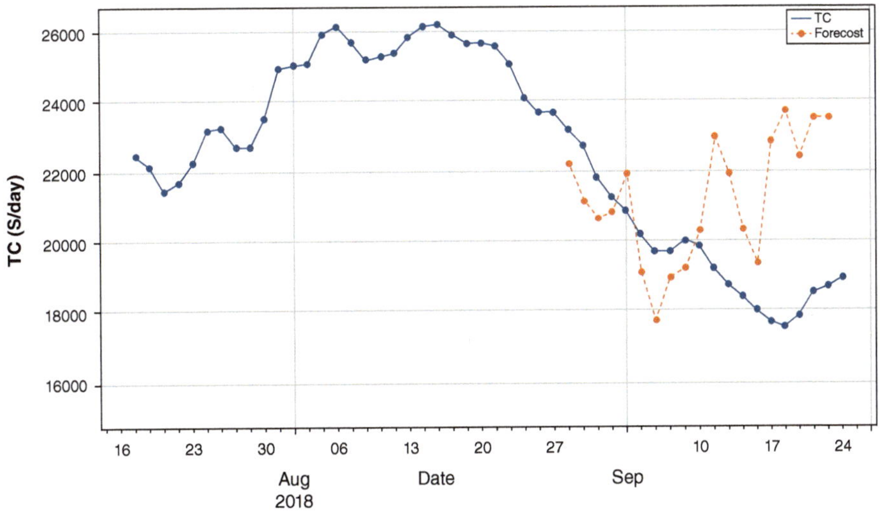

FIGURE 2.34 Variation of TC and 20-day forecasted TC using Gradient Boosting algorithm.

Figure 2.36 represents the TC and forecasted TC graphs utilizing the Polynomial Regression algorithm. The degree of the polynomial features is set to four. The forecasted TC values are far from the actual TC values, and the forecasted TC graph is primarily up to the TC graph. The maximum forecasting error is about 75%, which is unacceptable.

Figure 2.37 exhibited the variation of TC and forecasted the TC graph of the validation data using ANN. The forecast graph is close to the real TC graph in the

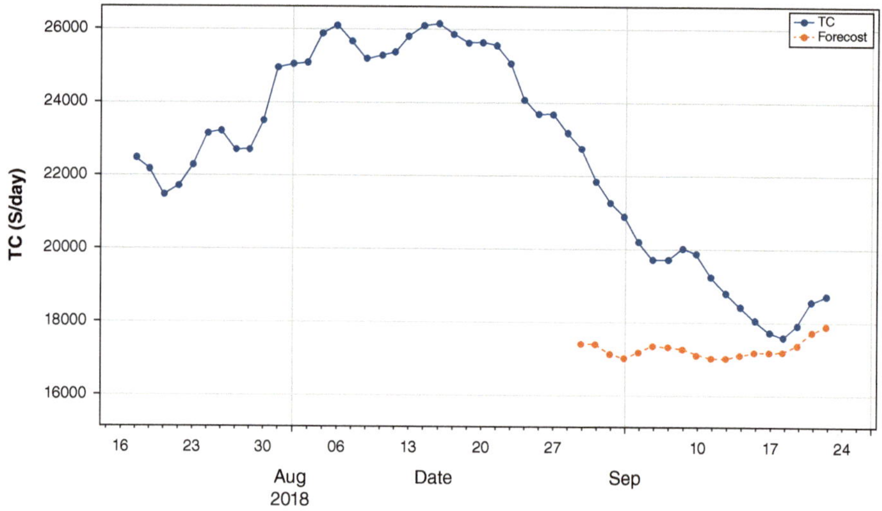

FIGURE 2.35 Variation of TC and 20-day forecasted TC using the SVM algorithm.

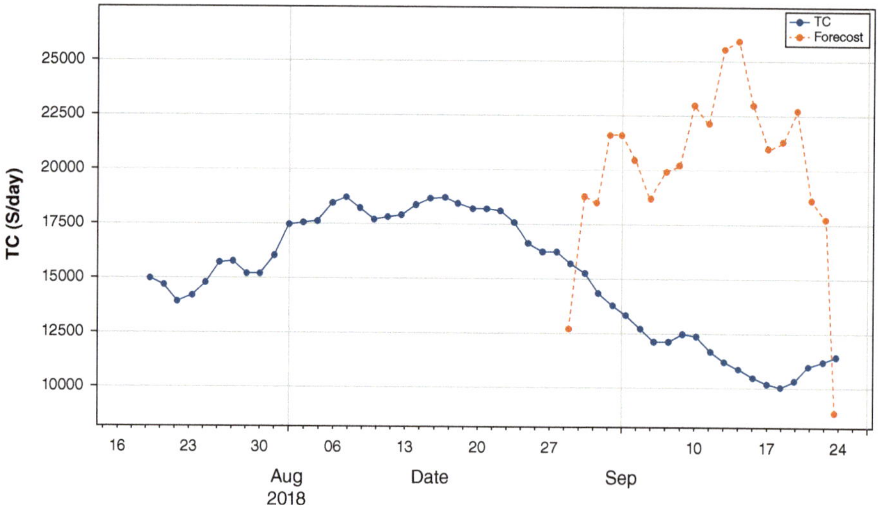

FIGURE 2.36 Variation of TC and 20-day forecasted TC using Polynomial Regression algorithm.

middle. But, it is very far from the real TC, and the end of the graph and the maximum error is about 50% on September 24.

In the next step, Stacking Regression is utilized to combine the best forecasting algorithms for the 20-day prediction of TC and examine the outcome. AdaBoost and Extremely Randomized Trees performed best among all other algorithms, according to Table 2.6. Using Stacking Regression and combining these algorithms for forecasting improved the model's accuracy significantly. The result is presented in

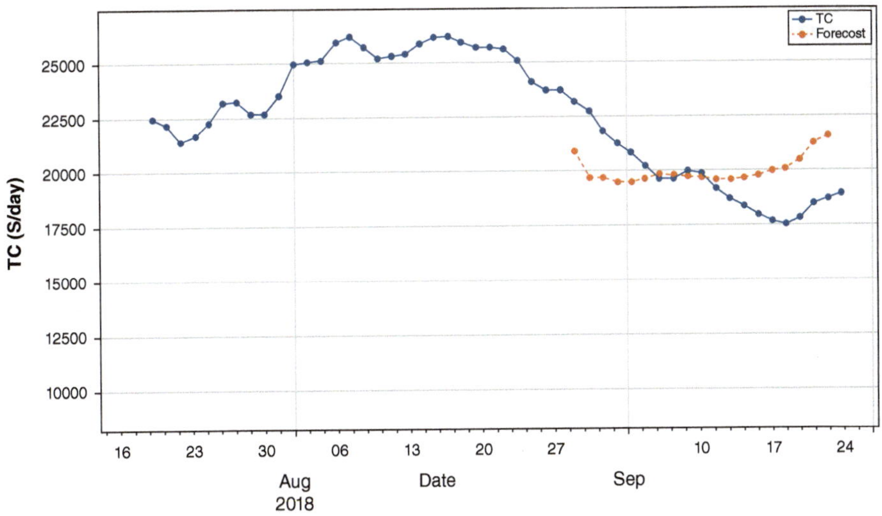

FIGURE 2.37 Variation of TC and 20-day forecasted TC using ANN algorithm.

TABLE 2.7
Stacking Regression Model Accuracy for 20-day Forecasting of TC (combining Bagging, Random Forest, AdaBoost, and Extremely Randomized Trees algorithms)

Model	R^2	MSE	MAE ($/day)
Stacking Regression	0.9736	750,215	572.01

Table 2.7. The R^2 of the 20-day predictive model has increased to 0.98, and the MAE is about 570 $/day. The 20-day forecasting model's result is satisfying.

The prediction result of the model for the validation data and actual TC values are illustrated in Figure 2.38. The forecasting value of TC is close to the real TC, especially in the first twelve days of the forecasting period. The forecasting value of TC is close to the real TC, especially in the first twelve days out of the twenty days of the forecasting period. The maximum error is about 3,000 $/day (17%) on September 18.

2.9.1.12 Summary

Data analysis is performed, and the desired parameters are selected to feed into the forecasting model.

Some advanced ML algorithms were utilized to develop the forecasting model. AdaBoost, Extremely Randomized Tree, and Stacking Regression (combining AdaBoost and Extremely Randomized Tree regressors) algorithms outperform the other algorithms for a 5-day (1 week) forecasting period. R^2 of 0.96 and MAE of 650 – 700 $/day are the results of these models to forecast TC quickly (5 days).

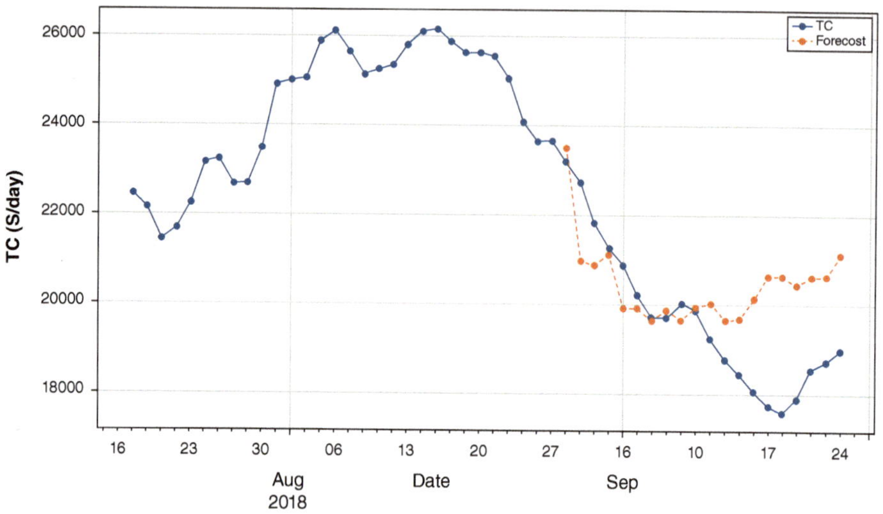

FIGURE 2.38 Variation of TC and 20-day forecasting model using Stacking Regression and combining Bagging, Random Forest, AdaBoost, and Extremely Randomized Trees algorithms.

AdaBoost, Extremely Randomized Tree, and Stacking Regression (combining AdaBoost, Extremely Randomized Tree, Random Forest, and Bagging regressors) algorithms perform better for mid-term forecasting of TC (20-day). R^2 of 0.96–0.97 and MAE of 570 – 600 $/day are the results of these models for forecasting TC in the mid-term period (20 days).

It is important to note that the 20-day forecasting models are more accurate than the 5-day forecasting models, according to Tables 2.4 and 2.6. Statistical investigations show a hidden pattern in the data, and its period is close to 20 days.

2.9.1.13 Recommendations

Advanced ML algorithms have been utilized to develop forecasting models. Other practical forecasting algorithms are based on time series analysis. Time series analysis is a subfield of statistics and econometrics. Because of the sequential nature of the data, time series analysis has particular goals, such as prediction, which aims to produce reasonable forecasts of the future. The most critical time series forecasting models are autoregressive integrated moving average (ARIMA) and ARIMAX.

The ARIMA model describes a univariate time series as a combination of autoregressive (AR) and moving average (MA) lags, which capture the autocorrelation within the time series. The order of integration denotes how many times the series has been differenced to obtain a stationary series.

Autoregressive integrated moving average models extend ARIMA models by including exogenous variables X.

In the next step, these models will be utilized to develop forecasting models and examine their accuracy compared to the advanced ML algorithms.

On the other hand, more analysis of the available data and parameters may be required. Plotting the statistical distribution of the parameter values can add some insights into the data and input selection for the forecasting model. Furthermore, analyzing the autocorrelation of the TC may help to find hidden patterns in the TC variation over time.

2.9.2 Case Study 2: Advanced Data Analytics—SAG Mill Process; Harnessing the Power of Data to Create Digital Mining Services

2.9.2.1 Introduction

This project aims to develop a customized, scalable AI model using the data collected by operations and P86 sensor systems to reduce mill variability and minimize the energy consumption of SAG Mills. Ultimately, this process will improve the operator's decision-making process.

This project's specific objective is to deliver a customized solution that leverages data and human insights augmented by AI to stabilize the client's processes, creating a competitive advantage for Molycop in the market by accelerating its digital transformation process. The following steps will accomplish this objective:

- Investigating data delivering the initial baseline assessment report;
- Making up the datasets and validating the data mining results with the physics;
- Testing and selecting the best prediction models (SAG Power, Variability);
- Developing an integrated optimization AI/ML models to minimize energy consumption, reduce variability, and maximize the throughput;
- Validating and testing the developed models; and
- Reporting and transferring the knowledge to the client.

While the project's scope investigates the available operational and sensor datasets the business provides, data scientists may use physics to supplement the datasets and validate the results. The client contact has validated this approach and detailed it in the corresponding status update or report.

2.9.2.2 Specific Phase 1: Scope—Initial Data and Insights

This phase's primary goal is to develop the AI prediction models for variability (pebble rate) and SAG Power. The structure of these models was presented in two prior reports. This report also contains the tested Gradient Boosting model for sensor simulation and the Extra tree model for MLP SAG Power and Pebbles rate prediction. These models have been selected based on a comprehensive investigation and data modeling during the prediction and data preparation phases.

Given the lack of MLP-specific sensor data, a dummy sensor dataset was created based on available Fortnum data to estimate sensor data according to the available operational data. This report also includes technical details related to the development of this model.

Several interviews were conducted with the MLP operational team to validate insights into the model and incorporate institutional knowledge in the day-to-day optimization and decision-making process.

2.9.2.3 Methodology

This project will use the CRISP-DM methodology. Figure 2.39 provides an overview of a data science project's life cycle.

The sequence of the phases is not rigid. Moving back and forth between different phases is always required. Each phase's outcome determines which phase, or task of a phase, must be performed next. The arrows indicate the most important and frequent dependencies between phases.

A high-level view of the activities in each of the phases in CRISP-DM methodology as it relates to the Molycop Advanced Analytics project is shared as follows:

Business Understanding: The initial phase focuses on understanding the project objectives and requirements from a business perspective. This section will present the results of the studies and technical discussions with the business teams.

Data Understanding: The data understanding phase starts with initial data collection. It proceeds with activities to familiarize themselves with the

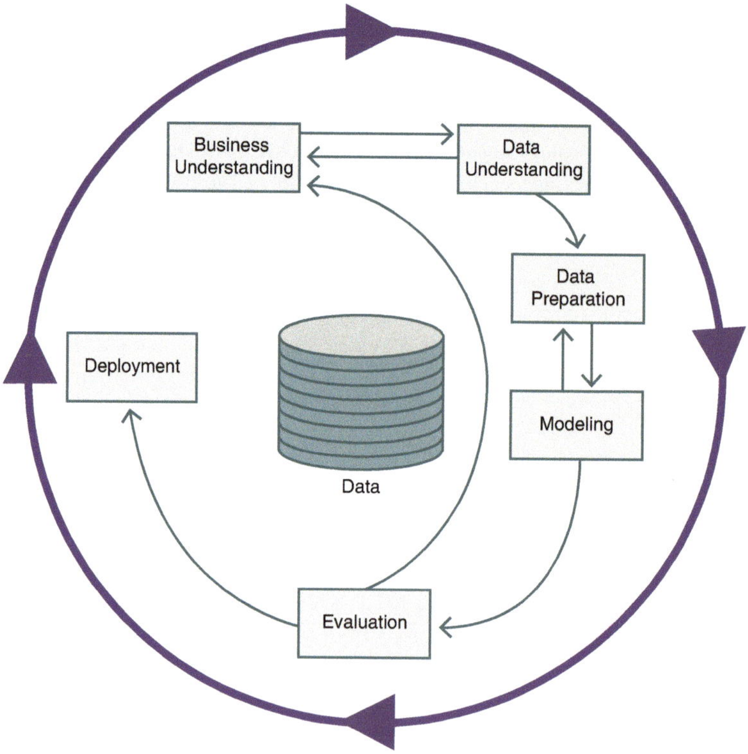

FIGURE 2.39 Cross-industry standard process for data mining.

data, identify data quality problems, and discover first insights into it. This section also includes the existing core understanding and refactoring.

Data Preparation: The data preparation phase covers all activities needed to construct the final dataset [data that will be fed into the modeling experiment(s)] from the initial raw data. Data preparation tasks will likely be performed multiple times and not in any prescribed order.

Modeling/Reporting: In this phase, appropriate experiments are designed and applied, and their parameters are calibrated to optimal values.

Evaluation: At this stage in the project, models have been built with high quality from a data analysis perspective. Before proceeding to the final deployment of the model, it is essential to thoroughly evaluate it and review the steps executed to create it to ensure the model properly achieves the business objectives. At the end of this phase, a decision on using the data mining results should be reached.

Deployment: In this phase, the model's production deployment has been described, including the data lake architecture, data engineering solution, model deployment, and visualization solution.

2.9.2.4 Business Understanding—Minera Los Pelambres—Chile

2.9.2.4.1 SAG Milling/Pebbles

The previous reports explained the details of the business process and operational procedures in Minera Los Pelambres (MLP). In Minera Los Pelambres, pebbles produced by the SAG Mill are not measured individually and could be returned to any of the three available SAG Mills (see Figure 2.40).

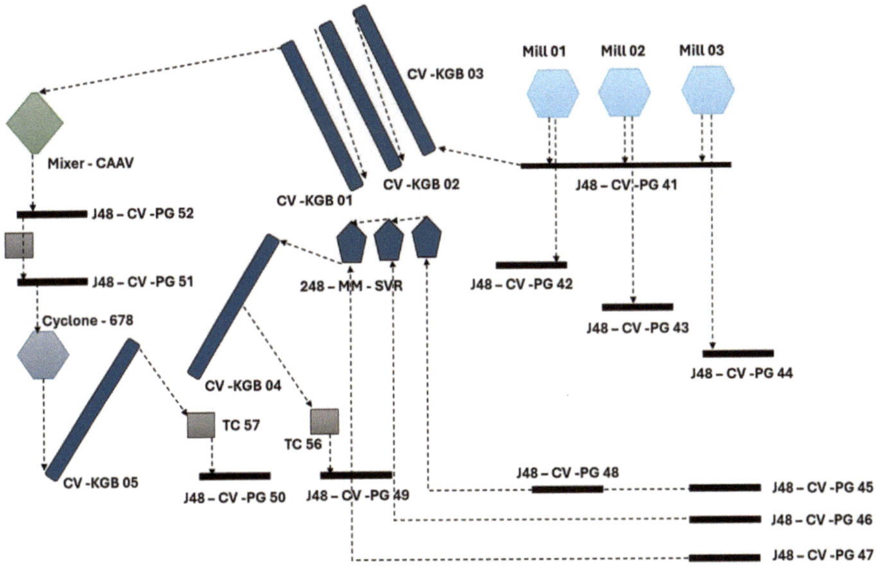

FIGURE 2.40 Pebble measurement points are marked.

Decreasing the pebble rate is the selected approach to minimize the materials' variability after milling by SAG Mill. In MLP, there is no measuring system to scale the pebble after each SAG Mill separately (see Figure 2.40). A statistical approach was applied to estimate the pebble rate discharged by SAG 2 using the information from the other SAG Mills and the overall measured pebbles.

2.9.2.4.2 SAG Mill Operator Role

The leading operator's role is to reach the planned throughput. This aim is usually achieved through load cell control. In a regular operation, the operators set ranges in which the expert system adjusts the control variables, including SAG speed, Solid percentage, and Fresh feed. Primarily, decision-making is based on the operator's experience. At present, no formal manual has been developed for MLP. In shift changes, basic information about SAG condition is exchanged; it is regular or restricted. There is general satisfaction in the number of tools available to achieve the planned objectives.

The following list presents gains, pains, and operators' tasks regarding their SAG operation role.

Gains and Positive aspects expected during the SAG operation:

- Global understanding of the plant and its process, broad vision;
- Capacity to improve the process in the medium and long term;
- Proactive and anticipated response to unforeseen events;
- Communication with vendors; and
- Participation in the continuous improvement process.

Pains and undesirable situations regarding SAG operation:

- Low confident measurements and uncalibrated sensors; Solid percentage and Freshwater;
- Non-intuitive differences between tools that aim to display similar information in control;
- Detentions and mechanical issues;
- Security incidents; and
- Reactive plant management.

Tasks operators have to make regarding SAG operation:

- Plant managing in restricted mode;
- Achieve production goals;
- Maintain SAG throughput;
- Lead unforeseen events management; and
- Be aware of the tendencies and types of minerals produced.

Table 2.7 shows the logic operator's decision-making regarding power consumption and pebble rate control.

TABLE 2.8
Variable Control

Parameter	Normal Range	How It Identified Range Situation	Control Variable
SAG Power	below14,500 kW	Control room	SAG speed and total media grinding
Pebbles Rate	700 tone per hour, depending on crusher capacity	On terrain visual inspection of total load cell tendency	Diminish freshwater and SAG speed, rise solid percentage

2.9.2.5 Data Understanding

2.9.2.5.1 Fortnum Data

The P86 sensor installation for Minera Los Pelambres has been delayed until January 2021. For this reason, the project team developed a model to create a required dummy sensor data set for MLP to train and validate the potential AI prediction algorithms. The client has already approved this approach. The following is a summary of the information collected supporting creating the sensors' dummy data sets for MLP.

As mentioned above, the project team must use the Fortnum mine sensor data to make an AI model to simulate the MLP sensor data. The Fortnum sensors collect peak magnitude and position information in each Zone once it passes the predetermined Zone limits in each section. Each section has two sensors, named by their section, Feed, Mid, and Discharge, and a distinctive A or B, separated by 180° (see Figure 2.41).

In this approach, the project team worked with raw signals of the peak magnitude, so no unit is specified. Later in production, this signal is transformed into acceleration units.

The simulation model tries to replicate how sensors record the data, so the developed model considers Zones and SAG mill sections, as indicated in Figure 2.41. The SAG Mill, a key component in the mining process, rotates counterclockwise.

This particular Zone and Section configuration is significant as it enables a detailed analysis of the SAG Mill operation. Using this methodology, it is possible to obtain a state of each zone across the SAG Mill, where zone 2 presents the toe, Zone 3 the kidney, and Zone 4 the shoulder. Only five sensors were occupied in the SAG Mill. The sixth sensor, identified as sensor B in the Feed section, did not present reliable information.

2.9.2.5.2 Additional MLP Requested Data

To estimate the pebble discharge from SAG Mills, all received data for SAG 1, SAG 2, and SAG 3 were investigated, and a summary of results is tabulated in Table 2.8.

Table 2.8 summarizes data for a specific period. During this period, SAG 2 had the lowest fresh feed rate; this is congruent because it is also, according to operators, the SAG processes the more significant amount of pebbles. This information has been used to estimate the modeling section's SAG 2 pebble generation rate.

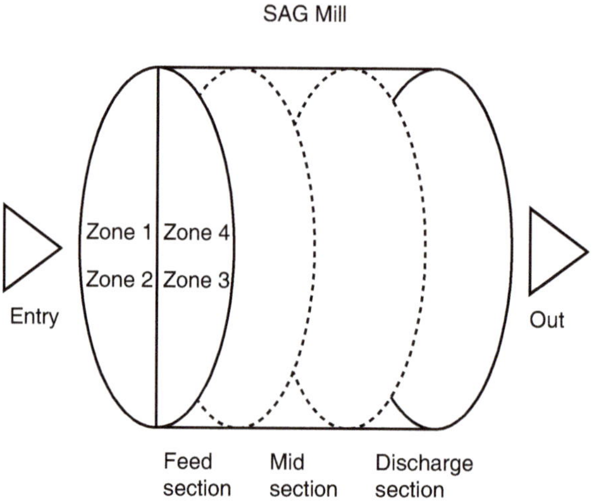

FIGURE 2.41 Generic sensor position placement in SAG Mill.

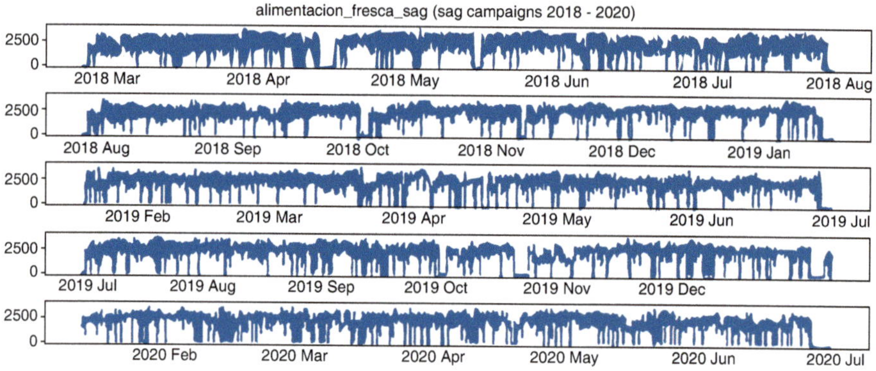

FIGURE 2.42 Fresh feed behavior of campaigns.

2.9.2.6 Data Preparation

2.9.2.6.1 MLP Data Preparation

Between 2018 and 2020, MLP carried out five campaigns (each lasting approximately six months). See Figure 2.42.

For the analysis, we meticulously created one file per campaign, aggregating data to 1 minute using the mean function. The data presented several gaps with outlier and null values (due to SAG stops or sensor reading failures), which we diligently managed.

A simple moving average operation of 5 minutes (5 periods) was applied to the entire dataset. This window was selected because the model needs to observe 5 minutes of past data to predict a future value. This future value should consider 5-minute observations to absorb the settling time of the SAG under new operating conditions

TABLE 2.9
MLP Pebble versus Fresh Feed Comparison

	Total SAG Pebbles	SAG 1 Fresh Feed	SAG 2 Fresh Feed	SAG 3 Fresh Feed
Mean	348.21	2794.71	2360.89	2561.20
Std	204.49	278.19	431.62	526.72
Min	3.43	0	0	0
25%	206.06	2702.64	2172.93	2425.35
50%	380.10	2805.81	2489.90	2710.93
75%	505.19	2941.59	2669.27	2893.19
Max	796.54	3099.26	3054.38	3317.39

and consider the residence time of the material that produces the data to be predicted. The selection of the 5-minute window was validated with various operators at MLP.

Two predictive models were developed to simulate and predict the behavior of the SAG:

1. SAG Power indicates the energy (kW) used during grinding.
2. Pebble Rate (SAG 2) indicates the rate of pebbles (tone per hour) discharged by the SAG.

2.9.2.7 Modeling

2.9.2.7.1 Fortnum Sensor Datasets Modeling

This modeling aims to create a data generation model based on available sensor data to simulate the MLP condition and make dummy datasets for MLP. Two data-generating models were developed to achieve the project's aim:

1. The first model predicts the position using operational data; and
2. The second model predicts the magnitude using operational data and the first model's output (Position).

The models mentioned above generated a dummy sensor dataset for MLP. Table 2.9 lists the input variables used to generate the sensors' parameters.

All collected and created data are scaled from 0 to 1, as described in Table 2.10.

After reviewing various models, the Gradient Boosting Model was selected. The results of the selected model (Gradient Boosting) are tabulated in Table 2.11.

Figures 2.43 and 2.44 illustrate the results of the sensor data generation model validation for Fortnum.

Figure 2.43 displays that the concentration around zone 2 has a more significant dispersion than the other three zones as it corresponds to the toe position where the heavy impacts occur, resulting in a noisier zone. Figure 2.44 shows a slightly larger dispersion for the greater Magnitudes. Most of them also correspond to the toe position of the SAG Mill.

TABLE 2.10
Input Parameters to the Sensor Data Generation Model

Position Model Input	Peak Magnitude Model Input
SAG Power	SAG Power
SAG Speed	SAG Speed
CV_02 (Fresh Feed)	CV_02 (Fresh Feed)
SAG Water Feed	SAG Water Feed
Zone	Position
Feed	Feed
Mid	Mid
Discharge	Discharge

TABLE 2.11
A Data Scaling Approach

Variable	Formula
CV02 (Fresh Feed)	X/(10*Vol)
SAG Water Feed	X/(10*Vol)
SAG Speed	X/Critical Speed
SAG Power	X/Max design SAG Power
Position	X/360

TABLE 2.12
Results of Position Model and Peak Magnitude Model (Gradient Boosting)

Parameter	Score (R^2)
Position	0.96
Peak Magnitude	0.86

This model can simulate dummy sensor data at other mine sites with appropriate scaling for a particular mine.

2.9.2.7.2 Minera Los Pelambres Operational Prediction Model

This project's comprehensive investigation has been completed to find the best prediction model. The Extra Trees model was selected for modeling because it provided the best results. The operational variables used to predict the SAG Power and SAG Pebble Rate were:

- Fresh feed SAG
- SAG Speed

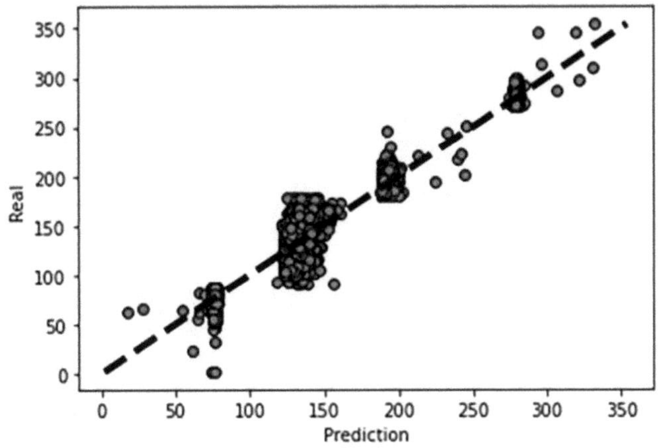

FIGURE 2.43 Peak Position model validation results.

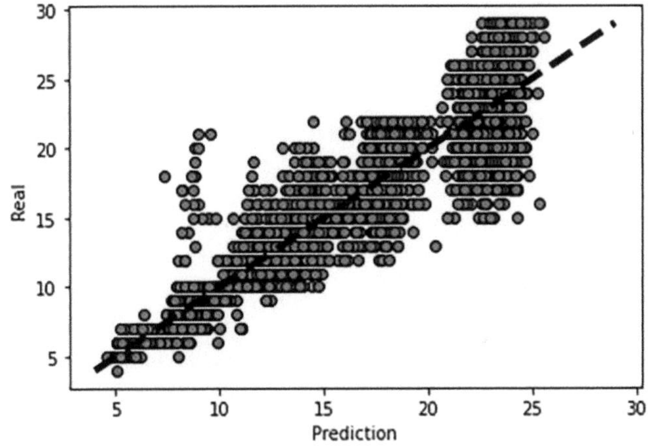

FIGURE 2.44 Peak Magnitude model validation results.

- Freshwater SAG
- Solid Percentage SAG
- Feed Size Reference to SAG
- Grinding Media Loading
- Total Load in Weight of the SAG Mill
- WI Day (daily Work Index)
- Pebbles to SAG2
- Lifter Age Day

The project team reached the high accuracy model, and the output score results are illustrated in Table 2.12, where the score means R^2.

TABLE 2.13
Model Results

Output	Score (R^2)
SAG Power	0.98
SAG Pebble Rate	0.96

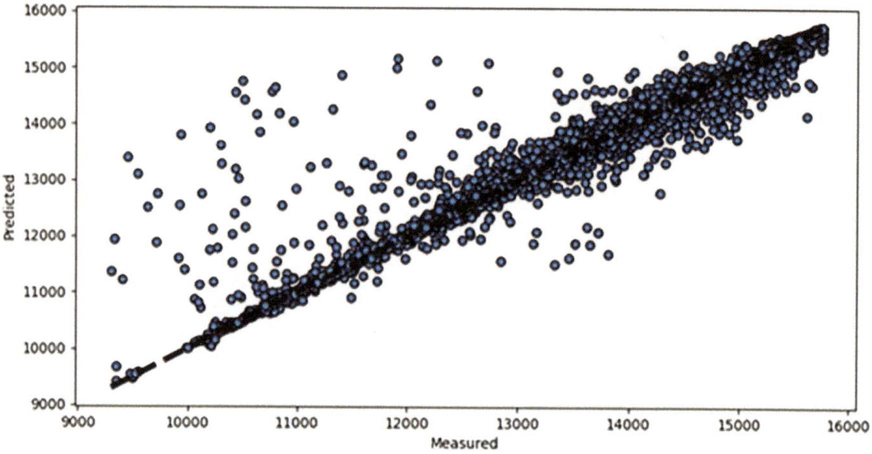

FIGURE 2.45 SAG Power prediction validation results.

Figure 2.45 illustrates the validation results of the SAG Power predictions model. This model's score (R^2) is 0.98, which shows the high accuracy of the developed prediction model. This model could be improved in the beta phase.

Figure 2.46 shows the validation results of the SAG Pebble Rate predictions model. This model's measured score is 0.96, an acceptable R^2 for this model type.

2.9.2.7.3 Minera Los Pelambres Integrated Predictive Model

An integrated model, combining the operational and sensor data, was created. The model simulated the behavior of the P86 sensors (which will be installed later in MLP) and was trained with Fortnum's actual sensor data.

Twenty-four sensor parameters were simulated and then joined with the operational MLP real dataset, replacing parameters already used to create the sensor model and obtaining the following input parameters.

Input parameters of the predictive model with the dummy sensor dataset are:

- Solid Percentage
- Feed Size Reference to SAG
- Total Load in Weight of the SAG Mill
- Work Index
- Pebbles to SAG

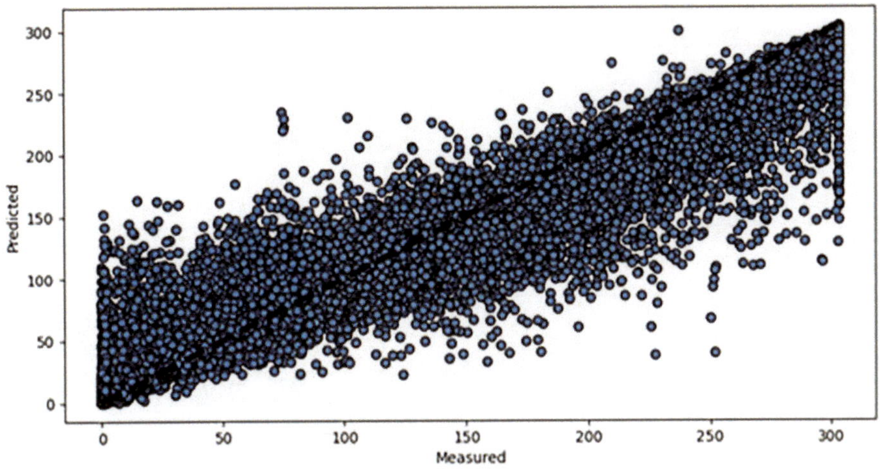

FIGURE 2.46 SAG Pebble Rate prediction validation results.

- Lifter age day
- Position zone 1 location feed
- Peak Magnitude zone 1 location feed
- Position zone 1 location mid
- Peak Magnitude zone 1 location mid
- Position zone 1 location discharge
- Peak Magnitude zone 1 location discharge
- Position zone 2 location feed
- Peak Magnitude zone 2 location feed
- Position zone 2 location mid
- Peak Magnitude zone 2 location mid
- Position zone 2 location discharge
- Peak Magnitude zone 2 location discharge
- Position zone 3 location feed
- Peak Magnitude zone 3 location feed
- Position zone 3 location mid
- Peak Magnitude zone 3 location mid
- Position zone 3 location discharge
- Peak Magnitude zone 3 location discharge
- Position zone 4 location feed
- Peak Magnitude zone 4 location feed
- Position zone 4 location mid
- Peak Magnitude zone 4 location mid
- Position zone 4 location discharge
- Peak Magnitude zone 4 location discharge

The simulation covers all four zones and three SAG Mill locations (Feed, Mid, and Discharge).

TABLE 2.14
MLP Integrated Predictive Model Accuracy

Output	Score (R^2)
SAG Pebble Rate	0.95
SAG Power	0.86

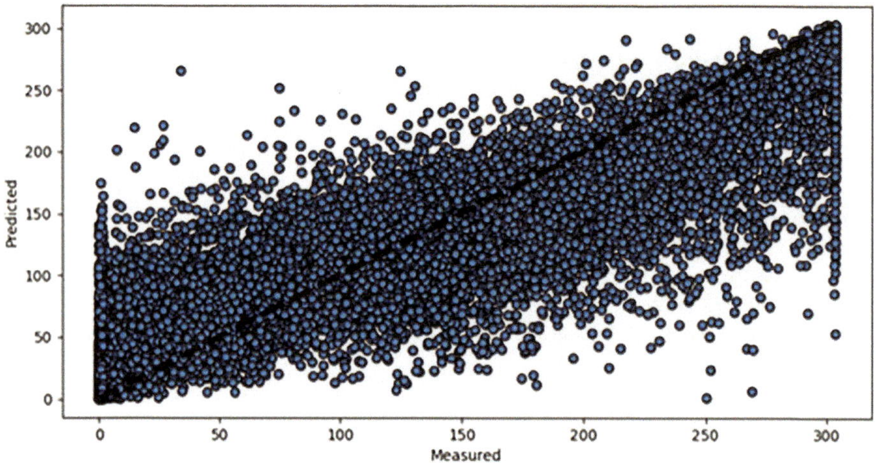

FIGURE 2.47 SAG Pebble rate model results.

After testing many different ML model types, the Extra trees model was selected for this prediction with the results, as shown in Table 2.13.

Figure 2.47 shows that the model tends to overestimate the Pebble rate for lower measurement values, and for higher measurements, it tends to underestimate it.

Figure 2.48 shows that power was overestimated in lower measurement values. This could be explained by the fact that most data concentrated over 10,000 kW, causing predictions in this range.

2.10 CONCLUSION: UNVEILING THE POTENTIAL OF ADVANCED ANALYTICS IN MINING 4.0

Integrating advanced analytics represents a transformative leap into the future of mining operations, where efficiency, safety, and sustainability are paramount. Throughout this chapter, we have delved into the intricate web of opportunities, challenges, and future directions that define the landscape of advanced analytics in Mining 4.0.

The unprecedented demand for natural resources has propelled mining companies to seek innovative solutions to optimize their business processes. Advanced analytics emerges as a beacon of hope, offering a pathway to significant improvements in

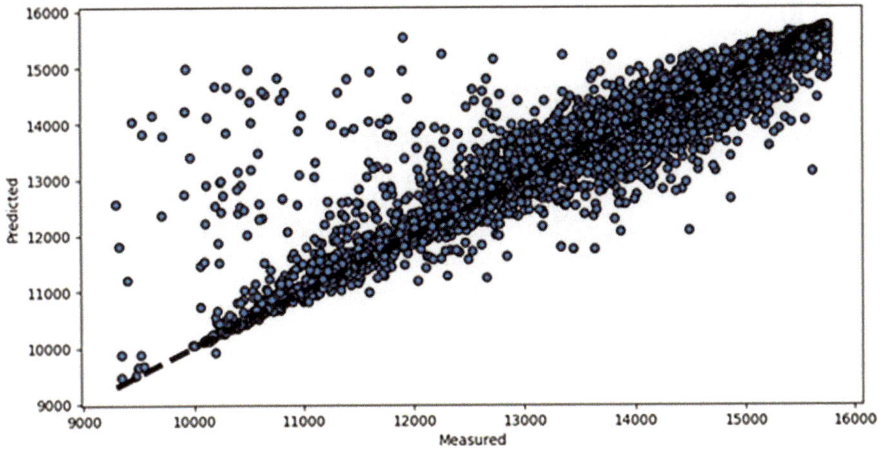

FIGURE 2.48 SAG Power model results.

strategic and operational decision-making. By harnessing the power of data, mining companies can transcend the limitations of traditional decision-making processes and unlock new levels of efficiency and productivity across the entire value chain.

However, the journey toward harnessing the full potential of advanced analytics in mining operations has its challenges. Challenges such as data integration, skill requirements, cultural barriers, and infrastructure constraints loom large, threatening to impede progress and stifle innovation. Yet, these challenges are manageable. Through strategic investments in talent development, technology infrastructure, and organizational change initiatives, mining companies can overcome barriers and pave the way for a data-driven future.

The future of advanced analytics in the mining industry is brimming with promise and potential. AI and ML technologies promise to revolutionize decision-making processes. In contrast, IoT and sensor technologies offer real-time insights into equipment performance and safety. Cloud-based analytics platforms provide scalable data management and analysis solutions. At the same time, predictive simulation techniques enable scenario analysis and risk management.

Moreover, the focus on sustainability and environmental stewardship drives the adoption of advanced analytics for environmental monitoring, compliance reporting, and resource optimization. By embracing advanced analytics as a strategic imperative, mining companies can navigate the complexities of the digital age and emerge as leaders in a rapidly evolving industry landscape.

In summary, advanced analytics in Mining 4.0 represents a paradigm shift in managing and executing mining operations. By leveraging data and analytics, mining companies can unlock new opportunities for optimization, efficiency, and sustainability. As we embark on this transformative journey, let us embrace the possibilities, overcome the challenges, and forge a path toward a brighter, data-driven future for the mining industry.

REFERENCES

[1] Alamdari, S., et al., Application of machine learning techniques to predict haul truck fuel consumption in open-pit mines. *Journal of Mining and Environment*, 2022. 13(1): p. 69–85.

[2] Bag, S., et al., The role of big data and predictive analytics in developing a resilient supply chain network in the South African mining industry against extreme weather events. *International Journal of Production Economics*, 2022. 251: p. 108541.

[3] de Moura, R.L., L.d.L.F. Ceotto, and A. Gonzalez, Industrial IoT and advanced analytics framework: an approach for the mining industry. In *2017 International conference on computational science and computational intelligence (CSCI)*. 2017: IEEE.

[4] Fan, C., et al., Unsupervised data analytics in mining big building operational data for energy efficiency enhancement: a review. *Energy and Buildings*, 2018. 159: p. 296–308.

[5] Martins, P. and A. Soofastaei, Process analytics. In *Data analytics applied to the mining industry*. 2020: CRC Press. p. 131–148.

[6] McGaughey, J., Artificial intelligence and big data analytics in mining geomechanics. *Journal of the Southern African Institute of Mining and Metallurgy*, 2020. 120(1): p. 15–21.

[7] Qi, C.-c., Big data management in the mining industry. *International Journal of Minerals, Metallurgy and Materials*, 2020. 27(2): p. 131–139.

[8] Soofastaei, A., *Development of an advanced data analytics model to improve the energy efficiency of haul trucks in surface mines* (Ph.D. thesis). 2016: The University of Queensland.

[9] Soofastaei, A., Intelligence predictive analysis to reduce production cost. *Australian Resources and Investment*, 2017. 11(1): p. 24–25.

[10] Soofastaei, A., The application of artificial intelligence to reduce greenhouse gas emissions in the mining industry. *Green Technologies to Improve the Environment on Earth*, 2018. 25: p. 234–245.

[11] Soofastaei, A., *Data analytics applied to the mining industry*. 2020: CRC Press.

[12] Soofastaei, A., Digital transformation of mining. In *Data analytics applied to the mining industry*. 2020: CRC Press. p. 1–29.

[13] Soofastaei, A., Advanced analytics for mining industry. In *Advanced analytics in mining engineering: leverage advanced analytics in mining industry to make better business decisions*. 2022: Springer. p. 1–22.

[14] Soofastaei, A., *Numerical simulation: advanced techniques for science and engineering*. 2023: BoD–Books on Demand.

[15] Soofastaei, A., et al., Development of a multi-layer perceptron artificial neural network model to determine haul trucks energy consumption. *International Journal of Mining Science and Technology*, 2016. 26(2): p. 285–293.

[16] Soofastaei, A., et al., Simulation of payload variance effects on truck bunching to minimize energy consumption and greenhouse gas emissions. *Coal Operators' Conference 2015*, 2015.

[17] Soofastaei, A., et al., Reducing fuel consumption of haul trucks in surface mines using artificial intelligence models. *Coal Operators' Conference*, 2016.

[18] Soofastaei, A. and M. Fouladgar, Energy efficiency improvement in surface mining. In *Latest research on energy recovery*. 2022: IntechOpen.

[19] Soofastaei, A., et al., Energy-efficient loading and hauling operations. In *Energy efficiency in the minerals industry: best practices and research directions*. 2018: AusIMM. p. 121–146.

[20] Soofastaei, A., P. Knights, and M. Kizil, Application of advanced data analytics to improve haul trucks energy efficiency in surface mines. *Extracting Innovations: Mining, Energy, and Technological Change in the Digital Age*, 2018. 2(5): p. 163–179.

3 Designing Intelligence
Harnessing Soft Sensors and Advanced Analytics in Petroleum Refining for Industry 4.0

Ajaya Kumar Pani and Ali Soofastaei

3.1 INTRODUCTION

The dawn of Industry 4.0 has ushered in an era of unprecedented digital transformation, reshaping the contours of various sectors, including the oil and gas industry. At the heart of this revolution lies the integration of advanced analytics and artificial intelligence (AI), which are pivotal in driving efficiencies, enhancing productivity, and fostering innovation within the sector. This introduction delves into the multifaceted applications of data analytics, AI, predictive modeling, and optimization in the oil and gas industry and highlights their transformative potential.

3.1.1 Embracing the Data Deluge

The oil and gas sector is inherently data-rich, with vast information from exploration, production, refining, and distribution processes. In the context of Industry 4.0, the industry's approach to this data has evolved from mere collection and storage to strategic utilization. Advanced analytics techniques enable the conversion of raw data into actionable insights, facilitating informed decision-making and strategic planning. Data sources range from geological surveys and drilling logs to sensor data from equipment and pipelines, encompassing structured and unstructured formats.

3.1.2 Analytical Prowess

Advanced oil and gas industry analytics encompasses various techniques, including statistical analysis, data mining, and machine learning (ML). These methodologies uncover patterns, trends, and correlations within complex datasets, offering a deeper understanding of operational dynamics. For instance, predictive analytics can forecast equipment failures or maintenance needs, minimizing downtime and optimizing resource allocation. Similarly, data mining can identify efficient drilling locations, reducing exploratory risks and enhancing yield potential.

DOI: 10.1201/9781003186823-3

3.1.3 The AI Revolution

AI has emerged as a cornerstone of innovation in the oil and gas sector, with applications ranging from automated drilling to intelligent asset management. ML algorithms, a subset of AI, are particularly adept at processing and analyzing large datasets. They learn from past data to make predictions or identify opportunities for efficiency gains. For example, AI-powered models can predict the lifespan of critical infrastructure, enabling proactive maintenance and reducing the likelihood of catastrophic failures.

3.1.4 Predictive Mastery

Predictive analytics and modeling stand at the forefront of the industry's strategic arsenal, offering foresight into future scenarios based on historical and real-time data. These predictive models can forecast market trends, demand fluctuations, and supply chain disruptions, allowing companies to adapt their strategies proactively. In exploration and production, predictive models analyze geological data to assess oil or gas presence probability, optimizing exploration efforts and reducing environmental impacts.

3.1.5 Optimization Strategies

Optimization in the oil and gas industry is a complex, multi-faceted endeavor to enhance efficiency, reduce costs, and maximize profitability. Advanced analytics and AI enable the optimization of various processes, from drilling and production to logistics and supply chain management. For instance, optimization algorithms can determine the ideal drilling paths, considering geological characteristics, safety parameters, and cost considerations. Similarly, supply chain optimization models ensure the timely and cost-effective delivery of materials, equipment, and products.

3.1.6 Navigating Challenges and Opportunities

Integrating advanced analytics and AI in the oil and gas industry is challenging. Data quality and integration, cybersecurity, and skilled personnel are significant hurdles. However, the opportunities for transformation and innovation far outweigh these challenges. Companies that successfully harness the power of data analytics and AI stand to gain a competitive edge, achieving operational excellence and sustainability in an increasingly complex and volatile market landscape.

3.1.7 Oil and Gas Industry 4.0

The ambitious target by most nations around the globe for increasing energy efficiency and reducing carbon emissions can be achieved by incorporating more intelligence into the manufacturing/process industries. Industry 4.0 aims to integrate information technology with the various aspects of industrial operations. An essential goal of Industry 4.0 is brilliant production in industries using AI or ML techniques. In industrial automation, ML techniques are efficient computational tools for feature (information) extraction, data pattern recognition, and prediction. Advancements in instrumentation hardware and data storage systems have resulted in the availability of a massive volume of data. The term “Big Data” refers to the large volume of data in the order of petabytes or exabytes, and in process industries, “Big Data” is characterized by the three Vs: volume, variety, and velocity (V^3). Industry 4.0 targets extracting helpful information from this Big Data to achieve better process monitoring and control, quality monitoring and control, decision support, sustainability, efficiency, etc. In process industries, the objective of intelligent production is accomplished by using process data and applying various AI/ML techniques to achieve improved process efficiency, profitability, reduced downtime, adherence to product quality and effluent quality norms, etc. Implementation of Industry 4.0 in the oil and gas industry has given rise to oil and gas 4.0. Intelligent manufacturing can be targeted at different levels in the oil and gas industry. Activities at different levels include exploration and drilling (upstream), storage and transport (mid-stream), and refining, sales, and management (downstream) [1]. Applying AI at these levels will achieve intelligent oilfields, pipelines, and refineries [1]. Among these different levels of oil and gas 4.0, this chapter is devoted to a discussion on intelligent refineries. The focus of this chapter in the context of Industry 4.0 is explained in Figure 3.1.

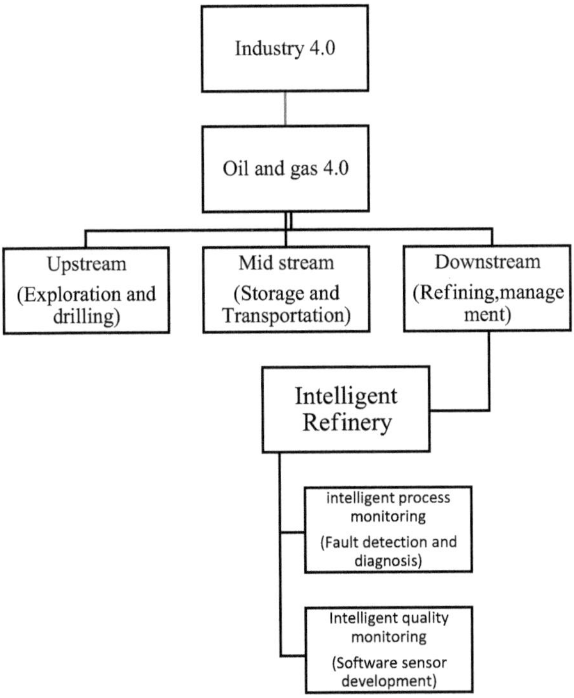

FIGURE 3.1 Soft sensor design in the context of Industry 4.0 in refinery.

A well-instrumented and automated petroleum refinery produces enormous amounts of data, 4 to 6 GB/day [1]. Efficient data mining and analytics processes achieve intelligent refining in the era of Big Data. This is accomplished by developing data-driven models using AI/ML techniques for efficient process monitoring (industrial fault detection and diagnosis) and prediction of critical unmeasured process outputs (soft sensing). Any unwanted deviation from normal operating conditions is known as a fault, and timely detection and diagnosis of industrial faults ensure reduced downtime, increased productivity, and process efficiency.

This chapter focuses on intelligent sensor development in petroleum refineries (the highlighted part in Figure 3.1). The need for such software-based intelligent sensors arises because hardware sensors are unavailable for real-time monitoring of specific process/quality parameters. Even if available, some hardware sensors are installed in such harsh industrial environments that they require frequent maintenance. Ensuring the desired product quality becomes challenging without continuous monitoring and control. These problems faced by process industries in general and petroleum refineries, in particular, are suitably addressed by the design and implementation of software sensors. Such intelligent sensors are designed using industrial data and necessary AI/ML-based computational tools to develop predictive models. These soft sensors may be steady-state process models or adaptive/dynamic process models. Irrespective of the type of soft sensor, there are two major activities in data-driven intelligent sensor development: (1) data preprocessing and (2) model development and validation. The application of AI/ML techniques is primarily significant during the modeling phase (and to a limited extent in some cases in the data preprocessing step).

The chapter is organized as follows. Section 3.2 briefly discusses the scope of soft sensor applications in monitoring various process/quality parameters in petroleum refineries. This is followed by discussing the need for data preprocessing and various commonly used data preprocessing techniques before soft sensor model development in Sections 3.3. Section 3.4 describes the model development and validation process for soft sensor design. Different AI/ML-based soft sensor models reported in the literature with applications in petroleum refineries have been surveyed in Section 3.5. The chapter ends with concluding remarks and future directions. Section 3.6 provides a case study regarding the application of data analytics in the oil and gas industry.

3.2 SCOPE OF SOFT SENSOR IMPLEMENTATION IN PETROLEUM REFINERY

In this section, the processing of crude in a petroleum refinery is briefly presented. During refining operations, various quality-related parameters are analyzed offline in laboratories with a much larger sampling time than monitored online parameters. Various quality parameters for which soft sensor models are reported in the literature are also mentioned.

Petroleum refineries process crude into valuable products. Typically, they consist of an atmospheric distillation unit (ADU) or crude distillation unit (CDU), vacuum distillation unit (VDU), and cracking unit.

An ADU consists of a crude distillation column, condenser, reboiler, and side strippers. After desalting (for impurity removal) and preheating, crude oil is sent to the ADU. Primary products of an ADU are unstabilized naphtha (which is subsequently processed in a naphtha splitter to obtain liquefied petroleum gas and gasoline), heavy naphtha, kerosene, light diesel oil, light gas oil, and heavy gas oil. Some essential quality parameters include 95% ASTM D86 temperature, density, viscosity, aniline point of various products, Reid vapor pressure of naphtha, flash point temperature of various light distillates, pour point of heavy fractions, the cetane number of diesel, freezing point, and cold filter plugging point of kerosene, and so on.

Heavy residues from CDU are further distilled at sub-atmospheric pressure conditions in a VDU. The quality of the VDU bottom product is assessed by a laboratory technique known as needle penetration. This method has a delay of approximately four hours in knowing the penetration quality of bitumen. Some process variables warranting the need for a soft sensor are a product purity prediction of a 95% point of distillation curve, viscosity index, and distillation temperature [2].

Heavy petroleum residues obtained from ADUs or VDUs are converted to more valuable lighter fractions such as gasoline or olefin by cracking. The cracking process is either thermal or catalytic. Catalytic cracking is more common in modern refineries, carried out in one separate refinery unit known as a fluid catalytic cracking (FCC) unit. The products from cracking units are light and heavy naphtha, diesel, and aviation turbine fuel (aviation kerosene). Some standard scopes of soft sensor development in this unit are for inference of product composition from dehumanizer and depolarizer column, naphtha IBP and EBP, and research octane number (RON) of gasoline. It may be noted that naphtha is obtained from different parts of the refinery. Depending on the processing unit, the product is known as straight run naphtha (obtained from CDU), cracked naphtha (obtained from cracking unit), and reformate naphtha (obtained from reforming unit).

Distillates from CDU and VDU and products from cracking units are further processed in the desulphurization unit, where the catalytic hydrogenation process removes compounds containing sulfur, nitrogen, oxygen, and other impurities. In hydrocracking, the heavy residues from CDU and VDU under high pressure and temperature undergo hydrogenation polymerization and cracking reactions in the presence of a catalyst. Hydrogenation reaction improves the input's H/C ratio, and cracking breaks the C-C bond to produce more valuable lighter components.

3.3 DATA PREPROCESSING

The first step in preparing a data-driven machine-learning model is extracting process data from the industrial database. During this activity, plant-operating conditions must be carefully examined. For example, suppose modeling develops a steady-state model for continuous monitoring. In that case, the plant should operate continuously and be stable during the data extraction.

After data extraction, data preprocessing is done to make the data suitable for ML model development. Data preprocessing consists of one or more activities: dimensionality reduction or variable selection, outlier detection and removal, missing value imputation, and finally, data normalization. The ML model will have good prediction accuracy only if the collected raw data is adequately preprocessed before being used for modeling [3].

In an adequately instrumented industrial process, real-time data are obtained for many process variables from installed hardware sensors. These sensors often need to be fixed, resulting in incorrect and sometimes absurd or abnormal measurement values. Such abnormal data values are known as outliers. A data-driven model is susceptible to the presence of outliers in the dataset. Therefore, these outliers must be detected correctly and removed from the dataset. Various univariate and multivariate outlier detection techniques available in the literature can be applied to this activity. Univariate outlier detection techniques include three σ edit rules, Hampel identifier, and box plot. Some popular multivariate outlier detection techniques include Hotelling's T^2 distance, minimum covariance determinant, smallest half volume, and k-means clustering [4]. Interested readers can refer to [5] for a detailed explanation and application procedure for various outlier detection techniques.

In addition to outliers, the dataset may have values that must be added. This arises because, often, the sensor is taken out of the process for maintenance, and during this period of sensor maintenance, information for that particular variable is unavailable—also, the deletion of an outlying observation results in missing values. Generally, three types of missing patterns are possible in an industrial dataset: missing completely at random, missing at random, and not missing at random. The simple approach to address the problem of missing values is to delete the samples with missing values. However, if the number of samples is not many, deletion will further reduce the number of objects, limiting our capability in the subsequent model development phase. To avoid deletion, missing values can be substituted with a mean value of the entire data or interpolated values of neighboring measurements. Other traditional methods include hot deck imputation, regression imputation, and multiple imputation. Besides these methods, ML techniques, such as clustering, decision trees, and neural networks, have been reported recently for missing value imputations [6].

After addressing the issue of outliers and missing values, the next step is variable selection (input selection). This is because information on all the measured variables may be optional for output prediction. Many inputs in the soft sensor model add to model complexity and increase the computational burden during the initial identification and simulation in real time. Therefore, only relevant variables should be considered as inputs. In the initial stage, operating personnel knowledge can be utilized to decide the set of influential variables. Such information is necessary to apply data-driven techniques for candidate variable selection. In the context of ML, methods for variable selection (feature selection) methods are categorized into the following classes: filter-based, wrapper-based, and embedded techniques. Correlation tests, mutual information, and chi-square tests are some of the filter-based techniques. Wrapper-based techniques include forward stepwise regression and backward stepwise regression. Least absolute shrinkage and selection operator (LASSO), elastic net, and ridge regression fall under embedded methods for variable selection. Recently a technique involving p-value computation based on permutation importance was proposed for performing the g variable selection.

After selecting relevant variables, input dimension reduction may have a greater scope. This is especially possible when some of the chosen inputs (and outputs) correlate. In such a scenario, the original variables are transformed into latent variables. This step is known as variable transformation or feature extraction. Principal component analysis (PCA) is one of this category's most popular techniques for input dimensionality reduction. Other techniques in this category include independent component analysis (ICA), multidimensional scaling, self-organizing map, ecomap, and locally linear embedding [7].

After outlier detection, missing value imputation, and variable selection comes the normalization of the data. This is required because some variable values may be several orders of magnitude higher than other variables. Data from a temperature sensor installed in a furnace may have a few hundred or thousand magnitudes. In contrast, numerical values recorded for the recycle ratio maintained in a distillation tower or pressure values (recorded in bar unit) will be much less than temperature values. While variables with low-magnitude values may be necessary, these variables may be overshadowed by high-magnitude variables during modeling if unscaled values are used. A few commonly used normalization or data scaling methods are *z*-score normalization and min-max normalization. In z-score normalization, the mean value of a particular variable is subtracted from each value, and the result is divided by the standard deviation of the observations of that variable. In this method, most of the values in an outlier-free dataset are between –3 and +3. Min-max normalization is performed by subtracting the minimum value of a particular variable from a particular observation of that variable and dividing the result by the range of that variable (maximum–minimum of that variable). All values in this method take values between 0 and 1.

Another problem encountered in data processing is input–output data dimension mismatch. This problem is encountered when the data-driven model is developed to predict unknown output (which is the focus of this chapter). In oil refineries, such process outputs are often quality parameters measured by offline laboratory analysis, performed hourly or once per shift. On the other hand, input data are obtained from online sensors and are available where values are available every few seconds or minutes. A dataset where both input–output values are available is known as labeled data, and a dataset involving only input data is known as unlabeled data. Data-driven predictive model development requires labeled data. However, the number of labeled data samples may be low (depending on the laboratory analysis frequency). Some researchers adopt some techniques to increase the labeled dataset size. Two types of methods that use some classifying techniques, i.e., self-training and co-training, are proposed to augment dataset size [8]. Novak et al. augmented the input–output dataset matrix by generating additional output data using the multivariate adaptive regression spline technique [9]. Mattos et al. proposed correlated data augmentation to expand the labeled dataset [10].

3.4 SOFT SENSOR MODEL DEVELOPMENT AND VALIDATION

After data preprocessing, the data is in the desired form for model development. After implementing the developed model online in the plant, the objective is to estimate product quality accurately when supplied with various model inputs (i.e., process parameter values received from installed online sensors). The input the model receives during online use differs from the input dataset, which will be used for model development. In other words, the developed model should possess good generalization capability (i.e., producing accurate enough outputs for unknown input data). The processed dataset is divided into training and testing data to achieve this objective. This division is usually done at 50:50, 70:30, or 80:20 (ratio of training set to testing set samples). The training data is used for model development, and testing data is used for model validation, as discussed later in this section. While dividing the total data into training and testing sets, it should be ensured that most of the total data characteristics are retained in the training subset. While random division is most widely used for this purpose, more is needed to capture the statistical characteristics of the total data in the training subset. Therefore, statistical methods for data division are preferable. Recently, Singh et al. have investigated the effect of data division on model generalization capability by developing regression neural network models on three benchmark datasets of the petroleum refining process [11].

The training dataset uses various AI/ML-based algorithms for model development. Various ML algorithms can be classified into four types: unsupervised, supervised, semi-supervised, and reinforced learning. Supervised, unsupervised, and, to some extent, semi-supervised algorithms have been investigated for innovative operations in process industries, and reinforced learning is applied in other areas such as robotics and gaming. [12] Unsupervised techniques are mainly used to identify abnormal process behavior (fault detection) by applying them to unlabeled datasets. Supervised and semi-supervised techniques are applied to labeled datasets (i.e., datasets containing input and output values) to predict key process parameters. Semi-supervised techniques are instrumental when few labeled samples are available for model development.

In the area of soft sensor development, where the purpose is to predict an output value (response variable) based on a set of inputs (predictor variables), supervised techniques are generally used. Supervised ML techniques commonly include linear regression (ordinary least square, partial least square, principal component, LASSO, etc.), support vector machine, and artificial neural network (ANN). ANN techniques in soft sensor design include feed-forward (such as backpropagation, radial basis function (RBF), and probabilistic) and recurrent neural network techniques.

Irrespective of the modeling technique used, the accuracy of the developed model is determined by simulating the model with unknown data (or testing data). This is the procedure of model validation. For this unknown dataset, which was not initially used for model development, the model-predicted output values are compared with actual output values in this dataset to compute model accuracy. Some standard statistical parameters that are used for model accuracy computation during the model validation stage are root mean squared error, mean absolute error, median absolute error, mean absolute percentage error, coefficient of determination (R^2), the standard deviation of residuals, final prediction error, etc.

3.5 SURVEY OF SOFT SENSOR DEVELOPMENT IN PETROLEUM REFINERIES

This section reviews various soft sensor models reported in the literature. It may be noted that the survey is limited only to literature mentioning applications in petroleum refineries.

Wang et al. combined partial least squares (PLS) and ANN to predict ASTM 90% distillation temperature [13]. Rogina et al. developed multilayer perceptron and RBF neural network models for continuously monitoring light naphtha Reid vapor pressure [14]. The inferential sensing technique using dynamic PLS is reported by Shang et al. for predicting the 100% cut point of naphtha [15]. Novak et al. reported a dynamic soft sensing approach using Hammerstein-Wiener and neuro-fuzzy modeling optimized with a genetic algorithm for soft sensing kerosene cold filter plugging point [9]. The sulfur content of treated gas oil must be maintained within the prescribed limit. Shokri et al. developed a soft sensor for the hydrodesulfurization (HDS) process by applying support vector regression to estimate the sulfur content of gas oil [16]. An adaptive soft sensor model employing a just-in-time learning mechanism and supporting vector regression local modeling technique is proposed by Vijayan et al. to monitor the initial and end boiling point of naphtha in a CDU [17]. Some researchers have used an ensemble approach to improve prediction accuracy, developing more than one local submodel instead of one model from the entire training data. Wnag et al. developed four regression-based submodels and integrated them to predict the 10%, 50%, and 90% boiling points of diesel produced in a hydrocracking unit.

Recently, researchers have started exploring advanced ML (deep learning [DL]) techniques for soft sensor design for petroleum refineries. Unlike traditional ML techniques, DL techniques do not require prior data preprocessing. These techniques comprise networks with stacked nonlinear layers, and the deep network model can feature extraction and output prediction, which comes with increased model complexity and more significant computational effort. Further, the multiple-layer structure of deep neural networks enables these models to extract more information from data than traditional ML techniques and are exceptionally well suited for application to massive and correlated process data. Some DL techniques that have been reported for industrial applications in recent times include stacked autoencoder (SAE), convolution neural network (CNN), and deep belief network. Yuan et al. reported a spatiotemporal attention-based long short-term memory (LSTM) network for soft sensor modeling in nonlinear dynamic processes [18]. During the hydrocracking process, the model monitored the initial boiling points of heavy naphthas and aviation kerosene. A DL model multirate stacked autoencoder is reported [19] for predicting 50% boiling point and cetane number of diesel. The model consists of a cascaded shared network and a parallel quality-specific network. Ou et al. proposed a quality-driven regularization-based stacked autoencoder (SAE) for monitoring aviation kerosene's 90% boiling point [20]. However, the availability of improved computational facilities should not drive researchers to replace traditional ML techniques with DL methods. As pointed out by Sun and Ge, DL techniques perform better than conventional data-driven ML techniques when we have large amounts of data [21]. The performance of conventional ML and DL techniques is comparable for low to moderate amounts of labeled data. A survey of intelligent sensors (soft sensors) developed for application in petroleum refineries is presented in Table 3.1. It may be noted that the survey is confined to the refinery-related soft sensor works reported in the past 15 years (2009–2024).

TABLE 3.1
Survey of Soft Sensors Reported for the Petroleum Refining Industry

Authors	Modeling Technique	Parameter Monitored
Wang et al. [13]	PLS + ANN	ASTM 90% distillation temperature
Rogina et al. [14]	MLP and RBF neural network	Reid vapor pressure of light naphtha
Napoli and Xibilia [22]	Bootstrap resampling, noise injection, and neural model stacking	The freezing point of kerosene
Novak et al. [9]	Hammerstein-Wiener and Neuro-Fuzzy	Cold filter plugging point of kerosene
Shang et al. [23]	Deep belief network	ASTM 95% cut temperature of diesel
Shang et al. [15]	Dynamic PLS	Naphtha 100% cut point
Shokri et al. [16]	Support vector regression	The sulfur content of treated gas oil in the HDS unit
Graziani and Xibilia [24]	Deep neural network	The research octane number of gasoline
Wang et al. [13]	Ensemble of regression models	10%, 50% and 90% boiling point of kerosene
Dias et al. [25]	Bagging and Random forest	Gasoline RON
Yuan et al. [26]	Dynamic convolutional neural network	90% recovery temperature of diesel
Steurtewagen and Van den Poel [27]	Random forest and XG Boost	Catalyst saturation level in FCC
Mojto et al. [4]	Regression	Depropanizer column bottom product composition Hydrogenated gas oil product purity
Vijayan et al. [17]	JITL SVR	Naphtha IBP and EBP
Yuan et al. [19]	Multi-rate stacked autoencoder	50% boiling point and cetane number of diesel
Yuan et al. [18]	LSTM network	IBP of heavy naphtha and aviation kerosene
Ferreira et al. [28]	Kaizen programming	Composition of C_4 hydrocarbon in the distillate
Yoon et al. [2]	Stacked recurrent neural network	Kinematic viscosity at 100°C (KV100) of distillates from VDU
Media et al. [29]	Random Forest and XG Boost	Diesel flash point temperature
Ou et al. [20]	Quality-driven regularization-based SAE	90% boiling point of aviation kerosene
Li et al. [30]	LASSO-particle swarm optimization-deep belief network	Drypoint of aviation kerosene
Mattos et al. [10]	Azure AutoML framework	ASTM 95% distillation temperature of heavy naphtha

3.6 DATA ANALYTICS IN THE OIL AND GAS INDUSTRY—CASE STUDY

3.6.1 Introduction

Recent technological advancements have ushered in an era where oil and gas exploration and production industries grapple daily with vast data. This surge in data generation has brought a pressing challenge for these industries: effective data management. Echoing this sentiment, Brule [31] highlights that a substantial portion of a petroleum engineer's and geoscientist's workload—over half, in fact—is consumed by the tedious tasks of data searching and assembly.

"Big Data" describes the innovative methodologies employed to manage, process, and make sense of these colossal data troves. These datasets are characterized by their variety, volume, and velocity, stemming from numerous facets of upstream and downstream operations within the oil and gas sector [32–40]. When harnessed correctly, this data holds the potential to unravel complex engineering enigmas by revealing the fundamental principles that govern them.

A notable insight from Mehta [41] sheds light on the industry's burgeoning interest in Big Data. A collaborative survey by General Electric and Accenture revealed that 81% of industry executives ranked Big Data as a critical priority for 2018, primarily driven by the imperative to enhance the efficiency of exploration and production activities. This enthusiasm for Big Data marks a significant shift in perspective when juxtaposed with Feblowitz's [42] Findings from 2013. An IDC Energy survey conducted in 2012 found that 70% of U.S. oil and gas company respondents were then largely unaware of Big Data's potential applications within petroleum engineering, underscoring a rapid evolution in industry sentiment from 2012 to 2018.

This study comprehensively reviews contemporary literature on Big Data analytics' application across the oil and gas industry's value chain. Initially, it delves into defining Big Data, outlining its characteristics, and introducing the tools available for data processing. Subsequently, the focus shifts to detailing how Big Data analytics is being leveraged within the oil and gas sector to drive innovation and efficiency. The concluding segment of the study addresses the myriad challenges that persist in integrating and optimizing Big Data analytics within industry practices, setting the stage for a discussion on potential pathways forward.

3.6.2 Big Data Analytics

3.6.2.1 Big Data Definition

Big Data encompasses diverse information, ranging from unstructured data, which lacks a predefined format and is often text-heavy, to multi-structured data, born from the interactions between humans and machines and manifesting in various formats [43]. The concept of Big Data, also known in some circles as Big Data analytics or business analytics, initially centers on the sheer scale of the data sets available for analysis. However, the essence of Big Data extends beyond mere size, encapsulating additional attributes that render it amenable to specialized analytical tools. These attributes have been succinctly captured by IBM's three Vs: volume, variety, and velocity [44]. The discourse has evolved, with scholars appending two more Vs—integrity and value—to provide a more holistic definition of Big Data [45].

Volume refers to the vast amounts of data amassed from myriad sources, such as sensors and data recording devices. The challenge lies in storing and analyzing such vast data reserves sustainably and efficiently. Many organizations are awash with data but need more tools to process and glean insights from it. Big Data technologies promise to bridge this gap by offering sophisticated processing and analytical capabilities [45].

In the oil and gas sector, the volume attribute of Big Data is vividly illustrated through various stages like exploration, drilling, and production. Seismic data acquisition during exploration, for instance, generates prodigious amounts of data, facilitating the construction of 2D and 3D subsurface models. Innovations such as narrow-azimuth towed streaming and comprehensive azimuth techniques in offshore seismic studies further amplify data generation, necessitating robust analytical tools to process and interpret this information. Similarly, advancements in drilling technologies, like logging while drilling (LWD) and measurement while drilling (MWD), contribute to the data deluge by transmitting real-time data to the surface. During production phases, integrating optical fibers and various sensors within well tubular enables the continuous monitoring of parameters like fluid pressure, temperature, and composition [42].

Velocity, another critical dimension of Big Data, refers to the rapidity with which data is generated, transmitted, and processed. The challenge with velocity lies in the disparity between the available processing capacities and the sheer volume of data produced. In today's digital era, the rate of data generation is staggering, with estimates suggesting that 5 exabytes of data are created every two days—a volume equivalent to all data generated by humanity until 2003 [46]. For the oil and gas industry, where operations are inherently complex, the ability to process vast datasets swiftly is not just a convenience but a necessity, especially in scenarios where real-time data interpretation can prevent catastrophic events like blowouts during drilling operations [42].

Variety refers to the myriad forms in which data presents itself, encompassing structured, semi-structured, and unstructured formats. This diversity stems from the devices and sensors employed, each contributing data in distinct sizes and formats, ranging from textual and numerical to multimedia types like images, audio, and video. While a significant portion of data in the oil and gas industry is structured, emanating from supervisory control and data acquisition systems, production logs, and project management reports, a considerable volume of unstructured data, such as daily drilling reports and computer-aided design drawings. Semi-structured data, often the output of various modeling and simulation exercises, occupies a middle ground, retaining some structural elements while eschewing a rigid format [42].

Veracity, the fourth V, concerns the integrity and reliability of data. In the context of Big Data, veracity involves distinguishing high-quality, actionable data from "dirty" data that may skew analysis and lead to erroneous conclusions. This distinction is crucial in fields like oil and gas, where data often originates from complex subsurface environments and is fraught with uncertainties. Ensuring the cleanliness and accuracy of this data is imperative for making informed decisions [42].

Value, the final V, underscores the importance of deriving tangible benefits from Big Data investments. Big Data analytics' ability to unearth trends, predict equipment performance, and preempt failures can give companies a competitive edge, transforming raw data into actionable insights that drive operational efficiency and strategic decision-making.

Beyond these five Vs, some scholars advocate including a sixth dimension: complexity. This attribute acknowledges the intricate nature of the problems Big Data seeks to address, especially in fields like oil and gas, where the stakes are high, and the data is often dense and nuanced. With their advanced algorithms and computational capabilities, Big Data tools are well suited to untangle these complexities, revealing patterns and insights that might otherwise remain obscured [47].

3.6.2.2 Big Data Methodology

Navigating the vast ocean of Big Data, particularly when addressing complex challenges, necessitates harnessing cutting-edge and potent technological solutions. The effectiveness of Big Data analytics relies heavily on the availability of sophisticated tools and technologies characterized by their speed, accuracy, and processing power. This chapter segment is dedicated to exploring and shedding light on the array of tools and technological innovations at the forefront of Big Data analytics (see Figures 3.2–3.5).

The cornerstone of practical Big Data analytics is utilizing advanced computational technologies. These include high-performance computing (HPC) systems, which provide the backbone for processing massive volumes of data in real time or near real time, ensuring that insights are derived promptly and efficiently. HPC systems leverage parallel processing techniques to break down enormous datasets into manageable chunks, facilitating faster analysis and decision-making processes.

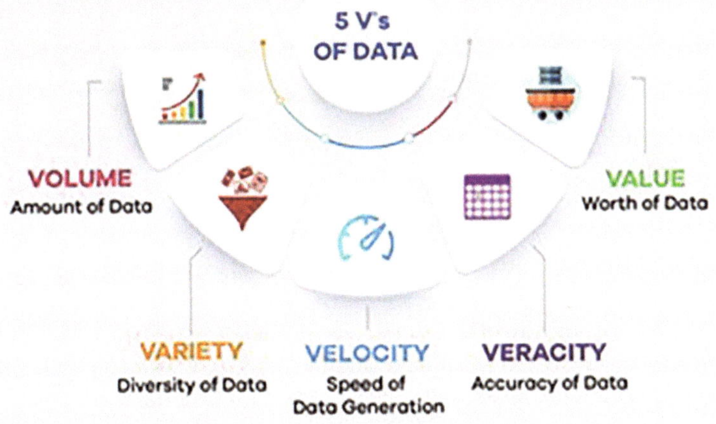

FIGURE 3.2 Big Data characteristics.

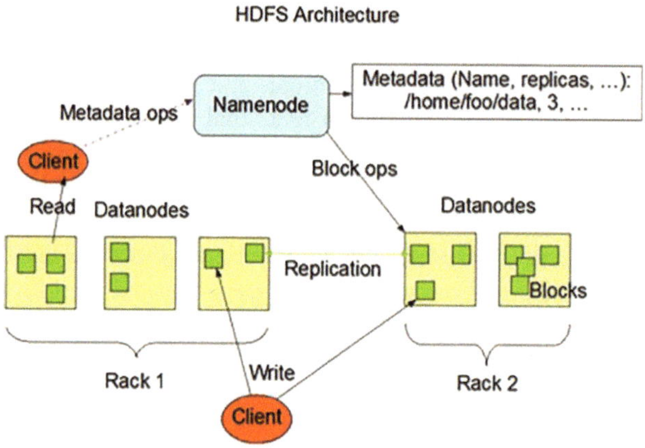

FIGURE 3.3 HDFS architecture with Namenode and Datanodes.

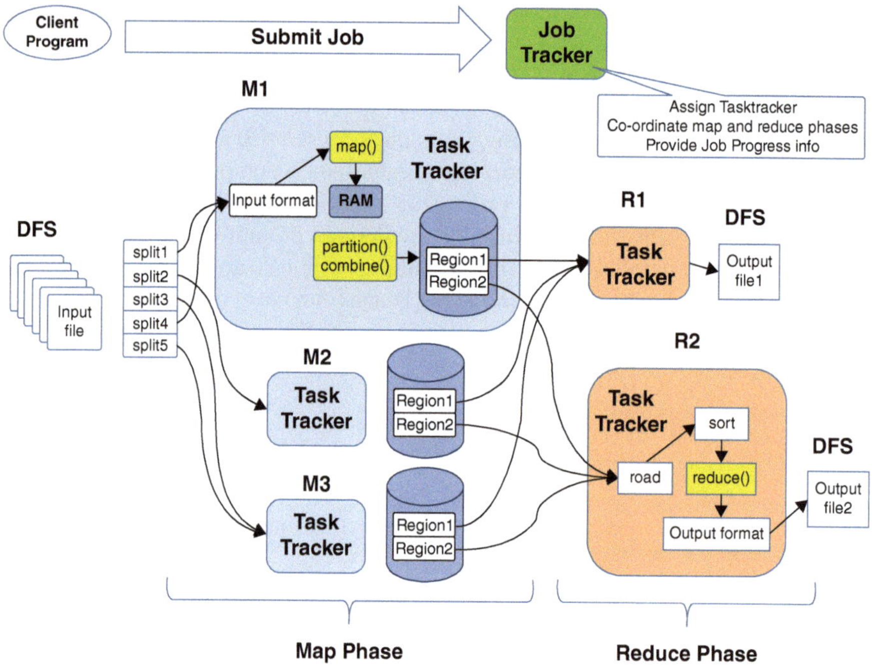

FIGURE 3.4 MapReduce architecture with Map and Reduce phases.

ML and AI are other pivotal technologies in Big Data analytics. These technologies empower computers to learn from and interpret data without explicit programming, adapt to new data inputs, and evolve in accuracy over time. ML algorithms and AI systems are particularly adept at identifying patterns and anomalies within large datasets, making them invaluable for predictive analytics and complex problem-solving within diverse sectors, including the oil and gas industry.

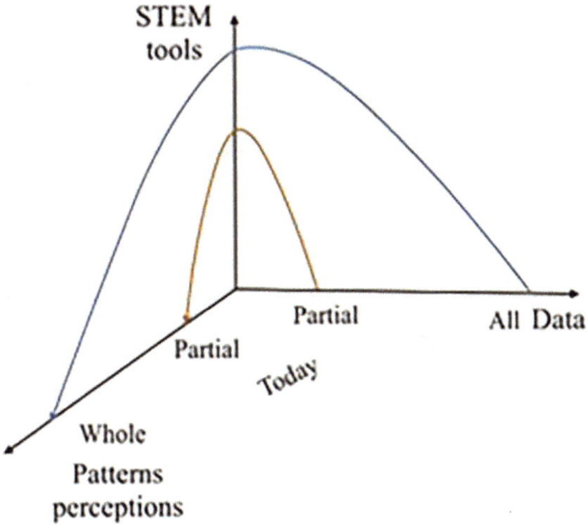

FIGURE 3.5 The relationship between data, STEM tools, and patterns perception.

Cloud computing platforms also play a critical role in the Big Data ecosystem. They offer scalable and flexible resources that can be tailored to the needs of Big Data projects. Cloud services provide access to vast amounts of computing power and storage, enabling organizations to process and analyze Big Data without significant upfront investments in physical infrastructure. This democratizes access to Big Data analytics, allowing even smaller entities to leverage these insights.

Data management and integration tools are essential for preparing and harmonizing the diverse datasets that comprise Big Data. Technologies such as data lakes and warehouses enable the storage and organization of vast quantities of structured and unstructured data, making it accessible and usable for analysis. Moreover, data integration tools facilitate merging data from disparate sources, ensuring a cohesive dataset primed for comprehensive analysis.

Visualization tools and software are indispensable for translating complex data insights into understandable and actionable information. These tools enable the creation of intuitive graphs, charts, and dashboards that succinctly convey the findings derived from Big Data analytics, allowing decision-makers to grasp critical trends and patterns at a glance.

Furthermore, edge computing has brought about a paradigm shift in how data is processed and analyzed. By processing data closer to the source, edge computing reduces latency and bandwidth, enabling real-time analytics and decision-making in environments where rapid responses are crucial.

In addition to these technologies, developing specialized analytical software and platforms tailored to specific industries and use cases has further expanded the capabilities of Big Data analytics. These platforms integrate various analytical tools and technologies, offering end-to-end solutions that span data ingestion, processing, analysis, and visualization.

In conclusion, the landscape of Big Data analytics is rich with diverse and influential technologies, each contributing unique capabilities that empower organizations to harness the full potential of their data. As these technologies continue to evolve and intersect, they promise to unlock even greater insights and efficiencies, driving innovation and progress across multiple domains.

3.6.2.2.1 Apache Hadoop

Developed by Doug Cutting and Mike Caferella in 2005, Hadoop is an open-source framework that owes its name to a toy elephant, symbolizing the project's strength and capacity to handle vast amounts of data [45]. Crafted primarily in Java, Hadoop is designed for distributed processing across extensive computer clusters, enabling it to efficiently manage and analyze large-scale data sets [44, 45]. At the heart of Apache Hadoop lies its ability to facilitate scalable computing through parallel data processing, a feature that is foundational to its widespread adoption and success.

The architecture of Apache Hadoop is built upon two foundational layers: the Hadoop Distributed File System (HDFS) and MapReduce, each serving distinct yet complementary roles in the data processing lifecycle [48]. HDFS, embodying an enslaver/enslaved person architecture, is the layer responsible for data storage. It is organized under the supervision of a central NameNode (the master) and numerous DataNodes (the enslaved people), as illustrated in the depicted HDFS architecture. This setup ensures efficient data distribution and storage across the cluster, laying the groundwork for resilient and scalable data management.

The second layer, MapReduce, handles task tracking and execution. It employs a similar controller/agent architecture, with JobTracker and TaskTracker as the controller and agent nodes [48]. Data processing within Hadoop unfolds through two distinct phases: the Map and Reduce phases, orchestrated by the MapReduce framework. This framework excels at processing vast data sets in parallel across multiple clusters and boasts scalability, flexibility, and fault tolerance.

During the Map phase, data is categorized into "Key" and "Value" pairs, where "Key" represents the node ID, and "Value" encapsulates the node's attributes. The essence of MapReduce lies in its ability to ingest data in these key-value pairs, with JobTracker delegating specific tasks to various TaskTrackers. These TaskTrackers then proceed to process the data further. Following the processing, the data output from the Map phase is temporarily sorted and stored locally, serving as a bridge to the subsequent Reduce phase. In the Reduce phase, this sorted data undergoes a consolidation process, where the segmented inputs are amalgamated to form cohesive outputs, as delineated in the MapReduce architecture diagram.

This dual-layered architecture, characterized by its robust data storage and efficient processing capabilities, underscores Hadoop's pivotal role in Big Data analytics. By enabling the distributed processing of massive data sets, Hadoop facilitates deeper insights and more informed decision-making, embodying a cornerstone technology in today's data-driven landscape.

3.6.2.2.2 MangoDB: An Overview of a Nonrelational Database Paradigm

In the realm of database technologies, MangoDB emerges as a quintessential example of NoSQL, a paradigm diverging from traditional relational database systems. Distinctively document-oriented, MangoDB is architected around the JSON (JavaScript Object Notation) format, a choice that underpins its flexibility and accessibility. JSON is a lightweight data-interchange format deeply rooted in JavaScript, designed to be easily understandable by humans and effortlessly parsed by machines. It organizes data into two fundamental structures: a collection of key/value pairs akin to objects in programming languages and an ordered list of values, which mirrors arrays.

MangoDB, crafted in the robust C++ programming language, symbolizes the NoSQL movement's adept management of unstructured data. This encompasses many data types, from sprawling documents and multimedia content to intricate social media interactions. Such versatility renders NoSQL databases adept at handling the voluminous and varied data generated in today's digital ecosystem.

A hallmark of MangoDB is its dynamic schema. This malleable framework permits databases to be tailored to the specific demands and workflows of diverse applications, a stark contrast to the rigid schemas of traditional relational databases. This adaptability facilitates a more organic development process, where data models can evolve with the application's requirements without requiring extensive reconfiguration or downtime.

Furthermore, MangoDB's document-oriented approach allows for a more intuitive mapping of real-world entities and relationships within the database, streamlining development and data retrieval processes. This is particularly advantageous in scenarios with complex and hierarchical data structures, as it enables a more natural and efficient representation of data entities and their interconnections.

In summary, MangoDB exemplifies the innovative capabilities of NoSQL databases in managing unstructured data through its document-oriented structure, JSON-based data format, and flexible schema. This technology provides a robust and scalable solution tailored to meet modern applications' diverse and evolving needs, underscoring its significance in the current database landscape as referenced in studies [43, 49–51].

3.6.2.2.3 *Cassandra: A Deep Dive into Key-Value and Column-Oriented Database Technology*

Cassandra stands out in the landscape of NoSQL database technologies with its unique blend of key-value and column-oriented data structures. Originally conceived within Facebook's innovative corridors to address specific scalability and performance challenges, Cassandra was released as an open-source project, significantly broadening its applicability and adoption across various industries.

At its core, Cassandra is designed to excel in environments that can afford the initial investment in mastering its complexities. Its substantial scalability, performance, and flexibility advantages offset this upfront learning curve. Unlike traditional relational databases that organize data in rows and tables, Cassandra's data model is predicated on keys and columns. This architecture enables efficient data storage and retrieval, mainly for handling vast volumes of data spread across many servers without compromising performance.

One of Cassandra's hallmark features is its distributed nature, allowing it to scale seamlessly across multiple data centers and cloud regions. This capability ensures high availability and fault tolerance, making it an ideal choice for applications that require uninterrupted access to large datasets. Furthermore, Cassandra's design permits incremental scalability, meaning it can grow alongside the application's needs by adding more nodes to the cluster without downtime or significant reconfiguration.

Cassandra's column-oriented structure facilitates flexible schema design. Columns can be added to any row without altering the entire table schema. This is particularly beneficial for applications that evolve and necessitate changes to the data model without extensive migration processes.

Cassandra's robustness and scalability have made it a preferred choice for various applications, from real-time analytics and data warehousing to Internet of Things (IoT) and e-commerce platforms. Its ability to handle large volumes of data across distributed environments, as well as its high performance and flexibility, underscores its value proposition in scenarios where traditional database systems might falter.

In conclusion, Cassandra's journey from a Facebook project to a widely adopted open-source NoSQL database highlights its efficacy in managing large-scale, distributed data with high throughput requirements. Its key-value and column-oriented data model offers a compelling blend of performance, scalability, and flexibility, making it an indispensable tool in the modern data management toolkit, as evidenced by literature and practical applications.

3.6.2.3 Big Data Processing: Unveiling Insights from Vast Datasets

The advent of Big Data has ushered in a new era where data is not only voluminous but also highly complex and varied. The imperative to harness this data for strategic insights has led to developing and adopting sophisticated processing tools. These tools are designed to distill vast and intricate data sets into actionable intelligence, facilitating informed decision-making across diverse domains.

Many processing tools have been developed to navigate the multifaceted landscape of Big Data, each tailored to specific aspects of data analysis. The following is an overview of some of the prominent tools that have become benchmarks in Big Data processing:

1. **Hadoop:** Emblematic of Big Data processing, Hadoop is a framework that allows for the distributed processing of large data sets across clusters of computers using simple programming models. It is designed to scale from single servers to thousands of machines, each offering local computation and storage.
2. **Spark:** Renowned for its speed and ease of use, Spark is a unified analytics engine for large-scale data processing. It can perform batch processing, real-time streaming, ML, and graph processing, making it a versatile tool for various data analysis tasks.
3. **Flink:** Focused on real-time data stream processing, Flink offers robust, scalable, and efficient handling of bounded and unbounded data streams. It is known for providing accurate real-time results, making it ideal for time-sensitive applications.
4. **Storm:** Tailored for real-time computation, Storm enables the processing of unbounded data streams. It benefits real-time analytics, online ML, and continuous computation, delivering fault tolerance and scalability.

5. **Kafka:** Originally designed as a messaging queue, Kafka is now widely used for real-time streaming data pipelines and applications. It provides high throughput, built-in partitioning, replication, and fault tolerance, making it suitable for large-scale data processing tasks.
6. **NoSQL Databases:** This category includes databases like Cassandra, MangoDB, and HBase. These databases are designed to handle various data types, including structured, semi-structured, and unstructured data. They offer scalability and flexibility for Big Data storage and real-time access.
7. **Data Lakes:** Platforms like Amazon S3 and Azure Data Lake enable the storage of vast amounts of raw data in their native format. These data lakes support the analysis and processing of Big Data using various analytical tools and frameworks.

These processing tools, among others, constitute the backbone of modern Big Data analytics, enabling organizations to extract meaningful patterns, trends, and insights from their data reservoirs. As data volume, velocity, and variety expand, these tools evolve and adapt, offering more sophisticated, efficient, and scalable solutions for Big Data challenges. The selection of the appropriate tool or combination of tools depends on the specific requirements of the task at hand, including the nature of the data, the desired outcomes, and the computational resources available.

3.6.2.3.1 R Programming: A Confluence of Statistical Computation and Graphic Excellence

R is a paradigm of modern programming languages tailored explicitly for statistical analysis and graphical representation. Conceived by Robert Gentleman and Ross Ihaka, its genesis was driven by the aspiration to offer a comprehensive data analysis toolkit incorporating functional and object-oriented programming paradigms. This dual approach equips R with the agility needed for the iterative exploration of data and the robustness required for systematic software development.

R's prowess is significantly amplified by its wide array of built-in functions, designed to streamline the processing of large data sets and the execution of complex statistical simulations. This intrinsic functionality and its extensive package ecosystem enable researchers and data scientists to apply sophisticated statistical techniques and models readily. The package system, a cornerstone of R's architecture, facilitates an open and collaborative environment where code and methodologies can easily be shared and replicated, fostering innovation and consistency within the community.

Moreover, R's capability to produce high-quality graphical outputs is unparalleled. This feature enhances the interpretability of data analyses and seamlessly integrates the workflow from data exploration and model building to the final stages of publication-ready output. This cohesive workflow capability makes R a preferred choice for academics, researchers, and professionals who seek an end-to-end solution within a single programming environment.

R's specialization in statistical computation is underscored by its comprehensive suite of modules and toolboxes, which are adept at handling a broad spectrum of data analysis tasks. From data ingestion and transformation to sophisticated statistical modeling and visualization, R provides a coherent and integrated platform for data analytics. Its ability to concisely express complex data operations and statistical computations and its graphical capabilities make R an invaluable tool for data-driven research and decision-making.

However, it is essential to acknowledge R's limitations, particularly its reliance on in-memory data storage. This constraint means that R's performance is inherently tied to the machine's memory capacity on which it operates, potentially limiting its utility in scenarios involving massive datasets that exceed a single machine's memory capacity. This challenge underscores the importance of considering the scale of data and computational resources in the planning stages of a project intended for R.

In summary, R embodies a sophisticated statistical analysis and graphical representation ecosystem, offering a rich set of features that cater to a wide range of data analysis needs. Its functional and object-oriented programming integration, comprehensive package system, and superior graphical capabilities position R as a tool of choice for statistical computation and data visualization. Nonetheless, awareness of its memory-dependent performance is crucial for optimizing its application across diverse data analytical projects.

3.6.2.3.2 Exploring Datameer: Enhancing Data Processing with Hadoop Integration

Datameer represents a significant stride in the democratization of data analytics. It embodies an intuitive programming platform that leverages Hadoop's power to elevate its data processing capabilities. Its design philosophy centers on simplifying the complex landscape of Big Data analytics, making advanced data processing accessible to a broader range of users, including those with limited technical expertise.

A standout feature of Datameer is its commitment to user accessibility. The platform offers tools to streamline the data import, enabling users to integrate data from various sources easily. This functionality, combined with its robust output visualization capabilities, empowers users to transform raw data into insightful, actionable information with relative ease.

Integrating Hadoop into Datameer's architecture is a pivotal aspect of its functionality. It allows it to harness Hadoop's scalable computing resources for enhanced data processing performance. This synergy enables Datameer to manage and analyze vast datasets more efficiently, making it a valuable tool for enterprises dealing with large data volumes.

Datameer's user-friendly interface plays a crucial role in its growing popularity. By abstracting the complexities inherent in Big Data technologies such as Hadoop and Datameer lowers the barrier to entry for conducting sophisticated data analysis. This approach facilitates a broader adoption of data analytics practices across various sectors. It enables users to focus more on deriving insights from data and less on navigating the intricacies of the underlying technology.

Furthermore, Datameer's emphasis on visualization tools underscores the importance of data presentation in the analytical process. By providing a range of visualization options, Datameer assists users in crafting compelling narratives with their data, enhancing the communicative power of their analytical findings.

Anticipated to capture increasing interest from the data analytics community, Datameer stands out for its blend of simplicity, power, and versatility. Its user-centric design and Hadoop's computational prowess position it as a formidable tool for a wide array of data processing tasks. As the platform continues to evolve, it is expected to offer even greater functionalities, further solidifying its role in simplifying and advancing the field of data analytics.

3.6.2.3.3 BigSheets: IBM's Web-Based Solution for Simplified Data Analysis

IBM's BigSheets emerged as a pioneering web application to democratize data analytics. It caters primarily to users with limited technical expertise or those not traditionally versed in data science disciplines. This innovative tool facilitates the collection, analysis, and visualization of unstructured data from various online and internal sources, bridging the gap between complex data analytics and non-technical user engagement.

At the heart of BigSheets' functionality is its adept use of Hadoop's distributed computing capabilities, which enables it to tackle vast datasets efficiently and effectively. This integration allows BigSheets to process and analyze large volumes of unstructured data, transforming it into actionable insights without requiring users to delve into the technical complexities of Hadoop's ecosystem.

Further enhancing its utility, BigSheets incorporates technologies like OpenCalais, a tool designed to extract structured information from unstructured datasets. This feature significantly streamlines the data analysis, enabling users to derive meaningful patterns and insights from diverse data sources.

One of BigSheets' key strengths is its intuitive, spreadsheet-like interface, which resonates with a broad user base familiar with conventional spreadsheet applications. This familiarity lowers the learning curve, making it more accessible for individuals to conduct sophisticated data analyses without requiring advanced technical skills.

Beyond mere data processing, BigSheets strongly emphasizes data visualization. Through its array of simple yet effective visualization tools, users can effortlessly present their analytical results visually engagingly. This not only aids in interpreting complex data but also enhances the communicability of insights derived from the analysis.

BigSheets is particularly suited for individual data analysis projects. Its ease of use and spreadsheet-like interface offer a seamless transition for those accustomed to traditional data management tools. Its easy handling of unstructured data, coupled with the power of Hadoop and the precision of tools like OpenCalais, makes BigSheets a valuable asset for users seeking to navigate the vast seas of data analytics with confidence and simplicity.

In summary, IBM's BigSheets stands as a testament to the evolving landscape of data analytics, where the power of Big Data is made accessible to a broader audience. By combining the robustness of Hadoop with user-friendly interfaces and visualization tools, BigSheets empowers users to harness the potential of their data, making it an indispensable tool in the repertoire of modern data analysis methodologies.

3.6.2.4 Big Data's Role in Transforming the Upstream Oil and Gas Sector

The transformative power of Big Data extends its reach far beyond traditional realms such as marketing, business analytics, and database management, making significant inroads into various engineering fields. Among these, the upstream oil and gas sector stands out as a domain where Big Data's impact is increasingly pronounced, driven by the industry's exponential growth in data generation. This surge in data availability is attributed to advancements in technologies such as seismic acquisition devices, enhanced channel counting, sophisticated fluid front monitoring geophones, as well as developments in carbon capture and sequestration sites, along with LWD and MWD tools. These innovations have collectively contributed to a data-rich environment ripe for applying Big Data analytics.

The relevance of Big Data in the upstream oil and gas industry is further elucidated by the work of Anand, who provides a compelling narrative on the potential of Big Data to uncover latent insights from the industry's extensive data repositories. Anand's discussion is anchored around a conceptual 3D plane that illustrates the interplay between combining data, science, technology, engineering, and mathematics (STEM) disciplines and pattern recognition capabilities. This model proposes that using essential STEM tools with a limited dataset might yield only rudimentary patterns, potentially needing more depth and carrying considerable uncertainty. Conversely, engaging a more voluminous dataset with advanced STEM tools opens the door to identifying more definitive patterns, thereby reducing uncertainty and aligning findings more closely with reality.

This dichotomy underscores a pivotal theme in applying Big Data within the oil and gas sector: the depth and reliability of insights are directly influenced by the quantity of data and the sophistication of the analytical tools employed. Big Data can facilitate the transition from sparse, simplistic data analysis to comprehensive, nuanced examinations, significantly enhancing decision-making processes, risk assessment, and operational efficiency in upstream oil and gas operations.

Moreover, integrating Big Data analytics in this sector is not just about handling voluminous data; it is about leveraging it to drive innovations in exploration, production optimization, and environmental sustainability. By harnessing the power of Big Data, the upstream oil and gas industry can achieve more accurate reservoir characterization, optimize drilling and production strategies, and enhance environmental impact monitoring.

In summary, the upstream oil and gas industry's embrace of Big Data signifies a paradigm shift toward more informed and strategic operational methodologies. The vast amounts of data generated by modern technological advancements, when analyzed with sophisticated STEM tools, have the potential to unveil patterns and insights previously obscured. This reduces uncertainty and propels the industry toward more efficient and sustainable practices, marking a new era of data-driven innovation in oil and gas exploration and production.

3.6.2.5 Big Data's Pivotal Role in Seismic Exploration

The interpretation of seismic data, a cornerstone of exploration in the oil and gas industry, has been profoundly transformed by Big Data analytics and advancements in computational technology. The surge in data volume attributable to state-of-the-art seismic acquisition devices necessitates a departure from traditional analysis methods toward more sophisticated, data-driven approaches. Utilizing Big Data in analyzing seismic datasets represents a significant leap forward, offering more profound insights into subsurface structures and facilitating more accurate resource identification.

ML, a subset of Big Data analytics, has emerged as a particularly effective tool for deciphering complex relationships within vast seismic datasets. The enhanced efficiency of ML algorithms in handling large volumes of data marks a pivotal advancement in seismic data analysis. An illustrative example of this approach is the research conducted by Roden, which integrates principal component analysis (PCA) with Self-Organizing Maps (SOM) to perform comprehensive multi-component seismic analyses. This research delineates a structured five-stage process, beginning with the precise definition of geological challenges, followed by PCA to identify crucial attributes, the application of SOM leveraging ML for predictive modeling, detailed 2D mapping to highlight key geological features, and culminating in a sensitivity analysis to fine-tune the findings through various attributes and training scenarios.

In another groundbreaking study by Joshi et al., the potential of Big Data to revolutionize the analysis of micro-seismic datasets was showcased. This study focused on modeling fracture propagation maps during hydraulic fracturing processes, employing the Hadoop platform to manage the extensive data generated. The researchers adeptly utilized datasets spanning exploration, drilling, and production phases to characterize reservoirs, enhancing predictive accuracy by identifying potential anomalies based on historical data.

Olneva et al. extended the application of Big Data in seismic exploration by studying the West Siberian Petroleum Basin. Their research employed an innovative dual-approach methodology, moving "from general to particulars" and "from particulars to general." The first approach leveraged drilling data and regional maps across 5,000 wells to establish broad geological patterns. In contrast, the second approach refined these insights using detailed seismic and geological data from over 40,000 km of exploration.

These studies collectively underscore the transformative impact of Big Data on seismic exploration within the oil and gas sector. By harnessing sophisticated data analytics and ML techniques, researchers and industry professionals can now explore previously untapped depths of seismic data, unveiling intricate geological structures with unprecedented precision. This paradigm shift enhances resource identification accuracy and propels the industry toward more efficient and sustainable exploration practices.

3.6.2.6 Leveraging Big Data in Enhancing Drilling Operations

The drilling sector is experiencing a data-driven revolution catalyzed by the influx of information from digital rig sites and human operators' manual inputs. The diversity of data sources, encompassing everything from real-time sensor readings to operational logs, provides a rich tapestry for analytical endeavors. Integrating novel data recording instruments and formats has further amplified the potential to harness Big Data technologies in drilling operations, with over 60 sensors routinely capturing myriad parameters throughout the drilling process.

Pioneering work by Duffy et al. illustrates the tangible benefits of applying Big Data analytics to drilling operations. Their study, which focused on optimizing weight-to-weight connection times during pad drilling in the Bakken formation, exemplifies the efficiency gains achievable through data-driven strategies. Implementing best-practice initiatives, informed by an automated drilling state detection service, yielded a significant reduction in non-productive time, culminating in a time saving of over 11.75 days across a nine-well pad, alongside a 45% improvement in overall non-drilling time efficiency.

Similarly, Maidla et al.'s research underscores the value of Big Data analytics in refining drilling performance by integrating diverse data sets, including electronic drilling recorder outputs and morning reports. Their emphasis on data quality control, filtering, and a robust understanding of drilling processes' underlying physics is critical for deriving actionable insights. This meticulous approach helps avoid misleading conclusions resulting in resource wastage and operational delays.

Yin et al. explored the utilization of Big Data to uncover invisible non-production time (INPT) using real-time logging data, showing how mathematical statistics, AI, and cloud computing can collectively optimize drilling operations. This optimization of INPT represents a forward leap in operational efficiency and cost-effectiveness.

In the context of risk mitigation, the study by Johnston and Guichard leveraged extensive drilling, well logging, and geological formation data from approximately 350 wells in the UK North Sea. Despite the heterogeneity of data formats, their work highlights the critical challenge of data aggregation and processing, emphasizing the foundational role of data management in leveraging Big Data analytics for drilling operations.

Hutchinson et al.'s study further exemplifies the innovative application of Big Data in drilling. It utilizes downhole vibration sensor data to analyze drill string dynamics. By merging actual sensor data with simulation outputs, their research facilitated the development of a drilling automation application that minimizes the risk of drilling failures and contributes to significant cost savings in drilling development.

These studies collectively illuminate the transformative impact of Big Data analytics on the drilling industry. Big Data stands at the forefront of driving operational excellence and innovation in drilling processes by enabling more precise and efficient operations, reducing downtime, and mitigating risks. The ability to navigate and analyze vast datasets, drawing from sensors and operational logs, is instrumental in enhancing the safety, efficiency, and cost-effectiveness of drilling operations worldwide.

3.6.2.7 Advancements in Reservoir Engineering through Big Data Analytics

The field of reservoir engineering is undergoing a significant transformation, driven by the integration of Big Data analytics and the proliferation of advanced downhole sensing technologies. Distributed sensors like distributed temperature sensors (DTS), distributed acoustic sensors (DAS), and permanent downhole gauges (PDG) are now pivotal in generating vast datasets that enhance reservoir characterization and management. The work by Bello et al. exemplifies this shift, showcasing a reservoir management application that leverages Big Data through components like data visualization, filtration, ML-based modeling, and application deployment on web platforms for improved user interaction.

Parallel to these developments, the reservoir simulation landscape is evolving with the infusion of AI and data mining into traditional methods. This synergy fosters a new era of reservoir modeling techniques characterized by Closed-Loop Reservoir Management (CLRM) and Integrated Asset Modeling (IAM). These data-driven approaches promise to capture complex reservoir behaviors that elude conventional theoretical models, thus offering a more nuanced understanding of reservoir dynamics.

In environmental stewardship, Haghighat et al.'s study on CO_2 sequestration epitomizes the potential of Big Data in mitigating climate change impacts. By employing ML algorithms to analyze pressure data from downhole gauges, they developed a sophisticated real-time detection system for CO_2 leakage, demonstrating Big Data's role in enhancing the safety and efficacy of sequestration projects.

Popa et al.'s exploration of optimizing heavy oil reservoir operations in Chevron's San Joaquin fields further underscores Big Data's utility. By analyzing a comprehensive array of data from over 14,200 wells, the study illustrates how Big Data can streamline operations ranging from steam-assisted gravity drainage (SAGD) to cyclic steam injection, optimizing resource recovery.

Beyond conventional resources, Big Data is also reshaping the exploration and exploitation of unconventional oil and gas reserves. Lin's work integrates analytical and physics-based models with Big Data to refine reservoir simulations, offering more profound insights into unconventional reservoir behaviors.

Moreover, the application of Big Data extends to enhancing hydraulic fracturing operations, as demonstrated by Udegbe et al. Through analyzing production data and applying pattern recognition techniques, their study reveals critical trends and parameters that influence fracturing efficacy, akin to advancements in facial recognition technology.

Finally, Big Data analytics is revolutionizing the strategic deployment of Enhanced Oil Recovery (EOR) methods. Xiao and Sun's research on optimizing EOR through hydrodynamic simulations exemplifies how Big Data can guide the selection and application of EOR techniques, maximizing recovery while minimizing costs and environmental impacts.

Collectively, these studies illuminate Big Data's transformative impact on reservoir engineering, from optimizing recovery processes and enhancing environmental safeguards to pioneering new approaches in reservoir modeling. As Big Data continues penetrating this field, it heralds a new age of efficiency, sustainability, and innovation in oil and gas exploration and production.

3.6.2.8 Harnessing Big Data in Production Engineering Innovations

The discipline of production engineering is witnessing a paradigm shift. Integrating Big Data analytics revolutionizes traditional methodologies and fosters predictive and optimized production strategies. For instance, Seemann et al.'s work at Saudi Aramco exemplifies this shift by developing an intelligent forecast and flow method to automate decline analysis to predict future production trends based on historical data patterns.

Similarly, Rollins et al.'s collaboration with Devon Energy shows the potential of Big Data in refining production allocation techniques. By leveraging public datasets alongside proprietary data and employing Hadoop as a processing backbone, they crafted a sophisticated allocation model that culminates in an intuitive, map-based visual representation of production data, enhancing interpretability and decision-making processes.

Another area in which Big Data has made significant inroads is the optimization of electric submersible pumps (ESPs). Sarapulov and Khabibullin's work, involving the analysis of vast logs from numerous wells, demonstrates how Big Data can pinpoint operational issues like overheating or startup failures, thereby improving ESP reliability and efficiency. Converting diverse data formats into a uniform comma-separated values format was critical in ensuring data consistency and analysis readiness.

Palmer and Turland extended the application of Big Data to the optimization of rod pump wells through a meticulously designed three-step workflow. This encompassed comprehensive data acquisition, automated analytical workflows for model development, and interactive data visualization to provide actionable insights, enhancing operational efficiency and performance.

Shale operators leverage Big Data to refine hydraulic fracturing techniques in unconventional resources. A notable project by Southwestern Energy demonstrated how variables such as proppant loading and fracturing stage spacing could significantly influence productivity, highlighting the critical role of data-driven strategies in optimizing hydraulic fracturing projects.

Ockree et al.'s study further exemplifies the innovative use of Big Data in developing AI-based production-type curves, integrating them with economic analyses for comprehensive field development planning. Their approach, beginning with an extensive data processing pipeline to filter, join, and prepare data for ML, employed the Robust Mahalonobis technique to eliminate outliers, ensuring the integrity and reliability of the analysis.

These pioneering efforts collectively underscore the transformative impact of Big Data on production engineering. From enhancing predictive analytics in production forecasting to optimizing equipment performance and refining hydraulic fracturing operations, Big Data drives efficiency, reliability, and sustainability in production engineering practices. As the industry continues to embrace these advanced analytical capabilities, the potential for innovation and improvement in production operations is boundless, heralding a new era of data-driven decision-making and operational excellence.

3.6.3 Leveraging Big Data in the Downstream Oil and Gas Sector

3.6.3.1 Refining Innovations

The downstream sector of the oil and gas industry is increasingly capitalizing on Big Data to enhance refining processes, asset management, and overall operational efficiency. The study by Plate exemplifies this trend by detailing the application of Big Data in refining through a focused analysis of a four-stage cracked gas compressor (CGC). By systematically analyzing current and historical operating data, the study predicts CGC performance, fine-tunes this prediction based on end-of-life and failure metrics, and culminates in generating comprehensive visual reports. These predictive analytics facilitate informed management decisions and play a crucial role in minimizing equipment downtime and reducing maintenance expenditures.

In a pioneering initiative by Repsol SA, Repsol is integrating Big Data analytics into refinery management, which shows the transformative potential of data-driven strategies in the refining domain. Collaborating with Google Cloud, Repsol aims to harness advanced data analytics and ML technologies to optimize operations at one of its flagship refineries in Spain. This partnership underscores the growing trend of leveraging cloud-based platforms and AI to drive efficiency and innovation in refinery management.

Khvostichenko and Makarychev-Mikhailov's research further demonstrates the application of Big Data in refining, particularly in evaluating the impact of completion parameters on good productivity. Their comprehensive study, encompassing data from 4,500 wells undergoing slickwater treatments, delves into various chemical treatments' effectiveness in enhancing sound output. By employing a statistical t-test approach to analyze data sourced from the IHS Energy database, the study offers valuable insights into optimizing chemical treatments for improved healthy productivity.

These instances underscore Big Data's pivotal role in revolutionizing the downstream sector, mainly refining operations. By enabling more precise predictive maintenance, optimizing refinery management, and enhancing well productivity through data-driven insights, Big Data is setting new benchmarks for operational excellence in the oil and gas industry. As these technologies evolve, their integration into refining operations promises to usher in unprecedented efficiency, sustainability, and competitiveness in the downstream sector.

3.6.3.2 Advancing Oil and Gas Transportation through Big Data Analytics

In oil and gas transportation, the strategic application of Big Data analytics is setting new paradigms for operational efficiency and environmental sustainability. Anagnostopoulos' research is a cornerstone in this evolving field, highlighting Big Data's potential to enhance maritime shipping performance. His study was mainly focused on optimizing ship propulsion power, with the dual objectives of elevating operational efficiency and reducing greenhouse gas emissions—a critical concern in the era of heightened environmental awareness.

The research utilized a comprehensive dataset collected over three months from an array of sensors deployed across a large car and truck carrier vessel, M/V. This rich dataset provided a granular view of the ship's operational dynamics, setting the stage for an in-depth analysis through advanced analytical techniques.

The employment of eXtreme Gradient Boosting (XGBoost) and multi-layer perceptron (MLP) neural networks in Anagnostopoulos' study represents a sophisticated approach to data analysis in the maritime transportation sector. XGBoost, known for its efficiency, scalability, and performance, is particularly adept at handling the complexities of large-scale and high-dimensional data typical in transportation systems. MLP neural networks, with their capability to model complex nonlinear relationships, offer a complementary tool for deciphering intricate patterns within the data, potentially uncovering insights into fuel consumption, vessel performance, and operational efficiencies.

By integrating these powerful analytical tools, the study aimed to predict optimal propulsion power and foster a deeper understanding of the multifaceted factors influencing ship performance. The insights from this analysis could inform strategic decisions around vessel operation, routing, maintenance, and fuel usage, contributing to more sustainable and cost-effective transportation practices.

This research exemplifies the transformative impact of Big Data analytics on the oil and gas transportation sector. By harnessing the vast amounts of data generated by modern vessels, coupled with cutting-edge analytical methodologies, the industry is poised to achieve significant advancements in ship performance optimization. The ripple effects of such innovations extend beyond operational efficiencies to encompass broader environmental benefits, marking a significant step toward greener and more sustainable maritime transportation solutions.

3.6.3.3 Empowering Health and Safety in the Energy Sector with Big Data

Integrating Big Data analytics into health and safety executive (HSE) practices is revolutionizing the approach to ensuring operational safety and environmental stewardship within the energy sector. Notably, the study by Park et al. shows the application of Big Data in enhancing energy efficiency in maritime operations. By leveraging Hadoop and Apache Spark, the researchers developed an energy efficiency model utilizing the energy efficiency operational indicator, derived from comprehensive datasets, including automatic identification system data and marine environmental metrics. This model provides critical insights into optimizing fuel consumption relative to operational parameters, contributing to safer and more sustainable shipping practices.

Similarly, Tarrahi and Shadravan's research underscores the pivotal role of Big Data in bolstering occupational safety within the oil and gas industry. By analyzing extensive injury data from the Bureau of Labor Statistics, covering a wide array of industries, the study employs sophisticated data processing and clustering techniques to unearth underlying trends in workplace injuries. Multidimensional statistical analysis further aids in translating these complex datasets into actionable insights, enhancing risk management and safety protocols.

Pettinger's insights into utilizing safety inspection data for predictive analytics further illuminate the potential of Big Data in preempting safety risks. Organizations can proactively address potential hazards by continuously monitoring and analyzing safety indicator data, such as behavioral assessments and compliance metrics, fostering a safer working environment.

Similarly, Cadei et al.'s innovative approach to hazard prediction through Big Data analytics represents a significant advancement in operational safety. Focusing on the chemical H_2S concentration as a critical hazard indicator, their study integrates diverse data sources, including real-time measurements, historical trends, and maintenance records, to develop predictive models using artificial neural networks and random forests. This comprehensive workflow facilitates accurate hazard forecasting and enhances operational readiness to mitigate risks, showcasing the transformative impact of Big Data on safety and operational integrity in oil and gas production.

These studies collectively highlight Big Data's transformative potential in advancing HSE objectives within the energy sector. The industry is poised to achieve unprecedented safety, efficiency, and environmental sustainability by harnessing vast datasets and employing advanced analytical techniques. As Big Data continues to permeate HSE practices, it paves the way for a more informed, predictive, and proactive approach to safety and operational management, heralding a new era of data-driven excellence in health and safety executive functions.

3.6.4 Navigating Big Data Challenges in the Oil and Gas Sector

The advent of Big Data in the oil and gas industry, like in many other sectors, brings many challenges, particularly in data management, analysis, and application. The costs associated with capturing, storing, and processing vast volumes of data are significant. Innovations like fog computing and the IoT offer promising solutions by decentralizing data storage and processing, alleviating latency issues, and data source mobility.

Cameron's exploration into the specific challenges oilfield service companies face highlights the gaps in personnel expertise and the complexities surrounding data ownership. His advocacy for a holistic approach to Big Data application in oil and gas underscores the need for an interdisciplinary collaboration that bridges computer science with petroleum engineering, ensuring that Big Data solutions are grounded in industry-specific knowledge and are presented through intuitive, user-friendly interfaces.

The digital transformation of oilfields, characterized by an extensive deployment of sensors and data acquisition systems, further complicates the Big Data landscape. The challenge extends beyond mere data collection to encompass the efficient transmission of this data to processing centers, necessitating robust protocols and infrastructure capable of handling diverse, voluminous, and continuous data streams.

Surveys, such as the one conducted by IDC Energy, reveal broader industry-wide challenges, including a prevalent lack of awareness and support for Big Data initiatives, the daunting task of discerning relevant data from the deluge, the scarcity of skilled personnel adept in Big Data technologies, and the prohibitive costs of establishing and maintaining Big Data infrastructure. Addressing these challenges necessitates a concerted effort to educate and engage stakeholders across all levels of the organization, ensuring alignment with business objectives and facilitating the integration of Big Data into core operational processes.

Moreover, as identified by Maidla et al., technical hurdles delve deeper into the nuances of data acquisition and quality. The limitations of current sensor technologies, the frequency and fidelity of data collection, and the imperative to thoroughly understand the underlying physical processes represent critical areas where the expertise of seasoned petroleum engineers becomes invaluable. Their collaboration with data scientists is crucial for crafting Big Data solutions that are not only technologically advanced but also profoundly attuned to the intricacies of petroleum engineering.

Preveral et al.'s recommendation that companies develop bespoke Big Data tools and infrastructures resonates as a strategic approach to surmounting these challenges. Tailoring solutions to specific operational needs can enhance data utility, reduce reliance on generic software solutions, and ultimately optimize investment returns in Big Data technologies.

In summary, while Big Data holds immense potential for transformative impacts across the oil and gas value chain, realizing this potential requires navigating a complex landscape of technical, organizational, and strategic challenges. Through targeted investments in technology, personnel, and cross-disciplinary collaboration, the industry can leverage Big Data not just as a tool for operational enhancement but as a catalyst for innovation and competitive advantage.

3.6.5 Synthesizing Big Data's Role in the Oil and Gas Industry: Insights and Prospects

This comprehensive examination sheds light on the expansive role of Big Data analytics within the oil and gas sector, revealing its multifaceted applications and inherent challenges. Big Data's essence lies in its sheer volume of data and its velocity, variety, veracity, value, and inherent complexity. These dimensions are crucial in understanding the full spectrum of Big Data's potential and its challenges.

The advent of advanced data recording technologies and the pressing need for more efficient exploration and production methodologies has catapulted Big Data to a critical position within the oil and gas industry. Particularly in exploration, seismic technology advancements have led to an exponential increase in data generation, necessitating sophisticated analytical methods like PCA and platforms like Hadoop for practical data interpretation.

In drilling engineering, automated monitoring services' use to analyze drilling data has significantly improved operational efficiency and safety. Similarly, integrating data from an array of sensors, including DTS, distributed dynamic temperature sensing, DAS, PDG, and downhole distributed sensing system, has revolutionized reservoir characterization and simulation, enhancing the precision and reliability of predictive models.

Big Data's influence extends into production engineering, where it has been instrumental in optimizing the performance of electric submersible pumps and refining production allocation methodologies. The downstream sector, too, has witnessed Big Data's transformative impact, with applications ranging from refining optimization to transportation efficiency and HSE advancements.

Despite the burgeoning interest from exploration and production (E&P) companies, the journey to fully leverage Big Data is fraught with challenges. Key among these are the pervasive need for industry-wide awareness and support for Big Data initiatives, concerns over data quality, and the complexities of translating vast data sets into actionable insights.

Addressing these challenges necessitates a concerted effort to enhance industry knowledge of Big Data's potential, invest in data quality improvement, and foster a deeper understanding of the intricate problems Big Data seeks to solve. As the oil and gas industry continues to navigate these challenges, the strategic application of Big Data analytics offers unprecedented opportunities for innovation, efficiency, and sustainability in evolving energy landscapes.

3.7 CONCLUSION

Unlike other industries, the oil and gas industry has recently been unstable. Dynamic supply-demand scenarios, increasing emphasis on green energy, etc., have resulted in stagnant or falling oil prices. This has reduced returns and increased unemployment in the oil and gas industry. Adopting advanced digital technologies in the era of Big Data will result in improved operational efficiency, risk minimization, and revenue maximization in the oil and gas industry. However, the pace of adoption of digital technologies could be faster, which may be attributed to factors such as the difficulty of integrating new technologies with existing ones and concerns about cybersecurity and data protection. Moreover, the application of ML and Big Data analytics techniques in the context of oil and gas industries is more challenging considering the time-varying and uncertain characteristics (such as changing crude composition, changing heat transfer rate due to heat exchanger fouling, varying conversion in reactors due to continuous change in catalyst activity).

The field of data-based soft sensor design and implementation in process industries (including the oil and gas industry) has witnessed massive interest among researchers and industrial practitioners in the past decade, and the field continues to evolve. In addition to classical data-preprocessing techniques, other methods are also increasingly investigated. A few popular ML algorithms for feature selection, such as random forest and gradient boosting, are being explored for input selection. Similarly, modified and improved versions of PCA, PLS, ICA, etc., are also applied to expand the applicability of soft sensors to processes possessing non-linearity and time-varying characteristics. Recent research in this domain also witnessed increasing application of deep learning techniques due to the availability of Big Data and improved computation facilities. Finally, ML techniques for industrial automation are not intended to replace human operators or first principle-based models. Instead, unlike other fields of AI/ML application, the techniques in the industrial context should be used along with existing first principle models as and when available. It must be capable of addressing the issue of uncertainties and time-varying characteristics associated with process data and should be understandable and interpretable by the plant operators and engineers.

REFERENCES

[1] Lu, H., et al., Oil and gas 4.0 era: a systematic review and outlook. *Computers in Industry*, 2019. 111: p. 68–90.

[2] Yoon, Y.S., et al., Development of inferential sensor and real-time optimizer for a vacuum distillation unit by recurrent neural network modeling of time series data. *Computers & Chemical Engineering*, 2022. 168: p. 108039.

[3] Pani, A.K. and H.K. Mohanta, A survey of data treatment techniques for soft sensor design. *Chemical Product and Process Modeling*, 2011. 6(1).

[4] Mojto, M., et al., Data-based design of inferential sensors for petrochemical industry. *Computers & Chemical Engineering*, 2021. 153: p. 107437.

[5] Rousseeuw, P.J. and A.M. Leroy, *Robust regression and outlier detection*. 2005: John Wiley & Sons.

[6] Xu, S., et al., Data cleaning in the process industries. *Reviews in Chemical Engineering*, 2015. 31(5): p. 453–490.
[7] Curreri, F., G. Fiumara, and M.G. Xibilia, Input selection methods for soft sensor design: a survey. *Future Internet*, 2020. 12(6): p. 97.
[8] Triguero, I., S. García, and F. Herrera, Self-labeled techniques for semi-supervised learning: taxonomy, software and empirical study. *Knowledge and Information systems*, 2015. 42: p. 245–284.
[9] Novak, M., et al., Continuous estimation of kerosene cold filter plugging point using soft sensors. *Fuel Processing Technology*, 2013. 113: p. 8–19.
[10] Mattos, J.G., et al., A framework for enhancing industrial soft sensor learning models. *Digital Chemical Engineering*, 2023. 8: p. 100112.
[11] Singh, H., A.K. Pani, and H.K. Mohanta, Quality monitoring in petroleum refinery with regression neural network: improving prediction accuracy with appropriate design of training set. *Measurement*, 2019. 134: p. 698–709.
[12] Ge, Z., et al., Data mining and analytics in the process industry: the role of machine learning. *IEEE Access*, 2017. 5: p. 20590–20616.
[13] Wang, D., J. Liu, and R. Srinivasan, Data-driven soft sensor approach for quality prediction in a refining process. *IEEE Transactions on Industrial Informatics*, 2009. 6(1): p. 11–17.
[14] Rogina, A., et al., Soft sensor for continuous product quality estimation (in crude distillation unit). *Chemical Engineering Research and Design*, 2011. 89(10): p. 2070–2077.
[15] Shang, C., et al., A comparative study on improved DPLS soft sensor models applied to a crude distillation unit. *IFAC-PapersOnLine*, 2015. 48(8): p. 234–239.
[16] Shokri, S., et al., Combination of data rectification techniques and soft sensor model for robust prediction of sulfur content in HDS process. *Journal of the Taiwan Institute of Chemical Engineers*, 2016. 58: p. 117–126.
[17] Vijayan, V.S., H.K. Mohanta, and A.K. Pani, Support vector regression modeling in recursive just-in-time learning framework for adaptive soft sensing of naphtha boiling point in crude distillation unit. *Petroleum Science*, 2021. 18(4): p. 1230–1239.
[18] Yuan, X., et al., Deep learning with spatiotemporal attention-based LSTM for industrial soft sensor model development. *IEEE Transactions on Industrial Electronics*, 2020. 68(5): p. 4404–4414.
[19] Yuan, X., et al., Deep learning for data modeling of multirate quality variables in industrial processes. *IEEE Transactions on Instrumentation and Measurement*, 2021. 70: p. 1–11.
[20] Ou, C., et al., Quality-driven regularization for deep learning networks and its application to industrial soft sensors. *IEEE Transactions on Neural Networks and Learning Systems*, 2022. p. 1–11.
[21] Sun, Q. and Z. Ge, A survey on deep learning for data-driven soft sensors. *IEEE Transactions on Industrial Informatics*, 2021. 17(9): p. 5853–5866.
[22] Napoli, G. and M.G. Xibilia, Soft Sensor design for a Topping process in the case of small datasets. *Computers & Chemical Engineering*, 2011. 35(11): p. 2447–2456.
[23] Shang, C., et al., Data-driven soft sensor development based on deep learning technique. *Journal of Process Control*, 2014. 24(3): p. 223–233.
[24] Graziani, S. and M.G. Xibilia, Deep structures for a reformer unit soft sensor. In *2018 IEEE 16th international conference on industrial informatics (INDIN)*. 2018: IEEE.
[25] Dias, T., et al., Predictive analytics in the petrochemical industry: research octane number (RON) forecasting and analysis in an industrial catalytic reforming unit. *Computers & Chemical Engineering*, 2020. 139: p. 106912.
[26] Yuan, X., et al., A dynamic CNN for nonlinear dynamic feature learning in soft sensor modeling of industrial process data. *Control Engineering Practice*, 2020. 104: p. 104614.
[27] Steurtewagen, B. and D. Van den Poel, Machine learning refinery sensor data to predict catalyst saturation levels. *Computers & Chemical Engineering*, 2020. 134: p. 106722.

[28] Ferreira, J., M. Pedemonte, and A.I. Torres, Development of a machine learning-based soft sensor for an oil refinery's distillation column. *Computers & Chemical Engineering*, 2022. 161: p. 107756.
[29] Mendia, I., et al., Machine learning based adaptive soft sensor for flash point inference in a refinery realtime process. *Results in Engineering*, 2022. 13: p. 100362.
[30] Li, Q., et al., A soft-sensing method for product quality monitoring based on particle swarm optimization deep belief networks. *Transactions of the Institute of Measurement and Control*, 2022. 44(15): p. 2900–2910.
[31] Brulé, M.R., The data reservoir: how Big Data technologies advance data management and analytics in E&P. In *SPE digital energy conference and exhibition*. 2015: SPE.
[32] Crockett, B. and K. Kurrey, "Smart decision making needs automated analysis" making sense out of big data in real-time. In *SPE intelligent energy international conference and exhibition*. 2014: SPE.
[33] Wu, W., et al., Retrieving information and discovering knowledge from unstructured data using Big Data mining technique: heavy oil fields example. In *International petroleum technology conference*. 2014: IPTC.
[34] Bin Mahfoodh, A., et al., Introducing a big data system for maintaining well data quality and integrity in a world of heterogeneous environment. In *SPE Kingdom of Saudi Arabia annual technical symposium and exhibition*. 2017: SPE.
[35] Perrons, R.K. and J. Jensen, The unfinished revolution: what is missing from the E&P industry's move to "big data". *Journal of Petroleum Technology*, 2014. 66(05): p. 20–22.
[36] Perrons, R.K. and J.W. Jensen, Data as an asset: what the upstream oil & gas industry can learn about "big data" from companies like social media. In *SPE annual technical conference and exhibition?* 2014: SPE.
[37] Akoum, M. and A. Mahjoub, A unified framework for implementing business intelligence, real-time operational intelligence and big data analytics for upstream oil industry operators. In *SPE Middle East intelligent oil and gas symposium*. 2013: SPE.
[38] Sousa, C., et al., Applying big data analytics to logistics processes of oil and gas exploration and production through a hybrid modeling and simulation approach. In *Offshore technology conference Brasil*. 2015: OTC.
[39] Hilgefort, K., Big data analysis using Bayesian network modeling: a case study with WG-ICDA of a gas storage field. In *NACE CORROSION*. 2018: NACE.
[40] Sukapradja, A., et al., Sisi nubi dashboard: implementation of business intelligence in reservoir modelling & synthesis: managing big data and streamlining the decision-making process. In *SPE Asia pacific oil and gas conference and exhibition*. 2017: SPE.
[41] Mehta, A., Tapping the value from big data analytics. *Journal of Petroleum Technology*, 2016. 68(12): p. 40–41.
[42] Feblowitz, J., Insights IDCE, Analytics in oil and gas: the big deal about big data. In *SPE digital energy conference and exhibition*. 2013: SPE.
[43] Trifu, M.R. and M.L. Ivan, Big data: present and future. *Database Systems Journal*, 2014. 5(1).
[44] Pence, H.E., What is big data and why is it important? *Journal of Educational Technology Systems*, 2014. 43(2): p. 159–171.
[45] Anuradha, J., A brief introduction on big data 5Vs characteristics and Hadoop technology. *Procedia Computer Science*, 2015. 48: p. 319–324.
[46] Sumbal, M.S., E. Tsui, and E.W. See-to, Interrelationship between big data and knowledge management: an exploratory study in the oil and gas sector. *Journal of Knowledge Management*, 2017. 21(1): p. 180–196.
[47] Khvostichenko, D. and S. Makarychev-Mikhailov, Effect of fracturing chemicals on well productivity: avoiding pitfalls in big data analysis. In *SPE international conference and exhibition on formation damage control*. 2018: SPE.
[48] Santoso, A.S. and L.A.M. Nelloh, Hasil Peer Reviewer Juril Elsevier2-LN. 2017.

[49] Győrödi, C., et al., A comparative study: MangoDB vs. MySQL. In *2015 13th international conference on Engineering of modern electric systems (EMES)*. 2015: IEEE.
[50] Mounir, N., et al., Integrating big data: simulation, predictive analytics, real-time monitoring, and data warehousing in a single cloud application. In *Offshore technology conference*. 2018: OTC.
[51] Kudo, T., et al., A proposal of a transaction processing method for MangoDB. *Procedia Computer Science*, 2016. 96: p. 801–810.

4 Harnessing the Convergence of Information Technology and Operational Technology for Digital Transformation

An Integrated Framework for Effective Project Management, Skill Development, Team Coordination, and Collaboration in Manufacturing Industry

Alper Ozpinar and Ali Soofastaei

4.1 INTRODUCTION

Based on Maslow's hierarchy of needs, humans initially concentrate on fulfilling primary physical needs such as sustenance and shelter. Once these necessities are met, individuals strive toward safety, belongingness, and love. Esteem needs to follow, eventually leading to the pursuit of self-actualization. While interpretations and expansions of these needs may evolve, they fundamentally revolve around similar core principles.

Driven by the intent to enhance living conditions and aspire to better lifestyles, relentless progress has been made in addressing these needs. This process accelerated in the previous century when collective human experiences were transformed into industrial systems.

Industrial revolutions are categorized into four principal phases or shifts in production techniques over time. Each phase is characterized by unique energy uses

DOI: 10.1201/9781003186823-4

(water, steam, electricity), technological evolutions, automation, and industrial communication strategies. Industry 1.0, or the First Industrial Revolution, witnessed a shift from production reliant on human, animal, and natural forces like wind and water to steam and water-powered mechanisms. This era notably catered to the basic tier of Maslow's hierarchy, with the introduction of the first industrial loom in 1784 being a prime example that allowed broader access to clothing.

Industry 2.0 encompassed a period where electricity generation, transmission, and distribution became pivotal, along with the emergence of electric motors. This era introduced assembly and production lines marked by distinct job roles and conveyor belt stations, ushering in the age of mass production. Notable examples from this period include the initial conveyor belt systems in meatpacking factories and Ford's renowned assembly line for Model T.

The Third Industrial Revolution, or Industry 3.0, marked a significant milestone in the 1970s. Substantial advancements and the extensive integration of electronics and information technology (IT) characterized it. This trend persists today, influencing various sectors across the globe. Network and internet technologies that began developing in the 1950s have become crucial components of the business landscape. These technologies continue to evolve, adapting to new advancements that emerge over time.

While progress in IT has been rapid and transformative, adopting these advancements in production and operational fields has been more measured and deliberate. Industry 3.0 brought significant progress in microprocessors and communication technologies, especially in automation, using programmable logic controllers (PLC), integrated circuits, and microcontrollers.

With the reduced size of integrated circuits, their capabilities and speeds have increased. This, coupled with the flexible and sustainable sharing of data across different platforms and systems, has shaped the unique dynamics of the business world along with software and hardware requirements. These dynamics have been grouped under IT.

Industry 4.0, synonymous with the fourth industrial revolution, signifies integrating modern innovative technology into manufacturing environments. It leverages critical elements such as automation, machine learning (ML), real-time data, and the Internet of Things (IoT) to create interconnected, efficient, and highly adaptive manufacturing ecosystems. However, Industry 4.0's influence extends beyond the production floor; it permeates various functions across the value chain, including product development, supply chain management, and customer service. This comprehensive transformation enables businesses to respond quickly and effectively to customer needs, providing a significantly superior value proposition to entities operating under outdated models.

Emerging on the horizon is Industry 4.0, a concept that encapsulates a technological vision characterized by automation and data exchange in manufacturing technologies. It includes cyber-physical systems, the IoT, cloud computing, and cognitive computing. This term was introduced during a presentation at the Hannover Fair 2011, marking a shift toward more integrated and intelligent systems.

Within the scope of Industry 4.0 and its associated technological goals, this chapter delves into themes about the harmonious operation of IT and operational technologies (OT), along with the structuring of the new era of OT systems for the Digital

Transformation of factories first into smart factories that leverages advanced digital technologies and data-driven processes to optimize and enhance production to the dark factories that extreme form of automation where manufacturing processes are entirely automated and little to no human intervention is required during production. The term "dark" refers to the absence of human workers on the factory floor, with the lights turned off.

This transformation illuminates the groundbreaking integration of cyber-physical systems, which offers unprecedented operational efficiency and productivity across different sectors, from manufacturing to services.

The constant progression of these dynamics has profound implications for businesses and individuals alike. It enables highly flexible, efficient, and scalable operations that adapt quickly to changing needs or market conditions. Furthermore, it fosters a more interconnected world where information can flow freely, facilitating incredible innovation and collaboration across geographical boundaries.

The chapter explores the synergistic relationship between IT and OT, unveiling opportunities for improved communication and data sharing and ultimately fostering more intelligent, efficient work environments across various industries. By leveraging these technological synergies, businesses can achieve higher performance levels, streamline processes, and create a more connected infrastructure.

The chapter also further explores how the concerted functioning of IT and OT can drive Digital Transformation efforts, providing a competitive edge in today's fast-paced business environment. It underscores how the fusion of these technologies can enable organizations to optimize their operations, enhance decision-making capabilities, and improve their responsiveness to market changes. By understanding these dynamics, businesses can better position themselves for success in an increasingly digitalized world.

Digital transformation embodies a fundamental shift in how organizations deliver value to their customers. This transformation integrates digital technology into all business areas, radically altering operational procedures and value delivery mechanisms. However, Digital Transformation extends beyond technological adoption; it is about restructuring organizations to be more agile, responsive, and customer-oriented. It is an essential strategy for businesses aiming to stay competitive in a market dominated by digital advancements. Within the provided context, Digital Transformation plays a critical role in assimilating various technological developments into business environments, ultimately boosting operational efficiency and catering to the dynamic needs of the marketplace.

IT-OT Convergence is a pivotal aspect of this digital evolution. It encompasses integrating IT systems, concentrated on data-centric computing, with OT systems responsible for overseeing and manipulating physical devices and processes. This synergistic convergence facilitates superior data exchange and analysis throughout an organization, culminating in more profound insights and elevated decision-making capabilities. The ensuing interconnectivity can manifest in varied ways; examples include harmonized supply chains, automated and optimized manufacturing procedures, or real-time analytical capabilities.

From a human resources standpoint, attaining IT-OT Convergence is not merely advantageous—it is crucial for organizations engaged in the manufacturing and production sectors. The flawless fusion of these technologies forms the bedrock of streamlined, efficient, and competitive operations, which are vital for success in the era of Industry 4.0. This convergence can enhance productivity, cut operational costs, improve product quality, and ensure faster responses to market changes. Thus, comprehending and leveraging IT-OT Convergence can provide a competitive edge for businesses tackling the intricacies of Digital Transformation in today's hyper-connected world.

Because of these advancements, new challenges and opportunities are emerging. The embrace of Digital Transformation, IT-OT Convergence, and Industry 4.0 necessitates organizations procuring new skills, introducing innovative technologies, and embracing new human resources management, team coordination, and role definition methodologies. It becomes essential to cultivate a culture of continuous learning and improve workforce adaptability as the required skills swiftly transform in response to technological progressions. Data and system security also emerge as significant concerns in this new landscape. The increased level of connectivity inherent in these trends exposes organizations to unprecedented risks and vulnerabilities. In this regard, cybersecurity measures must be robust and adaptive to protect data integrity, assure its confidentiality, and ensure its availability for authorized users.

Furthermore, the rise of Industry 4.0 also offers businesses opportunities to streamline operations, improve efficiency, and enhance their competitive edge. By leveraging technologies like artificial intelligence (AI) and big data analytics, organizations can gain invaluable insights to make informed decisions and optimize performance. The increased connectivity via IoT devices also allows remote monitoring of operations and predictive equipment maintenance. Thus, understanding and embracing these shifts are crucial for businesses aiming to thrive in the era of Industry 4.0.

Furthermore, the organizational strategy must align seamlessly with these digital initiatives. Leadership must envision and comprehend how Digital Transformation, IT-OT Convergence, and Industry 4.0 intersect with the organization's goals and objectives. This alignment ensures that digital endeavors are not only technologically advanced but also contribute positively to the strategic trajectory of the business, enhance customer value, and bolster competitiveness in the marketplace.

As organizations delve further into the digital sphere, ethical considerations surrounding the use of technology, data privacy, and broader societal and employment implications become increasingly pertinent. The ethical application of technology and responsible business practices should form the bedrock of strategies and operations for organizations navigating through Digital Transformation, IT-OT Convergence, and Industry 4.0.

These ethical considerations extend beyond merely complying with legal requirements. They are crucial in cultivating trust and goodwill among customers, employees, and other stakeholders. This trust is invaluable in a business environment increasingly driven by digital technologies. It forms the foundation for long-term customer loyalty, employee engagement, and positive stakeholder relationships.

Moreover, organizations must remain aware of their socio-economic impact in this era of rapid technological advancement. The rise of automation and AI technologies may lead to significant shifts in job markets and labor requirements, which businesses must address proactively to maintain a sustainable socio-economic balance. Ultimately, thriving in Industry 4.0 requires businesses to be technologically progressive, strategically aligned, ethically responsible, and socially conscious.

As the digital era continues to evolve, organizations find themselves at the intersection of technological advancement and business innovation. This chapter explores the complex maze of IT-OT Convergence, Digital Transformation, and Industry 4.0. The aim is to provide a comprehensive and actionable framework that guides organizations in incorporating these advanced concepts into their operational contexts.

At its core, this chapter is dedicated to bridging the gap between theoretical understanding and real-world application. It transcends the abstract domain, simplifying the intricate aspects of Digital Transformation into clear, practical steps that organizations can interpret, adapt, and tailor according to their distinct objectives and necessities.

Understanding the "what" and "why" behind Industry 4.0 and the "how" is crucial in this transformative era. This journey necessitates thoroughly exploring essential elements such as effective project management strategies, skill enhancement, efficient team coordination, and fostering collaborative practices. With new roles like the Edge-Fog, Big Data, and AI Integrator; Sustainability Expert; and Net Zero and Carbon Management Specialist, the pathway to integration and innovation in IT-OT Convergence is apparent. These roles epitomize the shift toward a future where operational efficiency, technological advancement, and environmental stewardship coexist harmoniously.

The overarching vision of this chapter is to arm businesses, managers, and professionals with the knowledge and tools required to navigate and lead the inevitable changes brought about by Digital Transformation and Industry 4.0. The ultimate goal is to create an environment that stimulates innovation, nurtures essential skills, and promotes productive team dynamics, aligning technological progress with sustainable practices. This narrative aims to demystify the complex aspects of Digital Transformation, providing a clear, practical roadmap for organizations to become more agile, adaptable, and environmentally conscious.

This chapter offers a comprehensive guide for organizations to thrive in the digital age by weaving together the technical and sustainable aspects of Industry 4.0. It encourages the adoption of advanced technologies and sustainable practices, ensuring businesses are well equipped to face the future's challenges and opportunities. In doing so, it empowers organizations to embark on a transformative journey toward a digital and sustainable future, where innovation, efficiency, and environmental responsibility drive success in the new industrial era.

4.2 FUNDAMENTALS OF IT-OT INTEGRATION

This segment delves into the foundational aspects of IT and OT, shedding light on the respective concepts, technologies, and their critical roles within modern organizational infrastructures. A nuanced comprehension of the differences and interplay between IT and OT is essential for delineating their respective responsibilities and ensuring coherent operational synergy [1].

As the term suggests, IT encompasses the spectrum of technologies dedicated to processing, storing, and transmitting information. It is the backbone of an organization's data management, communication, and computational services. On the other hand, OT refers to the array of hardware and software solutions that directly influence physical devices and processes—controlling, monitoring, and ensuring the smooth operation of machinery and industrial tasks [2].

To elucidate these concepts further, let us employ a more relatable analogy: envision a bustling restaurant. In this analogy, the kitchen, with its chefs, sous-chefs, and line cooks bustling about to prepare and cook dishes, epitomizes OT. The OT domain is akin to the kitchen's dynamic environment, where various elements—from ovens, stovetops, and refrigeration systems to sophisticated culinary gadgets—are meticulously orchestrated to deliver delectable dishes. This setting underscores the essence of OT: a realm where control systems, machinery, and operational processes converge to produce tangible outcomes.

Transitioning from the kitchen to the restaurant's front-of-house, we encounter the realm of IT. Here, in the restaurant's reception, dining area, and administrative offices, the focus shifts to managing reservations, processing orders, overseeing financial transactions, and nurturing customer relationships. Analogous to a restaurant's front-of-house operations, IT systems encompass the digital and communicative frameworks that facilitate data management, customer interaction, transaction processing, and overall administrative efficiency.

This culinary analogy demystifies the distinct functionalities and spheres of influence of IT and OT and highlights the critical importance of their integration within any organizational context. Just as seamless interaction between a restaurant's kitchen and front-of-house operations is pivotal for delivering an exceptional dining experience, harmonious integration of IT and OT systems is crucial for achieving operational excellence and strategic agility in today's digitally driven business landscape [3].

Building on this understanding, it becomes evident that while IT and OT have traditionally operated in separate domains, the evolving technological landscape and the push toward Digital Transformation necessitate a more integrated approach. The convergence of IT and OT opens up new avenues for innovation, operational efficiency, and enhanced decision-making, paving the way for more resilient, agile, and competitive business models.

In the subsequent sections, we will explore the implications of IT-OT integration in greater detail, examining the challenges, opportunities, and strategic considerations that businesses must navigate to harness the full potential of this convergence in the era of Industry 4.0. This exploration will give readers insights into the theoretical underpinnings, practical applications, and benefits of effectively merging IT and OT systems to drive Digital Transformation and sustainable growth [4].

4.2.1 IT in Modern Enterprises

It is intricately woven into every facet of organizational operations in the contemporary business ecosystem. IT transcends mere technical support to become a core element driving business strategies, innovation, and customer engagement. IT encompasses a broad spectrum of technologies and processes dedicated to the collection, processing, storage, and dissemination of information, thereby facilitating efficient business operations and enhancing decision-making capabilities.

With the advent of digitalization, IT's role has expanded significantly, becoming synonymous with business enablement and transformation. It is not just about managing data or supporting infrastructure; it is about leveraging technology to create value, optimize workflows, and deliver superior customer experiences. In this context, IT is often integrated with communication technologies, giving rise to information and communication technologies. This integration underscores the critical role of seamless communication in amplifying the impact of information technologies on business outcomes [5].

Given this discourse's focus on the synergies and interplay between IT and OT, it is essential to delineate IT's domain within this convergence. IT's realm in this context is broad, encompassing everything from enterprise resource planning (ERP) systems and customer relationship management (CRM) software to cloud computing and data analytics platforms. These technologies and systems collectively form the digital backbone of modern enterprises, enabling them to navigate the complexities of today's market dynamics [6, 7].

The following subsections will delve into the specific dimensions of IT within the framework of IT-OT Convergence, highlighting its transformative impact on various business operations:

1. **Data Management and Analytics:** At the heart of IT is the ability to manage vast volumes of data, turning raw data into actionable insights. Advanced analytics, AI, and ML algorithms are leveraged to predict trends, optimize operations, and personalize customer experiences.
2. **Digital Communication and Collaboration:** IT facilitates seamless communication and collaboration within and with external stakeholders. Tools and platforms that support instant messaging, video conferencing, and real-time document sharing have become indispensable in today's digital workplace.
3. **Cybersecurity and Risk Management:** As businesses increasingly rely on digital infrastructures, cybersecurity's importance within the IT domain cannot be overstated. Protecting data integrity, ensuring privacy, and mitigating cyber threats are paramount for maintaining trust and operational continuity.
4. **Cloud Computing and Infrastructure:** Cloud technologies offer scalable and flexible IT infrastructures, enabling businesses to adapt quickly to changing market demands. IT encompasses the management of these cloud resources, ensuring optimal performance and cost-efficiency.
5. **Application Development and Management:** IT involves developing, deploying, and managing applications supporting business processes. This includes custom software for internal use, customer-facing applications, and the integration of third-party solutions.
6. **Technology Strategy and Governance:** IT strategically aligns technology initiatives with business goals. This includes technology planning,

investment decisions, and governance frameworks that ensure alignment with regulatory requirements and industry standards.

This section comprehensively explores these facets to understand IT's critical role in modern enterprises. It sets the stage for discussing the dynamic interplay between IT and OT, which is fundamental to achieving Digital Transformation and operational excellence in Industry 4.0.

4.2.1.1 Role of IT in Facilitating Digital Transformation

IT is the foundational layer in Management Information Systems that orchestrates many business functions and processes. This encompasses various activities, from initiating and processing orders and production directives to managing sales, inventory movements, financial transactions, customer interactions, and comprehensive planning and organizational tasks. The scope of IT extends to include specialized processes and document flow such as invoices, dispatch notes, transactions with official institutions, executive reports, and strategic enterprise process management. It also covers critical operational domains such as human resources, production planning, material resource planning, and sophisticated decision support systems integral to management information systems [8].

At its core, IT is constituted by an intricate ecosystem of software solutions designed to fulfill these diverse business needs. These solutions are supported by robust hardware and network infrastructures that ensure seamless operation and connectivity. In a corporate setting, these systems are meticulously structured within a network, governed by stringent hierarchies and security protocols to safeguard data integrity and operational continuity [9].

The breadth of IT encompasses the systems housed within corporate server rooms and data centers. It extends to the broader infrastructure, facilitating internet connectivity, cloud computing, big data analysis, and comprehensive data management. This includes extensive database systems, network and security hardware, cutting-edge cybersecurity solutions, and communication services like email, web, and mobile platforms.

In the contemporary corporate landscape, software systems are layered and integrated across various operational levels, enabling businesses to achieve high automation and efficiency. These systems include but are not limited to, ERP systems that integrate core business processes, material requirements planning and manufacturing resource planning (MRP) systems that optimize manufacturing and production processes, and manufacturing execution systems (MES) that provide real-time monitoring and control of factory floor operations [10, 11].

Moreover, IT encompasses specialized applications and platforms for finance and accounting, human resources, product lifecycle management (PLM), and other functions such as fleet management and performance monitoring systems. The deployment of management information system applications, expert systems, and tailored vertical solutions further exemplifies the depth and diversity of IT's role in modern businesses [12].

These multifaceted layers of IT infrastructure and applications are pivotal in driving Digital Transformation, enabling businesses to navigate the complexities of the digital era with agility and strategic foresight. By leveraging the full spectrum of IT capabilities, organizations can streamline operations, enhance decision-making, and deliver unparalleled value to customers, securing a competitive edge in an increasingly digitalized market landscape.

This section delineates IT's comprehensive role in underpinning Digital Transformation initiatives. It highlights its critical function in integrating and optimizing business processes, enhancing data-driven insights, and fostering a culture of innovation and continuous improvement. Through this exploration, readers will understand how IT is the backbone of Digital Transformation. It facilitates the seamless convergence of technology and business strategy to drive growth and innovation in the digital age.

4.2.1.1.1 Understanding OT in Industrial Contexts

OT forms the cornerstone of industrial and manufacturing processes, acting as the nerve center akin to the bustling kitchen in a restaurant. OT's essence lies in its primary focus on controlling, monitoring, and automating the physical processes that are pivotal to production and operational efficiency. Unlike IT, which is oriented toward data management and information flow, OT is deeply rooted in directly interacting with and manipulating physical machinery, equipment, and operational environments [13].

At its core, OT involves various specialized equipment, tools, hardware, and software systems integral to executing technical operations within industrial settings. These technologies range from sensor networks, control systems, and robotics to more complex automated assembly lines and process control systems. They are designed to ensure precision, efficiency, and safety in the physical tasks and processes that underpin the production and delivery of goods and services.

OT extends beyond mere machinery and equipment; it embodies sophisticated software systems that provide critical real-time data and analytics, enabling operators to make informed decisions and adjustments to optimize performance and mitigate risks. These systems include supervisory control and data acquisition (SCADA) systems, Distributed Control Systems, and PLC, which collectively facilitate a high degree of automation and operational control [14, 15].

In the contemporary industrial landscape, the role of OT has evolved significantly, driven by technological advancements and the increasing demand for greater efficiency, sustainability, and adaptability in production processes. Integrating IoT devices, advanced robotics, and AI into OT systems has expanded their capabilities, allowing for more sophisticated monitoring, predictive maintenance, and autonomous operation.

The convergence of OT with IT systems—often called IT-OT Convergence—marks a significant shift in how industries approach production and operational challenges. This integration enables a seamless flow of information between the operational floor and strategic business functions, enhancing visibility, agility, and coordination across the entire value chain. As a result, businesses can achieve higher operational efficiency, product quality, and customer satisfaction while embracing innovation and driving Digital Transformation [16].

This section aims to delve deeper into the intricate world of OT, exploring their critical functions, the technologies that underpin them, and their evolving role in the modern industrial ecosystem. Through this exploration, readers will understand how OT is the bedrock of industrial operations, driving productivity, safety, and innovation in an increasingly interconnected and digitalized global economy.

4.2.1.1.2 OT and Their Role in Digital Transformation

OT is integral to the functioning and efficiency of critical industrial and infrastructural operations. These systems encompass the control and management of processes and equipment in sectors as diverse as water management, energy generation and distribution, and the manufacturing industry. OT systems are designed to be robust and reliable, capable of operating continuously under the demanding conditions of industrial environments. This reliability is paramount, as OT systems often underpin large-scale operations essential for societal well-being and economic stability, such as the distribution networks for electricity, natural gas, and petroleum [17].

The scope of OT systems is extensive, covering everything from industrial computing devices and edge computing systems to specialized production machinery such as computer numerical control machines and production workbenches. These include critical infrastructure components like boilers, heating, ventilation, and air conditioning systems, energy analyzers, building automation systems, and safety-critical elements such as security and alarm circuits. Unlike IT infrastructures subject to frequent updates and technological refreshes, OT systems are characterized by longevity and stability. It is not uncommon to find OT equipment operating on legacy platforms, such as Windows 95 or XP, due to their proven reliability and the specific requirements of industrial operations.

The user base of OT systems is predominantly composed of individuals with technical expertise or specialized skills, including line operators, engineers, maintenance personnel, and production workers. These users interact with OT systems through interfaces and control mechanisms tailored to the demands of industrial processes, such as PLCs, SCADA systems, and human–machine interfaces (HMIs) [18, 19].

In terms of communication, OT systems employ diverse protocols and technologies that extend beyond the standard transmission control protocol/internet protocol (TCP/IP) model prevalent in IT environments. This includes various industrial communication standards and physical media like Ethernet, fiber optics, and serial connections (RS-485, RS-232), which are chosen based on the specific needs of the operational environment and the requirements for reliability, speed, and security.

The networking architecture within OT environments also differs significantly from traditional IT networks. To ensure optimal performance and resilience, OT networks incorporate a variety of topologies, such as ring, bus, and star configurations. This diversity in communication strategies and network designs reflects operational technology's unique challenges and priorities, where the focus is on real-time performance, reliability, and the safety of both processes and personnel.

As industries embrace Digital Transformation, integrating OT with advanced IT systems and data analytics is becoming increasingly critical. This convergence can unlock new efficiency, agility, and innovation levels by leveraging real-time data and insights from operational processes. However, this integration also presents significant challenges, particularly in ensuring interoperability, maintaining system security, and managing the transition from legacy systems to more modern, interconnected platforms.

This section explores the complexities and opportunities OT systems present in Digital Transformation, aiming to comprehensively understand OT's pivotal role in the industrial ecosystem. It highlights the need for a strategic approach to integrating OT with IT systems, ensuring that Digital Transformation initiatives are grounded in a deep understanding of industrial environments' operational realities and requirements.

4.2.2 Deep Dive into the Impact of Digital Transformation and Industry 4.0 on Modern Business Landscapes

The profound and far-reaching importance and pertinence of Digital Transformation and Industry 4.0 in today's rapidly evolving business landscape are profound. As we navigate an age characterized by technological progress and continuously transforming market conditions, these paradigms emerge as critical navigational beacons. They offer strategic blueprints for organizations aiming to chart courses toward innovation, resilience, and sustainable development amid the complexities of a digital-first global economy [20].

Digital transformation encapsulates the comprehensive integration of digital technologies into all business areas, fundamentally altering how organizations operate and deliver customer value. It transcends mere technological adoption, embodying a cultural and operational shift that requires organizations to continually challenge the status quo, experiment, and adapt to changes with agility.

Industry 4.0, often synonymous with the Fourth Industrial Revolution, further advances this transformation, emphasizing the fusion of advanced digital technologies such as the IoT, AI, robotics, and big data analytics with traditional industrial practices. This convergence facilitates the creation of intelligent, autonomous systems that enhance manufacturing processes, supply chain management, and product development, fostering greater efficiency, customization, and quality.

The synergistic relationship between IT and OT lies at the heart of these transformative movements. With its focus on data, networks, and systems for information processing and communication, IT is the backbone for Digital Transformation strategies. It enables data collection, analysis, and strategic use to drive decision-making, innovation, and customer engagement.

Conversely, OT, centered on the physical devices, machinery, and processes critical to production and industrial operations, becomes increasingly interconnected and intelligent in Industry 4.0. This intelligence allows for more responsive, flexible, and efficient operational environments where real-time data and insights can lead to optimized performance and reduced downtime.

The confluence of IT and OT within the Digital Transformation and Industry 4.0 framework represents a paradigm shift in how businesses view and leverage technology. It underscores a move from siloed IT and operational functions toward an integrated, collaborative approach that blurs the lines between digital and physical domains. This integration has challenges, including cybersecurity risks, the need for cultural change, and the complexities of managing legacy systems alongside cutting-edge technologies. However, the potential benefits of operational excellence, innovation, and competitive advantage are significant.

As organizations strive to adapt to and thrive in this new era, a critical understanding of the principles, technologies, and strategic implications of Digital Transformation and Industry 4.0 becomes indispensable. This knowledge equips business leaders, technologists, and policymakers with the insights needed to make informed decisions, embrace change, and capitalize on the opportunities a digitized, interconnected world presents.

This section delves into these topics to elucidate the multifaceted impact of Digital Transformation and Industry 4.0 on modern businesses. It seeks to provide a comprehensive overview that highlights the transformative potential of these movements and addresses the practical considerations and strategic approaches necessary for successful implementation and sustained growth in the digital age.

4.2.2.1 Significance and Impact

In the current business climate, marked by rapid technological progress and dynamic market changes, Digital Transformation and Industry 4.0 concepts have become pivotal. They act as critical drivers for organizational innovation, efficiency, and sustainability. Digital Transformation represents the strategic adoption of digital technologies across all business operations. This process is crucial for cultivating flexible and resilient business models that swiftly adapt to evolving market trends and customer expectations, securing a competitive edge in an increasingly digitalized marketplace.

Digital Transformation is not merely about adopting new technologies; it is about rethinking existing business processes, structures, and strategies to leverage digital advancements. This transformative journey enables businesses to enhance operational efficiencies, improve customer experiences, and innovate products and services. It encourages a culture of continuous improvement and agility, enabling organizations to anticipate market shifts and respond with speed and precision.

Industry 4.0 complements and extends this transformation within the manufacturing sector and beyond, signifying the fusion of advanced digital technologies with traditional industrial practices. Integrating IoT, AI, advanced robotics, and cloud computing heralds the advent of smart factories and industrial setups. These environments are characterized by their self-monitoring, analytical, and autonomous decision-making capabilities. The application of these technologies enables unprecedented levels of process optimization, resource efficiency, and product customization, driving the next wave of industrial productivity and growth.

The synergy of Digital Transformation and Industry 4.0 facilitates seamless interplay between the virtual and physical realms, creating intelligent networks along the entire value chain. This connectivity enhances operational efficiencies and opens up new opportunities for innovation and value creation regarding products and business models. For instance, predictive maintenance, powered by AI and IoT, can drastically reduce downtime and maintenance costs. At the same time, big data analytics can uncover insights that lead to better product designs and customer experiences [21, 22].

Moreover, these transformative trends have far-reaching implications beyond operational improvements. They are critical in addressing broader economic, environmental, and societal challenges. For example, advanced analytics and IoT can optimize energy use and reduce waste in manufacturing, contributing to more sustainable industrial practices. Similarly, digitalizing supply chains can enhance transparency and resilience, making them more responsive to disruptions.

In sum, the importance and relevance of Digital Transformation and Industry 4.0 lie in their ability to empower organizations to navigate the complexities of the modern business environment. Embracing these paradigms allows businesses to achieve operational excellence, drive innovation, foster sustainable practices, and create value in new and exciting ways. As such, understanding and implementing the principles of Digital Transformation and Industry 4.0 is crucial for any organization looking to thrive in the digital era. This section aims to delve deeper into these themes, exploring the strategic implications, technological underpinnings, and practical approaches to harnessing the full potential of these transformative trends [23].

4.2.2.2 Foundational Pillars for Understanding Digital Transformation and Industry 4.0

A thorough grasp of certain core concepts and principles is vital to effectively navigating the complex terrain of Digital Transformation and Industry 4.0. These foundational elements equip individuals and organizations with the necessary insights to explore the vast potential of these transformative trends, enabling them to harness opportunities and tackle the inherent challenges in today's technology-infused business ecosystem.

Digital Maturity Assessment: This concept measures the extent to which an organization has embedded digital technologies across its operations and culture. Evaluating an organization's digital maturity is critical for understanding its current capabilities and identifying strategic areas for digital enhancement, ensuring a smooth and impactful transformation journey [24].

Agility and Lean Thinking: Central to Digital Transformation and Industry 4.0, these frameworks advocate for flexibility, efficiency, and customer focus. They encourage organizations to adopt adaptive structures and processes, facilitating rapid response to evolving market demands and enhancing customer value delivery.

Data-Driven Culture: The power of data lies at the heart of digital and industrial revolutions. Cultivating a data-driven mindset is imperative, enabling organizations to harness analytics for strategic insights, foster evidence-based decision-making, and align actions with overarching business goals and customer expectations [25].

Cybersecurity Vigilance: As digital footprints expand, safeguarding data integrity and system security becomes paramount. A comprehensive understanding of cybersecurity practices and adherence to privacy regulations are essential for protecting organizational and customer data against cyber threats, ensuring trust and compliance in a digital world [26].

System Interoperability: The essence of Industry 4.0 lies in the harmonious integration of diverse technologies and systems. Facilitating seamless communication and data flow among interconnected devices and platforms is critical to unlocking efficiencies and enabling intelligent, autonomous operations.

User-Centric Innovation: Despite the heavy reliance on technological advancements, the significance of the human aspect cannot be understated. Embracing a human-centric approach in design and strategy ensures that technological solutions enhance human work and life, promoting accessibility and user engagement.

Commitment to Lifelong Learning: The rapid pace of technological advancements necessitates a culture of ongoing education and skill enhancement. Organizations must prioritize learning initiatives and skill development opportunities, empowering their workforce to remain proficient in the latest technological tools and methodologies [27].

Armed with a solid understanding of these fundamental concepts, individuals and businesses are well equipped to explore the multifaceted implications of Digital Transformation and Industry 4.0. This section aims to detail each of these principles, providing a comprehensive framework for readers to understand and apply these concepts in driving successful transformation initiatives within their respective domains.

4.2.3 Bridging IT and OT for a Unified Digital Ecosystem

The fusion of cutting-edge digital technologies such as cloud computing, the IoT, Big Data, AI, ML, Unified Communications, and Business Intelligence within IT has elevated these tools from mere operational aids to strategic imperatives. This evolution in IT necessitates a corresponding transformation in OT, compelling them to integrate more closely with IT frameworks to achieve comprehensive and seamless synergy, encapsulating 360-degree digital integration.

This paradigm, often encapsulated under the banner of Industry 4.0 or the Next-Generation Digital Factory, heralds a transformative shift toward embracing advanced digital solutions. This shift encompasses the adoption of industrial cloud platforms, advanced analytics, Industrial IoT (IIoT) applications, Big Data analytics in manufacturing processes, Digital Twins, augmented reality for enhanced operational visibility, sophisticated system and process simulations, innovative virtual designs, additive manufacturing through 3D printing, predictive maintenance protocols, and the development of autonomous quality control and production systems.

Such a profound transformation within OT domains necessitates a forward-thinking approach, transitioning from traditional electromechanical maintenance paradigms to dynamic, technology-driven operational models. In this new era, production systems are not just mechanical workflows but are imbued with the full spectrum of technological advancements, mirroring the innovation-driven ethos of IT systems. Global logistics and distribution network demands accentuate the imperative for industrial entities to swiftly integrate competitive and innovative methodologies into their operations. Organizations must comply more with antiquated production methodologies, or failing to enhance their industrial processes risks falling behind in cost efficiency, productivity, operational transparency, quality control, and overall competitiveness.

In this transformative landscape, OT systems are pivoting from their conventional standpoints, striving for greater alignment with IT systems to foster enhanced collaboration and data exchange capabilities. The interaction between IT and OT, once limited to specific applications such as PLM, MES, and ERP, is now expanding into broader and more intricate dimensions of integration and communication.

This convergence facilitates the transformation of raw industrial data into actionable corporate insights that are accessible and analyzable corporate insights across all organizational levels. It enables precise interventions in production parameters and control mechanisms underpinned by structured access hierarchies. The traditional industrial control networks rooted in OT evolve into sensor-driven IIoT ecosystems. As many devices, from sensors to machinery, begin to generate voluminous and multifaceted data, this information is channeled through IT systems to corporate data centers and big data platforms, where it is processed and analyzed for a wide array of strategic applications beyond mere process control. These applications range from comprehensive asset management and supply chain optimization to sophisticated cost analysis and integrating real-time operational data with financial metrics.

For instance, in optimizing facility operations, the Facilities Management division must integrate energy analyzers with IT systems, which are pivotal for monitoring and controlling the electrical infrastructure. This integration mandates that OT hardware seamlessly connects with IT networking solutions via TCP/IP within the framework of corporate security protocols. This intersection raises critical considerations around cybersecurity and network integrity, as traditional OT communications and protocols might introduce vulnerabilities or conflict with established IT security and firewall policies.

Navigating this intricate landscape of IT-OT Convergence necessitates a strategic and holistic approach. This approach ensures that the integration enhances operational efficiency and innovation and aligns with stringent cybersecurity standards to safeguard the unified digital ecosystem.

4.2.3.1 Advantages of Integrating IT and OT Systems

The fusion of IT and OT represents a transformative shift for asset-intensive organizations, as highlighted by Gartner. This integration transcends the traditional separation of IT and OT into distinct domains, fostering a unified environment where processes and information flow seamlessly across the organizational spectrum. The amalgamation of OT with IT unlocks many substantial benefits, pivotal among which is enhancing decision-making capabilities. By tapping into a broader and richer data pool, stakeholders can make well-informed decisions, leveraging the operational intelligence and real-time insights previously siloed within the OT domain [1].

Consider the example of a vast petroleum pipeline network, sprawling over 10,000 miles, equipped with thousands of PLCs and an extensive array of devices interconnected by miles of cabling. In a conventional setup, only a fraction of the operational data—critical for monitoring and ensuring safe pipeline operation—is utilized, leaving most of the data untapped at the field level. Integrating this wealth of pipeline data into the broader business analytics framework can significantly enhance operational decisions, ranging from logistical considerations like dispatching repair units to strategic decisions informed by a comprehensive understanding of the health and performance of the entire network.

The synergy between IT and OT extends beyond improved decision-making, encompassing a range of benefits such as:

- **Cost Efficiency:** Integrating OT with the corporate IT infrastructure can lead to substantial cost savings, offsetting the initial investment through enhanced maintainability, reduced licensing fees, and streamlined operational expenses.
- **Process Optimization:** The convergence facilitates the optimization of business processes by ensuring that data from both domains is readily accessible, fostering a more cohesive and efficient operational framework.
- **Risk Mitigation:** A unified IT-OT environment enhances risk management by providing a holistic view of organizational operations. This enables proactive identification and mitigation of potential issues before they escalate.
- **Accelerated Development:** The integration shortens development and integration timelines thanks to standardized communication protocols and control mechanisms.
- **Standardization and Flexibility:** With IT-OT Convergence, organizations can leverage standard technologies such as structured query language databases, Java, and secure sockets layer, moving away from proprietary solutions. This standardization fosters the rapid development and deployment of secure, scalable solutions, enhancing operational flexibility.

This integrated approach also diminishes dependency on proprietary HMI/SCADA systems, often associated with high costs and restrictive environments. Instead, organizations gain the ability to monitor and control diverse systems more efficiently, enhancing their agility and responsiveness to market dynamics and operational challenges.

In essence, the convergence of IT and OT systems is not merely a technical upgrade but a strategic reorientation that equips companies to navigate the complexities of the modern industrial landscape more effectively. It transforms how information is managed and utilized, driving innovation, enhancing operational efficiency, and fostering a more agile and responsive organizational culture.

4.3 TEAM MANAGEMENT AND COLLABORATION OF THE OT DEPARTMENT FOR IT-OT CONVERGENCE

4.3.1 Reimagining Organizational Structures for IT-OT Synergy

Integrating OT within the organizational framework in the evolving Digital Transformation landscape necessitates a strategic approach to structure and collaboration. The OT Unit, the pivotal link between the IT Directorate and the broader Structural General Directorate's activities, is critical in bridging the gap between digital and physical operational realms. Despite the geographical proximity of OT functions to production, manufacturing, assembly, and maintenance operations, there is a compelling rationale for aligning the OT Unit closely with the IT Directorate [17].

This alignment is underpinned by the unity in job functions, processes, and responsibilities that OT shares with IT. Such an organizational positioning fosters functional synergy and ensures adherence to compliance standards and best practices traditionally within the IT domain's purview. Furthermore, integrating OT within the IT framework enhances the strategic oversight of technology investments and cybersecurity measures, where IT departments typically excel.

To institutionalize this alignment, it is prudent to formalize the presence of OT teams within the organizational hierarchy, specifically within the realms of Structural and Industrial Systems. This formalization ensures that OT functions are not isolated but part of a cohesive, interdisciplinary effort that aligns with the organization's goals and strategic imperatives.

The proposed structure facilitates a harmonious and efficient work environment characterized by seamless communication and coordinated efforts between OT and IT teams. This collaborative ethos is essential for optimizing technological operations, safeguarding against security vulnerabilities, and ensuring that the organization's infrastructure is both robust and flexible enough to adapt to evolving operational demands.

Moreover, this integrated organizational model promotes a culture of continuous improvement and innovation, where insights and best practices from the IT domain can inform and enhance OT strategies and vice versa. It encourages sharing knowledge and skills across departments, leading to a more informed, agile, and responsive organizational tech landscape.

In summary, the strategic positioning of the OT Unit within the IT Directorate, coupled with a clear delineation of OT roles within the structural framework, is not merely a structural adjustment. It represents a forward-thinking approach to organizational design that recognizes the intertwined nature of digital and OT in driving business success. This model lays the groundwork for a dynamic, integrated, and resilient technological ecosystem that can navigate the complexities of the digital age, aligning operational practices with the organization's strategic vision.

4.3.2 Enhancing Collaboration Between the OT Unit and Organizational Ecosystem

The OT Unit is poised to play a pivotal role within the organizational ecosystem, particularly in areas where industrial processes are integral. This unit's effectiveness hinges on fostering robust coordination mechanisms with the organization's structural and industrial segments. To this end, establishing clear communication channels, shared protocols, and collaborative frameworks is essential to seamlessly integrate the OT Unit's initiatives with the broader organizational operations [28].

Moreover, the scope of collaboration extends beyond the organization's internal confines to encompass external stakeholders and suppliers engaged in OT-related activities. Representatives from the OT Unit must be actively involved in all interactions with these external entities. This involvement ensures the partnerships are built on mutual understanding and shared objectives, particularly establishing, executing, monitoring, and maintaining agreed-upon standards.

This collaborative ethos is crucial for aligning OT efforts with the overarching goals and operational benchmarks set by both internal divisions and external collaborators. This alignment will ensure industrial processes are optimized for peak performance and adhere to the stringent compliance and regulatory frameworks governing these activities. The engagement of all relevant stakeholders, from the conceptual phase to the execution and maintenance stages, fosters a collective dedication to maintaining the highest standards of integrity, security, and operational efficiency in the OT systems and the industrial workflows they support.

Consistency in maintaining these standards and practices across the board is instrumental in achieving a cohesive operational environment. Such uniformity eliminates potential discrepancies and inefficiencies, paving the way for a more streamlined and integrated operational framework. This enhances the productivity and effectiveness of the OT initiatives and reinforces the organizational resilience and adaptability to evolving industrial landscapes.

In essence, the strategic coordination between the OT Unit and the broader organizational and external ecosystem is not just about facilitating smoother operations. It is about building a collaborative culture that values proactive engagement, shared responsibility, and a unified approach to achieving excellence in OT and industrial processes. This collective endeavor is pivotal in driving the organization's mission forward, leveraging technological advancements to optimize industrial operations, and sustaining competitive advantage in an increasingly complex and interconnected world.

4.3.3 General Duties and Tasks for OT Unit and Department

This section provides general definitions and information about the tasks to be performed by the employees of the OT Unit.

4.3.3.1 Clarifying Roles, Authority, and Communication Within the OT Unit

Integrating the OT Unit within the broader organizational structure necessitates meticulous coordination with various departments to ensure operational harmony. A well-structured framework delineating the responsibilities, authority levels, and communication protocols is necessary for the organization to avoid facing challenges such as overlapping jurisdictions, communication breakdowns, and unwarranted interventions, all of which can impede efficiency and productivity.

To mitigate such risks, it is imperative to establish well-defined boundaries delineating the responsibilities and authority of the OT Unit over other departments. This clarity is essential across a range of operational domains, including but not limited to:

- **Communication Networks:** Establish clear protocols for inter-unit communication and data sharing to ensure the seamless flow of information across IT and OT networks.
- **Software and Hardware Management:** This involves defining the ownership and management responsibilities for various software packages (from

engineering applications to Level 3 and 4 systems) and hardware components (including servers, workstations, and mobile devices).

- **System Infrastructure:** This section outlines responsibilities for managing operating systems, peripheral devices, radio-frequency identification (RFID) and barcode readers, and physical networking infrastructure.
- **Connectivity and Protocols:** Specifying the authority over network configurations, protocol support, conversions, routing, and switching to ensure robust and secure connectivity across IT and OT systems.
- **Procurement and Vendor Relations:** Clarifying roles in procurement, maintenance contracts, and interactions with supplier companies to ensure alignment with organizational procurement policies and standards.
- **Backup and Data Management:** Establish protocols for system and data backups at various levels, from PLC program backups to database and big data environment backups, to ensure comprehensive risk management and data integrity.
- **Technical Specifications and Installations:** Assigning responsibility for drafting technical specifications, overseeing installations and deliveries, and conducting forensic investigations as needed.

Given the diversity and complexity of OT systems, ranging from turn-key solutions to custom-built applications, it is crucial to develop specific process scenarios for each system, supported by thorough documentation. This documentation should detail the operational procedures, backup strategies, and contingency plans tailored to each system's unique requirements and configurations.

For instance, the backup protocols must encompass multiple layers, including the backup of PLC programs, the data they generate, associated application programs, databases, and the overarching data storage environments. Each layer requires a distinct approach to backup and recovery, underscoring the need for detailed planning and documentation.

Establishing clear guidelines and protocols for the OT Unit's interaction with other organizational divisions is foundational to maintaining operational efficiency, ensuring data security, and fostering a collaborative work environment. By delineating the roles, responsibilities, and communication channels, the organization can create a cohesive framework that supports its operational goals while mitigating potential risks associated with system overlaps and miscommunications.

4.3.3.2 Optimizing Physical Operations in Industrial Settings

Orchestrating physical tasks such as installing, relocating, and maintaining devices within OT Systems is critical to ensuring seamless industrial operations. These tasks necessitate a synergistic approach, particularly close collaboration between the OT and Maintenance Unit. Such collaboration ensures that the expertise and responsibilities of both units are leveraged effectively, optimizing the efficiency and reliability of industrial systems.

For foundational industrial components categorized as Level 1 equipment, planning and executing installations, relocations, and maintenance routines demands a joint effort. The collaborative planning phase involves the OT Unit, which has a deep understanding of OT requirements and configurations, and the Maintenance Unit, which brings expertise in the physical upkeep and technical servicing of industrial equipment.

While the planning stage is a collaborative endeavor, the execution of installation and termination tasks for these systems is primarily the domain of the OT Unit. This delineation of responsibilities ensures that the specialized technical requirements of OT systems are met with precision and according to industry best practices. The OT Unit's role in this context includes overseeing the technical aspects of equipment setup, ensuring proper integration with existing OT systems, and verifying that installations comply with organizational standards and operational requirements.

The Maintenance Unit, on the other hand, plays a pivotal role in providing ongoing support for these systems. It addresses physical and technical maintenance needs to ensure optimal performance and longevity of the equipment. This includes routine inspections, repairs, adjustments, and responding to emergency maintenance requirements to minimize downtime and maintain operational continuity.

Clear communication channels and protocols between the OT and Maintenance Units are essential to facilitate these processes. This includes establishing joint procedures for scheduling maintenance activities, sharing technical information and documentation related to the equipment, and coordinating efforts during complex installation or relocation projects.

Moreover, integrating modern maintenance management systems and tools can enhance the efficiency and effectiveness of these collaborative efforts. These systems can provide real-time visibility into equipment status, maintenance schedules, and inventory levels, enabling proactive planning and execution of maintenance and installation activities.

In summary, the effective collaboration between the OT and Maintenance Units in managing the physical aspects of industrial equipment installation, relocation, and maintenance is crucial for maintaining the operational integrity and efficiency of industrial systems. By combining the technical expertise of the OT Unit with the maintenance capabilities of the Maintenance Unit, organizations can ensure that their industrial operations are supported by robust, well-maintained, and optimally functioning OT systems [29].

4.3.3.3 Optimizing Network Infrastructure Through Active Device Management

Managing active network devices forms the backbone of a robust and secure OT infrastructure. This involves meticulous processes to ensure efficient installation, configuration, and maintenance of these devices. The tasks encompass a range of activities, each critical for the seamless operation and security of the network system [30, 31]:

- **Designing Network Topologies and Configuring Internet Protocol (IP) Settings:** Crafting an optimal network topology and configuring IP settings, including virtual local area network (VLAN) setups and virtual private network (VPN) configurations, are foundational steps. These actions ensure the network infrastructure is tailored to meet the operational environment's specific communication needs and structured to facilitate secure and efficient data flow across devices.

- **Implementing and Managing Firewall Security:** Establishing and rigorously managing firewall security protocols is paramount. This includes configuring firewall settings to regulate network traffic, authorizing active devices, and making precise adjustments to ports and connections to safeguard the network against unauthorized access and potential threats.
- **Adjusting Protocols for Enhanced Security and Functionality:** The dynamic nature of network environments necessitates occasionally altering communication protocols. Such adjustments are crucial for adapting to evolving security requirements and ensuring network devices' continued functionality and interoperability.
- **Monitoring Network Protocols and Managing Permissions:** Continuous monitoring of network protocols is essential for identifying and promptly addressing potential issues. Permission planning and management also play a crucial role in defining and controlling access rights within the network, ensuring that only authorized users and devices can perform specific actions.
- **Conducting Authority Violation Checks and Log Analysis:** Regular checks for authority violations are vital for detecting and mitigating unauthorized attempts to access the network or manipulate device functionalities. Log reviews complement this process by providing a detailed audit trail of network activities, enabling the identification of irregular patterns or security breaches.

While technically challenging, these tasks underscore the importance of a strategic approach to network device management within OT systems. By ensuring that active devices are correctly installed, configured, and maintained, organizations can enhance the resilience and efficiency of their operational technologies. This involves technical expertise and a keen understanding of the operational context and security landscape.

Furthermore, integrating advanced network management tools and technologies can significantly streamline these processes. From automated network monitoring solutions to sophisticated cybersecurity platforms, leveraging the right technologies can enhance the effectiveness of network management practices, ensuring that the OT infrastructure remains robust, secure, and aligned with the organization's operational objectives.

The meticulous management of active network devices is a critical component of a well-functioning OT system. It demands a comprehensive strategy encompassing technical proficiency, strategic planning, and the judicious use of technology to ensure the network infrastructure's integrity, security, and efficiency.

4.3.3.4 Streamlining Infrastructure through Passive Device Management

The establishment and upkeep of passive devices, including foundational cabling and connections within OT systems, form the bedrock for reliable and efficient network infrastructure. Though not actively involved in data processing or network traffic management, these elements are crucial for the physical framework that supports active components. Their installation and maintenance encompass several vital activities [32]:

- **Network Cabling Infrastructure:** This activity entails laying down the physical wiring that forms the backbone of the OT network. It involves planning the routing, installing the cables, and ensuring they are protected and organized to facilitate smooth operation and ease of maintenance. This foundational layer supports the entire gamut of OT systems, providing the necessary connectivity for seamless data flow.
- **Termination in Active Device Cabinets:** A critical aspect of network setup involves making precise connections within cabinets that house active networking devices. The termination process ensures that cables are correctly connected to these devices, adhering to technical standards and specifications to guarantee optimal performance and secure data transmission.
- **Fiber Optic Cabling Implementation:** Fiber optic cabling is indispensable for high-speed data transmission requirements, especially over longer distances or in environments with high electromagnetic interference. Implementing and maintaining these cables demands specialized skills to handle the delicate fibers, connectors, and precise termination techniques required for efficient light transmission.

These tasks are fundamental to ensuring the robustness and reliability of both active and passive components within OT systems. The effectiveness with which these devices are installed, configured, and maintained directly impacts the operational efficiency and resilience of the organization's technology infrastructure.

A well-coordinated approach involving both the OT and Maintenance Units is essential to achieve this. This collaboration ensures that the physical infrastructure is technically sound and aligned with the organization's broader operational strategies and requirements. Proper planning, execution, and maintenance protocols must be established to ensure that the passive network components support the dynamic needs of the OT environment.

Moreover, advancements in cabling technology and installation practices offer new opportunities to enhance network performance and reliability. Structured cabling systems and advanced cable management solutions can significantly improve network scalability, flexibility, and maintenance efficiency.

In conclusion, managing passive devices, particularly the network cabling infrastructure, is a critical yet often overlooked component of a comprehensive OT strategy. Organizations can enhance their operational technologies' integrity and performance by meticulously planning, installing, and maintaining these foundational elements. This, in turn, supports the seamless functioning of active devices and systems, underpinning the organization's operational capabilities and resilience in the face of evolving technological and operational demands.

4.3.3.5 Establishing Standards and Protocols for Network Integration

Establishing clear standards and communication protocols early ensures seamless integration and communication within network-connected devices. This foundational step is essential for the organization's operational technologies' internal coherence and facilitates smooth interactions with external entities such as stakeholders, suppliers, and partners. From the onset of project planning, it is imperative that the designated unit responsible for network integration is equipped with comprehensive guidelines and receives robust support to navigate the complexities of integrating diverse systems and technologies.

This preparation involves meticulously defining the technical standards and communication protocols that will govern the interoperability and functionality of network-connected devices. These standards serve as a blueprint, guiding device configuration, installation, and maintenance to ensure they can communicate effectively within the organization's network infrastructure and with external systems. By establishing these parameters early in the project-planning phase, organizations can mitigate potential integration challenges, streamline the deployment of new systems, and enhance overall system compatibility and performance.

Moreover, clearly defined standards and protocols are pivotal for aligning expectations, ensuring compliance, and facilitating effective collaboration with external stakeholders and suppliers. They provide a common language and expectations that guide the selection, design, and implementation of network-connected devices and systems, ensuring they meet the organization's operational requirements and security standards.

Integrating newly acquired systems, often involving a blend of legacy and cutting-edge technologies, further underscores the importance of well-defined standards and protocols. These guidelines enable a structured approach to system integration, ensuring new technologies can be seamlessly incorporated into the existing infrastructure without disrupting operational continuity or compromising network security.

Continuous communication and knowledge sharing between the responsible unit and external parties are crucial to support these integration efforts. Regular updates, meetings, and workshops can help maintain alignment, address emerging challenges, and share best practices. In addition, leveraging documentation, such as integration manuals and protocol specifications, can provide valuable reference points throughout the integration process.

The early definition and identification of standards and communication protocols for network-connected devices lay the groundwork for successful system integration and collaboration. This strategic approach facilitates the technical alignment of internal and external systems and fosters a collaborative ecosystem conducive to innovation, efficiency, and operational excellence [33].

4.3.3.6 Optimizing OT Software Management Through Collaborative Practices

Effective management of OT software, encompassing automation platforms and other specialized applications, is pivotal for maintaining the operational integrity and efficiency of industrial systems. Due to OT software's unique requirements and critical nature, differentiating these processes from general software system management is crucial. To this end, the OT Unit, in close collaboration with the Maintenance Unit, plays a central role in overseeing the lifecycle of OT software—from installation and relocation to maintenance and backup operations [34, 35].

Software Installation and Relocation: The initial installation and any subsequent relocations of OT software demand meticulous planning and execution to ensure seamless integration with existing systems and minimal disruption to ongoing operations. This involves technical installation and a thorough assessment of system compatibilities, network configurations, and operational workflows to ensure the software aligns with the specific needs of the OT environment.

Routine Maintenance and Updates: Regular maintenance of OT software is essential to ensure that the systems remain functional, secure, and efficient. This includes applying updates and patches to address vulnerabilities, improve functionality, and adapt to evolving operational requirements. The OT and Maintenance Units must establish a proactive maintenance schedule that balances system reliability with the organization's operational demands.

Robust Backup Strategies: Given the critical role of OT software in industrial operations, implementing comprehensive backup strategies is paramount. This ensures minimal impact on operational continuity in the event of system failures, data corruption, or other unforeseen issues. Backup protocols should cover the software applications, configuration settings, operational data, and custom scripts or algorithms integral to the OT systems.

Collaboration between the OT and Maintenance Units is critical to effectively managing these software-related tasks. By combining the OT Unit's specialized knowledge of OT with the Maintenance Unit's expertise in system upkeep and troubleshooting, organizations can foster a synergistic approach to software management. This collaboration ensures that OT software is technically sound and aligned with the organization's broader operational strategies and compliance standards.

Moreover, leveraging advanced software management tools and technologies can enhance the efficiency and effectiveness of these processes. Technology can be crucial in streamlining software installation, maintenance, and backup operations, from automated update and patch management systems to sophisticated backup and recovery solutions.

In summary, the strategic management of OT software, characterized by a collaborative approach and supported by advanced tools and methodologies, is fundamental to the resilience and performance of industrial systems. By prioritizing the specialized needs of OT software and fostering a culture of proactive maintenance and risk management, organizations can ensure that their OT remains robust, secure, and capable of supporting the dynamic needs of industrial operations.

4.3.3.7 Strategic Software Acquisition and Management

The acquisition and management of software for Level 2 and Level 3 systems, including HMI, SCADA, and Historian applications, demand a collaborative and strategic approach within the organization. The unit responsible for OT plays a pivotal role in this process, orchestrating the procurement activities in close coordination with various internal departments to ensure alignment with organizational objectives and technological standards [36, 37].

Pre-approval and Procurement Coordination: The journey begins with the pre-approval phase, where the OT Unit liaises with relevant departments, such as Procurement and Facilities, to outline the specific requirements and expectations for the software. This collaborative effort ensures the software procurement process is in synchronization with the organization's broader procurement policies and budgetary considerations, facilitating a streamlined and efficient approval process.

Monitoring of Purchasing and Maintenance Agreements: Post-procurement, the focus shifts to actively monitoring purchasing and maintenance agreements. This oversight is crucial to ensure that the software and its associated services meet the agreed-upon standards and deliverables, providing the organization with the necessary tools and support to maintain operational efficiency and system integrity.

Collaboration with Facilities and Engineering Units: The OT Unit also engages in a continuous dialogue with Facilities Engineering units, assessing application requests and proposals through the dual lenses of IT/OT integration and Industry 4.0 principles. This evaluation process is vital for identifying software solutions that meet the immediate operational needs and align with the organization's long-term technological roadmap and Industry 4.0 aspirations.

Evaluating Software within the IT/OT Framework: The critical task of evaluating and selecting software extends beyond functional capabilities to include considerations of interoperability, scalability, and compliance with industry standards. With its deep understanding of IT and OT ecosystems, the OT Unit is uniquely positioned to assess software options based on its ability to integrate seamlessly within the existing technological infrastructure and support the organization's Digital Transformation goals.

Industry 4.0 Alignment: In keeping with Industry 4.0's principles, the software procurement process strongly emphasizes solutions that facilitate excellent connectivity, data integration, and analytics capabilities. This forward-looking approach ensures the organization can leverage advanced technologies such as IoT, AI, and big data analytics, enhancing operational visibility, predictive maintenance, and decision-making processes.

In essence, procuring and managing Level 2 and Level 3 software are collaborative endeavors that require careful planning, coordination, and evaluation. By involving relevant departments and adhering to strategic procurement practices, organizations can ensure that their software investments are cost-effective and aligned with their operational requirements and long-term Digital Transformation objectives. This holistic approach to software procurement and management underscores the importance of IT/OT convergence in driving innovation, efficiency, and competitiveness in the era of Industry 4.0.

4.3.3.8 Streamlining Configuration Management through Specialized Teams

Configuring hardware and software components in OT systems is a critical task that demands precision, expertise, and a deep understanding of the system requirements. To ensure the highest levels of efficiency and efficacy in this process, responsibilities are strategically allocated among specialized teams. The Maintenance team takes charge of hardware configuration, leveraging their technical knowledge and hands-on experience with physical devices and infrastructure. Meanwhile, the OT team assumes responsibility for software configuration, applying their expertise in software applications, integration, and operational functionality [38].

Hardware Configuration by the Maintenance Team: The Maintenance team's role in hardware configuration encompasses various tasks, from setting up network devices and industrial controllers to configuring sensor arrays and other field devices. Their hands-on approach ensures that all hardware components are installed, configured, and optimized to meet the rigorous demands of industrial operations. Their responsibilities include ensuring the hardware is aligned with safety standards, operational efficiency benchmarks, and compatibility with existing systems.

Software Configuration by the OT Team: The OT team's responsibilities on the software side involve configuring application settings, network parameters, and user interfaces to meet specific operational requirements. This includes setting up HMI displays, defining SCADA system parameters, and customizing software applications to ensure seamless integration with hardware components and alignment with the organization's workflow and data management practices.

Collaborative Approach for Integrated Systems: This division of labor is designed to capitalize on each team's specialized skills and knowledge. It fosters an environment where each aspect of the system configuration is handled by professionals best suited to the task. This enhances the quality and reliability of the configurations and streamlines the setup and integration process, reducing the potential for errors and inconsistencies.

Aligning Configurations with Organizational Standards: Both teams work within a framework of established organizational standards and best practices, ensuring that all configurations—hardware and software alike—adhere to the required specifications, security protocols, and performance criteria. This standardized approach facilitates a unified and cohesive OT environment that supports the organization's operational objectives and technological strategies [39].

Ensuring Secure and Reliable OT Environments: The collaborative effort between the Maintenance and OT teams is crucial for maintaining a secure and reliable OT environment. By meticulously planning and executing hardware and software configurations, the organization can achieve optimized system performance, enhanced security, and greater resilience against operational disruptions.

In conclusion, the strategic division of configuration responsibilities between the Maintenance and OT teams exemplifies a targeted approach to managing complex OT systems. By leveraging the specialized expertise of each team and fostering a collaborative working environment, organizations can ensure that their OT infrastructure is not only technically sound but also aligned with broader operational goals and industry standards. This synergy is critical to achieving a robust, secure, and efficient OT environment that supports the dynamic needs of modern industrial operations.

4.3.3.9 Managing Domain Controllers and Operating System Compliance

Effective management of domain controllers for operating systems within OT environments necessitates a meticulous approach, particularly when PCs are interfaced with OT devices. To maintain system integrity, security, and compliance, it is imperative to implement a comprehensive inventory and compliance management framework. This framework should catalog all PCs within the OT landscape, detailing their respective operating systems, configurations, and compliance statuses with organizational standards [40].

Comprehensive Inventory Management: Instituting a robust inventory management system ensures that all PCs connected to OT devices are accounted for and assessed. This inventory should include detailed specifications, operating system versions, patch levels, and other relevant information contributing to a comprehensive overview of the OT network's technological ecosystem.

Operating System Compliance Matrix: Alongside the inventory, a dynamic compliance matrix is essential for monitoring and ensuring that each PC adheres to the defined security protocols, software requirements, and configuration settings. This matrix is vital for identifying compliance gaps and guiding remediation efforts to align with best practices and regulatory standards.

Isolated Domain Management for Industrial Systems: Implementing isolated domain management is crucial for recognizing the potential need for users to access OT and IT networks from the same PCs. This approach segregates the OT and IT environments at a domain level, enhancing security and minimizing the risk of cross-contamination between the two networks.

Automated System Management: Automated PC management is vital to accommodate the unique operational dynamics of production environments, particularly those operating across multiple shifts. This includes automatic startup procedures, software updates, and enforcing authorization restrictions. Such automation ensures that PCs at production stations remain secure, up-to-date, and configured according to predefined policies, regardless of the current shift or user.

Authorization Restrictions and Access Control: Implementing stringent authorization restrictions is paramount, especially in production environments where shift work is standard. Access controls should be defined and enforced to limit user privileges based on roles, ensuring that individuals can only access information and functionalities pertinent to their responsibilities. This minimizes the risk of unauthorized access or actions that could compromise system security or integrity.

In summary, managing domain controllers and operating system compliance within OT environments requires a structured and proactive approach. Organizations can enhance their OT systems' security, reliability, and efficiency by establishing a detailed inventory, maintaining a compliance matrix, segregating OT and IT domains, automating system management, and enforcing strict access controls. This comprehensive management strategy supports the seamless operation of OT environments, ensuring they remain resilient against threats while accommodating the organization's operational needs.

4.3.3.10 Enhancing Data Integration in OT Environments

The capability to efficiently collect, process, and transfer data is crucial in the OT domain, particularly within Level 1 systems encompassing the foundational layers of industrial automation. This encompasses various devices and systems, including PLCs, IIoT devices, and robotic systems, generating valuable data that drives operational insights and decision-making [41].

Streamlined Data Collection and Retrieval: The collection process from these diverse sources must be streamlined and automated to ensure that production parameters and operational data are accurately captured and made available for analysis. This involves developing or deploying specialized software solutions that facilitate seamless data extraction from OT devices and systems.

Data Transfer and Integration: Beyond mere collection, the data must be effectively transferred to platforms that can be analyzed and utilized, such as IT systems, Edge servers within the OT infrastructure, or cloud-based analytics platforms. This transfer process should ensure data integrity and security, employing protocols and formats that enable efficient data integration and accessibility for downstream applications.

Support for Research and Development (R&D) and Analysis Departments: The collected data plays a vital role in supporting the activities of the R&D and analysis departments. These departments rely on operational data to drive innovation, optimize processes, and develop new solutions. Hence, the OT systems must be equipped to deliver the requisite data in formats conducive to analysis and research applications.

Configuration and Integration of New Devices: As industrial environments evolve, new machinery, robotic systems, sensors, and IIoT devices are constantly introduced. Setting up the configurations and system definitions for these new additions is essential to ensure seamless integration into the existing OT landscape. This includes defining network parameters, communication protocols, and data collection points to ensure the new devices can communicate effectively within the OT system and contribute to the overall data pool.

Collaborative Framework for System Expansion: Expanding OT systems with new devices and technologies necessitates a collaborative approach involving the OT team and IT, R&D, and maintenance departments. This framework ensures that new devices are integrated with a holistic view of system functionality, data coherence, and operational objectives.

In summary, the processes related to data collection, retrieval, and transfer in OT environments are foundational to leveraging the full potential of industrial automation and innovative manufacturing initiatives. By establishing robust mechanisms for data integration, supporting the needs of R&D and analysis functions, and facilitating the seamless addition of new devices and systems, organizations can enhance their operational intelligence, drive innovation, and maintain a competitive edge in the rapidly evolving industrial landscape [42].

4.3.3.11 Enhancing Support for OT and Engineering Software Applications

In the dynamic landscape of OT, comprehensive support for the myriad of software applications that drive device functionality, data analysis, and operational efficiency is crucial. The dedicated team responsible for OT systems ensures that all software components, from device drivers and connectivity solutions to advanced analytical and optimization tools, receive the requisite installation, maintenance, and user support services. This multifaceted support framework encompasses several key areas [43]:

Comprehensive Software Support: The OT support team oversees the full spectrum of engineering and operational software applications within the OT environment. This includes but is not limited to connection software that facilitates communication between devices, testing and recording software that captures and logs operational data, and quality control applications that ensure product and process integrity.

Advanced Analytical and Optimization Tools: In addition to foundational software applications, the support extends to more sophisticated tools such as computational engineering software, which aids in the design and analysis of complex systems, and optimization software that enhances operational efficiency. The team also supports applications focused on overall equipment effectiveness and energy analysis, which are vital for monitoring performance metrics and identifying opportunities for improvement.

User Assistance and Training: Besides technical support, the team provides user assistance and training to ensure operators and engineers can effectively utilize the software applications. This includes developing user manuals, conducting training sessions, and offering on-demand assistance to resolve operational queries or issues.

Software Maintenance and Updates: Regular maintenance and timely updates of all OT and engineering software are essential to ensure these tools remain effective, secure, and compatible with the evolving OT environment. The support team manages software updates, patches, and upgrades, coordinating with vendors to address emerging issues or integrate new functionalities.

Integration and Interoperability Support: The OT environment has diverse software applications, so ensuring seamless integration and interoperability between different systems is crucial. The support team ensures that data flows smoothly across applications and that the software ecosystem is cohesive and aligned with the broader operational goals.

Responsive Support Mechanisms: Implementing responsive support mechanisms, such as help desks, ticketing systems, and emergency response protocols, ensures software-related issues can be swiftly addressed, minimizing downtime and operational disruptions.

In essence, supporting existing OT and engineering software within the organizational framework is a comprehensive endeavor that spans installation, maintenance, user assistance, and continuous improvement. By providing robust support for these critical software applications, the OT team enhances the OT landscape's reliability, efficiency, and innovation capacity, supporting the organization's broader objectives and ensuring a competitive edge in the fast-evolving industrial sector.

4.3.3.12 Seamless Integration of OT Systems with Enterprise Software

Synchronizing OT systems with broader enterprise software platforms, such as PLM, ERP, manufacturing resource planning (MRP), and MES, is pivotal for achieving a holistic and efficient operational framework. This integration facilitates the seamless flow of data across different levels of the organization, enabling informed decision-making and streamlined processes. Achieving this integration involves a collaborative approach with various internal units, ensuring that the OT system software is fully compatible and interconnected with these essential enterprise systems [44].

Strategic Collaboration for Integration: The process begins with strategic collaboration between the OT Unit and departments responsible for managing PLM, ERP, MRP, and MES platforms. This collaborative effort aims to identify integration points, define data exchange protocols, and establish a unified data model that supports seamless information sharing across these systems.

Direct Connection and Data Synchronization: Establishing direct connections between OT systems and enterprise software ensures real-time data synchronization, which is crucial for maintaining up-to-date information across the organizational spectrum. This direct linkage facilitates instant updates on production metrics, inventory levels, quality control parameters, and other critical operational data, enhancing the enterprise's responsiveness and agility.

Supporting Digital Factory Initiatives: The integration extends to advanced applications such as Digital Twins and the management of Design Material Lists bills of materials (BOMs) within the scope of the Digital Factory concept. Digital Twins, virtual replicas of physical systems, rely on real-time data from OT systems to simulate and analyze operations, enabling predictive maintenance, process optimization, and product development insights. Similarly, the integration ensures that BOMs are accurately reflected in both the design and production stages, supporting efficient material planning and procurement processes.

Ensuring Interoperability and Compliance: A vital aspect of this integration effort is ensuring that all systems are interoperable and adhere to industry standards and compliance requirements. This involves regularly reviewing and updating system configurations, software versions, and communication protocols to ensure compatibility and compliance with regulatory standards.

Facilitating Continuous Improvement: The integrated OT-enterprise software ecosystem is designed to support continuous improvement initiatives. It leverages data analytics and insights to drive operational enhancements, cost reductions, and quality improvements. By providing a comprehensive view of the entire product lifecycle and manufacturing operations, the integrated system enables organizations to identify bottlenecks, uncover inefficiencies, and implement targeted improvements.

In summary, integrating OT systems with PLM, ERP, MRP, and MES platforms is a strategic endeavor that enhances the coherence and efficiency of organizational operations. Through collaborative planning, direct system connections, and a commitment to interoperability and continuous improvement, organizations can harness the full potential of their technological investments, driving innovation and competitive advantage in an increasingly digital industrial landscape.

4.3.3.13 Strategizing User and Access Management in OT Environments

OT systems, designed for reliability and uninterrupted operation, embody a unique user landscape that significantly diverges from conventional IT systems. In OT environments, users are not limited to human operators; they can also include automated entities such as robots and collective units like shift teams and leadership roles, each requiring specific access rights and capabilities. This distinctive nature of OT systems necessitates a tailored approach to user and authority process planning, ensuring that permissions and access controls are meticulously defined and managed to support operational efficiency and system security [45].

User Definition and Access Permissions: Effective management of user roles and access permissions in OT systems involves thoroughly analyzing operational requirements and user interactions. This includes defining roles for human operators, automated entities, and team units, each with tailored access rights that align with their operational functions and responsibilities. The planning process should identify critical stations and interfaces where user interactions occur, establishing clear guidelines for role-based access controls.

Peripheral Device Communication: Integrating and managing peripheral devices, such as RFID readers, industrial cameras, and document scanners, is crucial for the seamless operation of OT systems. This requires a strategic approach to device communication management, ensuring that ports and interfaces, such as USB connections, are appropriately configured to support device functionality without compromising system security.

Balancing Operational Needs and Security Risks: The configuration of peripheral device interfaces exemplifies the delicate balance between operational requirements and security considerations. While turning off USB ports can mitigate the risk of unauthorized data transfers via portable disks, it may also impede the functionality of essential USB-operated devices. Therefore, the planning process must consider alternative security measures, such as device allowlisting, encryption, and activity monitoring, to safeguard the system while maintaining operational integrity.

Collaborative Approach to Access Management: Developing and implementing user and authority processes in OT systems should involve input from operational leaders, cybersecurity experts, and system users. This approach ensures that access management strategies are effective in supporting operational workflows and robust in mitigating security vulnerabilities.

Continuous Review and Adaptation: Given the dynamic nature of OT environments and the evolving threat landscape, user and authority process planning must be an ongoing effort. Regular reviews and updates to access controls, user roles, and device configurations are essential to adapt to changes in operational requirements, technological advancements, and emerging security threats.

Essentially, the strategic planning of user and authority processes in OT systems is critical to ensuring operational efficiency and security. By adopting a tailored approach that accounts for the unique characteristics of OT environments, organizations can establish a secure and resilient operational framework that supports human and automated users while effectively managing access to critical systems and devices.

4.3.3.14 Enhancing Disaster Recovery and Response in OT Environments

OT systems, characterized by their criticality to continuous production and manufacturing processes, present unique challenges regarding disaster recovery and immediate response mechanisms. Unlike IT systems, where data backup and recovery procedures can often be executed relatively easily and quickly, OT systems require a more nuanced approach due to the specialized software and development environments utilized for programming devices such as PLCs, robots, HMIs, and SCADA systems [46].

Specialized Backup Procedures: Backing up OT systems often necessitates using proprietary software provided by equipment manufacturers. For instance, backing up Mitsubishi PLCs may involve tools like MXComponent and GX Developer, while Siemens PLCs might use the TIA Portal. This dependency on specialized software underscores the need for tailored backup strategies that accommodate each system's requirements and capabilities.

Centralizing Backup Data: Historically, maintenance and repair teams have often been responsible for backups in OT environments. However, this can lead to backup data scattered across various storage mediums, from network-shared spaces to portable hard disks and personal computers. This dispersion complicates disaster recovery efforts and increases the risk of data loss or corruption.

Implementing a Comprehensive Backup and Versioning System: To mitigate these risks, it is imperative to establish a centralized and systematic approach to backing up OT systems. This includes consolidating backup data in secure, accessible locations and implementing version control for all software components. Versioning facilitates easier tracking of changes and updates and ensures that recovery processes can revert systems to the most stable and recent configurations [47].

Documenting Changes and Maintaining Clear Records: Beyond backup and versioning, maintaining detailed documentation of all changes, updates, and modifications to OT systems is crucial. This documentation should include clear records of what changes were made, who made them, and the reasons behind each change. Such meticulous recordkeeping is vital for diagnosing issues, understanding the impact of modifications, and executing effective disaster recovery strategies.

Enhancing Immediate Response Capabilities: In addition to robust backup and documentation practices, enhancing the immediate response capabilities of OT systems is essential for minimizing downtime and mitigating the impact of system failures. This includes establishing rapid response teams, implementing automated alert systems, and developing clear emergency protocols.

Finally, regular review and testing of disaster recovery plans are essential to ensure they remain practical and current with the evolving OT landscape. Simulated disaster scenarios and recovery drills can help identify potential weaknesses in the plans and provide opportunities for continuous improvement.

In conclusion, addressing the unique challenges of disaster recovery and immediate response in OT environments requires a multifaceted approach encompassing specialized backup procedures, centralized data management, comprehensive documentation, and enhanced response capabilities. By adopting these strategies, organizations can ensure the resilience and reliability of their OT systems, safeguarding critical industrial processes against disruptions and minimizing the impact of unforeseen events.

4.3.3.15 Streamlining OT Systems with Enhanced Inventory and Documentation Practices

A prevalent challenge within OT systems is the need for more thorough documentation and inventory management, often attributed to their decentralized management and multifaceted nature. A structured approach to documenting and inventorying critical system components and configurations is essential to address this. This approach not only aids in the effective management and operation of OT systems but also enhances security, compliance, and disaster recovery efforts [48].

Comprehensive Device Inventory: A detailed inventory of all devices within the OT network, including PLCs, HMIs, SCADA systems, sensors, and network communication devices, is foundational. This inventory should encompass hardware specifications, firmware versions, configuration settings, and network interfaces (ports), providing a complete overview of the OT environment's physical and logical layout.

System User Management and Access Controls: Documenting system users, their assigned roles, and corresponding access permissions is essential. This information is crucial for maintaining system security and ensuring users have appropriate access levels to perform their duties without compromising sensitive system functionalities or data.

Network Configuration and Segmentation Details: Documenting network configurations, including IP address assignments, subnet divisions, VLAN configurations, and routing protocols, is vital for maintaining network integrity and performance. Precise mapping of these configurations supports network troubleshooting, security zoning, and compliance with network architecture standards.

Closed System Connections and Interdependencies: Understanding the connections and interdependencies within closed systems, such as proprietary or isolated networks, is essential for system maintenance and change management. Documenting these internal connections aids in assessing the impact of system modifications and ensures the stability of critical processes.

Communication Infrastructure Documentation: Detailed records of the network and communication infrastructure, including communication cards, media converters, and other network devices, are necessary for managing data flow and ensuring robust communication links within the OT environment.

Implementing a centralized documentation and inventory management system is recommended to overcome the challenges of disorganization and multiple custodianship. This centralized approach ensures consistency, accessibility, and reliability of information and facilitates better coordination among various stakeholders involved in managing OT systems.

Regular Updates and Audits: Regular audits and updates of the documentation and inventory are crucial to ensure they remain accurate and reflective of the current state of the OT systems. This ongoing maintenance process should be integrated into the operational routines, with clear responsibilities assigned to relevant teams or individuals.

Integration with IT Documentation Practices: Where possible, aligning OT documentation and inventory practices with existing IT management frameworks can leverage established processes and tools, fostering a more integrated approach to technology management across the organization.

In summary, establishing rigorous documentation and inventory practices for OT systems is critical for overcoming the challenges posed by these environments' complex and decentralized nature. By adopting a structured and centralized approach to managing this information, organizations can enhance their OT systems' operational efficiency, security, and resilience, supporting the broader objectives of reliability and continuous improvement in industrial operations.

4.3.3.16 Implementing Certification and Audits for Enhanced OT System Integrity

Implementing rigorous certification and audit processes is essential to upholding OT systems' integrity, security, and compliance. Unlike traditional IT environments, which are often concentrated within corporate office settings, OT systems span various industrial and operational contexts. This diversity necessitates a tailored approach to certification and auditing that accommodates OT environments' unique challenges and requirements [49].

Bridging IT and OT Certification Standards: Establishing mutual certification criteria that bridge IT and OT systems is crucial for maintaining a cohesive security and compliance posture across the organization. Given the distinct operational dynamics of OT environments, certification processes must be adapted to address these systems' specific risks, technologies, and operational priorities while aligning with broader corporate IT standards.

Adapting Personnel and Processes: OT systems' specialization often means that personnel and processes familiar in corporate IT settings may not directly translate to the industrial context. Therefore, developing OT-specific certification and audit protocols requires input from professionals with expertise in industrial systems, cybersecurity, and compliance frameworks relevant to OT environments.

Incorporating Industry-Specific Standards: Besides general IT security standards, it is vital to incorporate industry-specific frameworks and guidelines, such as those provided by the National Institute of Standards and Technology (NIST) for industrial control systems. These standards offer detailed guidance on securing OT systems, addressing network architecture, device security, and incident response tailored to the OT context.

Regular Audits and Control Mechanisms: Regular audit and control mechanisms are essential for continuously assessing OT systems' compliance and security posture. These audits should be comprehensive, covering hardware, software, network configurations, and operational procedures to identify potential vulnerabilities and areas for improvement.

Collaborative Approach to Certification: Effective certification of OT systems requires collaboration among stakeholders, including IT and OT teams, compliance officers, and external certification bodies. This collaborative approach ensures that certification efforts align with internal standards and external regulatory requirements, facilitating a unified risk management and compliance approach.

Continuous Improvement and Adaptation: The dynamic nature of technology and the evolving threat landscape necessitate an ongoing commitment to certification and audit processes. Continual review and adaptation of certification criteria and auditing practices are essential to responding to new challenges and ensuring that OT systems remain secure, compliant, and aligned with best practices.

In conclusion, the certification and auditing of OT systems represent critical components of a comprehensive security and compliance strategy. Organizations can enhance the resilience and integrity of their OT systems by developing tailored certification standards, adapting processes to the unique requirements of OT environments, and fostering a collaborative approach to compliance. This proactive approach safeguards critical industrial operations and supports the organization's broader objectives of operational excellence and risk management.

4.3.3.17 Navigating Audit and Sustainability Challenges in OT Environments

Critical to the industrial operations landscape, OT systems often face unique challenges in audit and sustainability, especially concerning software management and system updates. While traditional electromechanical audits and controls are routine, the landscape for software audits—including operating system updates, patches, and version management—presents a complex scenario. System compatibility constraints and legacy operating systems often compound this complexity [50].

Addressing Legacy System Challenges: A significant hurdle in maintaining OT systems is the reliance on outdated operating systems no longer supported by manufacturers, such as Windows 95, 98, or Windows 7. The discontinuation of support for these platforms poses security risks and limits the system's ability to accommodate modern software solutions, including necessary updates and patches.

Compatibility and Sustainability Issues: The technological gap between legacy and current systems introduces several issues, from the incompatibility of new 64-bit software with older 32-bit operating systems to challenges integrating contemporary technologies and libraries, such as those used in Java-based applications with outdated platforms. These compatibility challenges hinder the sustainable evolution and secure operation of OT systems.

Security Vulnerabilities in Legacy Systems: A critical risk is the inability to install current software on older operating systems due to known security vulnerabilities. This limitation impacts the functionality and efficiency of OT systems and exposes industrial operations to potential cyber threats, compromising the integrity and reliability of critical infrastructure.

Strategic Audit and Update Framework: A strategic framework for auditing and updating OT systems is essential to address these challenges. This framework should include a comprehensive assessment of existing systems, identification of critical vulnerabilities, and a roadmap for system updates that considers compatibility, operational continuity, and security requirements.

Gradual System Modernization: Given the operational criticality of OT systems, a gradual and strategic approach to modernization is advisable. This approach involves phased updates, where possible, and the implementation of intermediary solutions that bridge the gap between legacy and modern systems, ensuring operational continuity while progressively enhancing system capabilities and security.

Collaboration with Manufacturers and Specialists: Equipment manufacturers and IT specialists can provide insights into alternative solutions for legacy systems, including custom patches, virtualization options, or secure gateways that isolate legacy systems from the broader network, mitigating security risks.

Investment in Sustainable Solutions: Long-term sustainability in OT environments necessitates investment in solutions that address immediate compatibility and security challenges and align with the industry's future technological direction. This includes exploring next-generation OT systems designed with interoperability, security, and upgradability in mind.

In summary, addressing the audit and sustainability challenges in OT systems, particularly those related to software audits and system updates, requires a multifaceted approach. By acknowledging the complexities of legacy systems, implementing a strategic audit and update framework, and investing in gradual modernization and sustainable solutions, organizations can enhance their OT environments' resilience, security, and efficiency, ensuring their readiness to meet the demands of contemporary industrial operations and future challenges.

4.4 DEFINING THE OT TEAM STRUCTURE: BALANCING IT AND OT EXPERTISE

Delineating team members' roles and responsibilities is crucial for maintaining a balanced and effective operational framework in the intricate OT ecosystem. The team is broadly segmented into two core areas of expertise: the IT Core and the OT Core. Each addresses different facets of the OT environment while ensuring synergy and cohesion in achieving the organization's operational objectives [51].

4.4.1 IT Core: Ensuring Technological Integrity and Security

The IT Core within the OT team is pivotal in establishing and maintaining the technological backbone that underpins operational processes. This segment of the team is tasked with:

- **Infrastructure Management:** Deploying and managing the network and computing infrastructure as the foundation for OT systems, ensuring connectivity, reliability, and scalability.

- **Cybersecurity Measures:** Implement robust security protocols, monitor systems for potential threats, and respond to cybersecurity incidents to protect sensitive operational data and systems from unauthorized access or sabotage.
- **Software Development and Integration:** Developing custom software solutions and integrating off-the-shelf applications to facilitate efficient data exchange and process automation, enhancing the overall functionality of OT systems.

Professionals in the IT Core may include network engineers, who architect and optimize the network infrastructure; cybersecurity analysts, who safeguard systems against digital threats; and software developers, who create and maintain bespoke solutions tailored to operational needs.

4.4.2 OT Core: Driving Operational Excellence and Process Optimization

The OT Core represents the operational heart of the organization, focusing on the direct management and optimization of production processes and machinery. Key responsibilities of this group include:

- **System Control and Automation:** Implementing and managing automation systems that control production machinery and processes, ensuring precise and efficient operation.
- **Process Monitoring:** Continuously monitoring operational processes to detect anomalies, optimize performance, and prevent downtime, utilizing advanced diagnostic tools and analytics.
- **Maintenance and Troubleshooting:** Conduct regular machinery maintenance and swiftly address any operational issues to minimize disruptions and maintain production continuity.

Key personnel within the OT Core might include automation engineers, who design and implement control systems; process technicians, who monitor and optimize production processes; and maintenance specialists, who ensure the ongoing reliability and efficiency of machinery and equipment.

4.4.3 Fostering Collaboration Between IT and OT Cores

While the IT and OT Cores have distinct focal areas, fostering collaboration and communication between these groups is essential for the holistic performance of OT systems. This collaborative dynamic ensures that technological advancements and cybersecurity measures seamlessly integrate with operational processes, machinery, and control systems.

4.4.4 Continuous Learning and Adaptation

Given the rapidly evolving landscape of technology and industrial processes, IT and OT Cores members are encouraged to engage in continuous learning and professional development. This commitment to ongoing education helps the team stay abreast of the latest technologies, methodologies, and best practices, ensuring the organization remains competitive and responsive to changing operational demands [52].

In conclusion, delineating the key roles and responsibilities within the OT team, with a clear distinction between the IT and OT Cores, provides a structured approach to managing and optimizing operational technologies. By leveraging the specialized expertise of each group and fostering a culture of collaboration and continuous improvement, organizations can achieve operational excellence, enhance system security, and drive innovation in their industrial processes [53].

4.4.5 Optimizing Organizational Synergy: The Role of the Enterprise Core

The Enterprise Core is the critical nexus within organizations, seamlessly merging IT's technological prowess with OT's operational acumen. This fusion is instrumental in harnessing the full potential of both domains to bolster organizational efficiency, drive strategic initiatives, and pave the way for comprehensive Digital Transformation. The Enterprise Core is the strategic architect of this integration, ensuring that technology deployment and operational processes are in lockstep with the overarching business objectives [54].

Strategic Alignment and Data-Driven Insights: Central to the Enterprise Core's mission is aligning IT and OT capabilities with the organization's goals. This involves a deep dive into data analytics, extracting actionable insights from the vast repository of information generated by IT and OT systems. Data analysts and business intelligence specialists play a pivotal role here, employing advanced analytics to inform decision-making, optimize operations, and identify new opportunities for innovation and growth.

Project Management and Digital Transformation: Project managers and strategic planners within the Enterprise Core spearhead initiatives that bridge the gap between technology and business. They manage cross-functional projects encompassing IT and OT elements, ensuring that Digital Transformation efforts are cohesive, well coordinated, and aligned with strategic business goals. Their work is crucial in orchestrating complex integrations, system upgrades, and process optimizations to enhance organizational agility and competitiveness.

Sustainability and Environmental Stewardship: Increasingly, the Enterprise Core is also taking on the mantle of sustainability, focusing on net zero initiatives and carbon footprint management. This expanded role requires a multidisciplinary approach, blending technical expertise with a commitment to environmental sustainability. Professionals in this domain are tasked with weaving sustainable practices into the fabric of IT and OT operations, advocating for energy-efficient technologies, waste reduction, and integrating renewable energy sources. They aim to ensure that the organization's technological and operational pursuits are practical, innovative, environmentally responsible, and aligned with global sustainability standards.

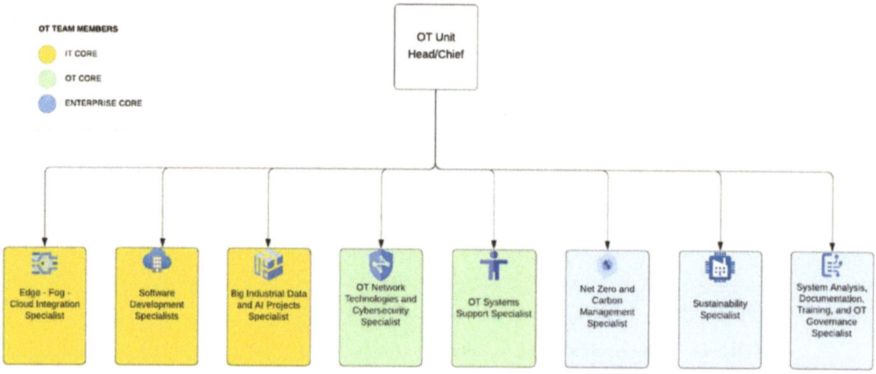

FIGURE 4.1 IT, OT, and Enterprise Core.

Developing Sustainability Strategies: The work of sustainability-focused personnel involves a holistic assessment of the organization's environmental impact, followed by developing and implementing strategies to minimize this footprint. This includes initiatives to reduce energy consumption, optimize resource utilization, and promote the adoption of green technologies. Moreover, they are responsible for ensuring that these sustainability efforts are transparent, measurable, and compliant with regulatory requirements, often involving collaboration with external stakeholders, regulatory bodies, and industry partners.

The Enterprise Core is the strategic linchpin in aligning technological innovation with business imperatives and sustainability goals. By fostering a culture of collaboration, continuous improvement, and environmental stewardship, this core group empowers organizations to navigate the complexities of the digital age, achieve operational excellence, and contribute positively to global sustainability efforts. Through their concerted efforts, the Enterprise Core ensures that the organization remains at the forefront of technological advancements, competitive in its operations, and responsible for its environmental impact (see Figure 4.1).

4.4.6 Leadership at the Helm: The Role of the OT Unit Chief

At the core of an organization's OT framework, the OT Unit Chief is the pivotal leader, orchestrating the unit's strategic and operational dimensions. This role demands a comprehensive oversight of all activities related to OT systems, encompassing the intricacies of project planning, intra-unit collaboration, and external stakeholder engagement. The OT Unit Chief is the linchpin ensuring that the unit's initiatives are technically sound and strategically aligned with the broader organizational goals.

Strategic Planning and Execution: The OT Unit Chief is entrusted with the critical responsibility of steering the unit through the complexities of operational technologies, from conceptualizing projects to overseeing their execution. This involves a meticulous approach to planning and prioritizing initiatives that drive efficiency, innovation, and resilience in the organization's operational processes.

Collaborative Leadership: In fostering a collaborative environment, the OT Unit Chief ensures seamless interaction within and across different organizational departments. This role involves bridging gaps, facilitating cross-functional teams, and ensuring that the OT Unit's projects and policies are integrated with the broader organizational objectives, enhancing synergy and operational coherence.

Transformational Leadership: The OT Unit Chief drives the unit's adaptive strategies and policies, keeping an eye on the ever-evolving technological landscape. They champion innovation by integrating cutting-edge technologies and industry best practices, ensuring the OT Unit remains at the forefront of operational excellence and technological advancement.

External Engagement and Partnership: Beyond internal coordination, the OT Unit Chief is vital in cultivating productive relationships with external stakeholders, including technology partners, suppliers, and industry peers. This external engagement is crucial for leveraging external expertise, staying informed about industry trends, and ensuring that a broad spectrum informs the OT Unit's strategies of insights and innovations.

Visionary Leadership and Continuous Improvement: The OT Unit chief leads with a strategic vision and is committed to the unit's continuous improvement and long-term success. They foster an environment where innovation is encouraged, challenges are addressed proactively, and team members are empowered to contribute to the unit's goals. They ensure the OT framework is robust, adaptable, and aligned with the organization's strategic direction.

Key Responsibilities

- Strategic oversight of OT-related processes from project inception to completion.
- Effective management of internal and external relationships, ensuring cohesive collaboration.
- Championing developing and revising policies and roadmaps to align with technological advancements and operational needs.
- Fostering a culture of innovation, continuous improvement, and strategic agility within the unit.

Essential Competencies

- Comprehensive understanding of OT and the industrial operational landscape.
- Proficient in strategic planning, project management, and execution.
- Skilled in fostering collaboration within the team and across organizational boundaries.
- Adept at leading transformational initiatives, with a keen eye on emerging technologies and industry shifts.
- Strong leadership qualities, including decision-making, problem-solving, and team empowerment.

In essence, the OT Unit Chief embodies the strategic, operational, and transformative leadership essential for navigating the complexities of modern operational technologies. Through their guidance, the OT Unit achieves its immediate objectives and contributes significantly to its broader innovation goals, efficiency, and competitive advantage in an increasingly digital industrial environment.

4.4.7 Empowering OT Security: The Role of the OT Network Technologies and Cybersecurity Specialist

The OT network technologies and cybersecurity specialist is a critical defender and architect in the intricate realm of OT. This specialist ensures that the operational network is robust and shielded against the ever-evolving landscape of cyber threats. This specialist melds expertise in network engineering with cybersecurity acumen to fortify the organization's OT systems, a cornerstone for maintaining uninterrupted and secure industrial operations [55, 56].

Strategic Network Design and Management: This role's heart is the strategic oversight of network configurations, including VLANs and security settings, tailored to meet the unique demands of OT environments. The specialist meticulously plans, establishes, and optimizes IP networks and systems, ensuring they are efficient and resilient against disruptions.

Enhancing Communication Security: With a deep understanding of communication protocols integral to OT systems, the specialist ensures secure and reliable data exchange between devices, systems, and networks. This involves both the selection of appropriate protocols and the continuous monitoring and adjustment of configurations to safeguard against vulnerabilities.

Proactive Cybersecurity Vigilance: Central to the role is a proactive stance on cybersecurity, where the specialist is constantly looking for potential threats and vulnerabilities. Through regular assessments and the implementation of advanced security measures, they work to preemptively address risks, preserving the integrity and confidentiality of critical operational data.

Collaborative Integration with IT Networks: Recognizing the interconnectedness of IT and OT networks, the specialist collaborates closely with IT units to ensure seamless and secure integrations. This collaboration is crucial for maintaining a unified defense strategy across the organization's technological landscape, particularly as the lines between IT and OT blur.

Implementing BT/OT Convergence: The specialist is pivotal in bridging Business Technology (BT) and OT networks, facilitating a harmonious integration that supports the organization's broader operational and business objectives. This includes managing BT/OT connections and systems to ensure continuity, security, and efficiency across all technological domains.

Staying Ahead of the Curve: In an ever-evolving field, staying abreast of the latest trends, technologies, and threats in network technologies and cybersecurity is essential. The specialist continuously seeks new knowledge and skills, applying innovative solutions and best practices to enhance the organization's OT security posture.

Key Responsibilities

- Architecting secure and efficient network infrastructures tailored to OT environments.
- Ensuring secure communication protocols and configurations across OT systems.
- Leading the charge against cybersecurity threats with proactive risk management strategies.
- Facilitating seamless IT/OT network integrations and managing BT/OT connections.
- Championing continuous improvement through the adoption of cutting-edge security technologies and practices.

Essential Competencies

- Comprehensive expertise in VLAN configurations, IP networking, and OT-specific communication protocols.
- Acute awareness of cybersecurity trends and the ability to translate this knowledge into effective security strategies.
- Collaborative spirit, with the ability to work closely with IT and other organizational units to achieve cohesive security objectives.
- Analytical prowess and problem-solving skills, essential for navigating complex network and security challenges.

The OT Network Technologies and Cybersecurity Specialist embodies the confluence of network engineering and cybersecurity, ensuring the OT environment is operationally sound and fortified against cyber threats. Through strategic planning, vigilant security practices, and collaborative integration efforts, this role is instrumental in safeguarding the organization's operational technologies, thereby supporting the continuity and security of critical industrial operations.

4.4.8 Advancing Integration: The Edge–Fog–Cloud Integration Specialist

In the complex interplay of modern IT infrastructures, the Edge-Fog–Cloud Integration Specialist emerges as a pivotal figure tasked with harmonizing the dynamic layers of IT systems, cloud services, and the emergent domain of fog computing. This role is instrumental in architecting a cohesive ecosystem where data seamlessly traverses from on-premise systems to the edge, through fog nodes, and into the cloud, enabling agile, scalable, and efficient data processing and analytics across the continuum [57].

Strategic Integration Across Computing Paradigms: This specialist's core responsibility is devising and executing sophisticated integration strategies that bind together disparate computing realms—IT, fog, and cloud—ensuring a fluid, secure, and efficient data journey. This involves understanding each layer's unique attributes and advantages and crafting integration solutions that leverage these characteristics to enhance overall system functionality and data utility.

Big Data Infrastructure and Analytics Enablement: A significant aspect of the role involves developing and managing a robust, extensive data infrastructure that spans these computing layers. The specialist ensures that this infrastructure is equipped to handle the volume, velocity, and variety of data, facilitating advanced analytics and insights that drive decision-making and innovation.

Optimizing Database Technologies for Scalability and Performance: This specialist configures and maintains advanced database systems, including NoSQL databases and object storage solutions. These efforts are geared toward optimizing data storage, retrieval, and processing capabilities, ensuring the infrastructure can scale flexibly to meet evolving data demands.

Collaboration for Operational and Analytical Excellence: Working with IT and OT teams, the Edge–Fog–Cloud Integration Specialist ensures the integrated systems meet the organization's nuanced data needs. This collaboration is vital for tailoring the integration to support operational efficiencies and unlock deep analytical insights that can inform strategy and operations.

Ensuring Data Security Across Layers: Data traverses multiple layers, so implementing stringent security measures to safeguard data integrity and privacy is paramount. The specialist deploys comprehensive security protocols and practices, ensuring that data is protected at every point in its journey from edge to cloud.

System Performance Monitoring and Optimization: Continuous integrated systems' fine-tuning is crucial for maintaining optimal performance and addressing any emergent issues related to compatibility, throughput, or latency. This ensures that the infrastructure remains robust and responsive.

Technological Vigilance and Continuous Improvement: Staying attuned to the latest advancements in cloud computing, fog computing, and big data technologies enable the specialist to continually refine and enhance the integration solutions, ensuring the organization remains at the cutting edge of data processing and analytics capabilities [58].

Key Competencies

- Deep understanding of the synergies and distinctions between cloud, fog, and edge computing paradigms.
- Expertise in big data infrastructure design, management, and optimization, with a firm grasp of modern database technologies and data warehousing.
- Proficiency in networking, data protocols, and security best practices specific to integrated IT, fog, and cloud environments.
- Collaborative acumen to align technological solutions with business objectives and operational needs, ensuring a holistic approach to integration.
- Strategic problem-solving skills, with the ability to navigate complex integration challenges, enhance data flow, and maximize system performance.

The Edge - Fog - Cloud Integration Specialist is at the forefront of driving seamless, secure, and efficient integration across the IT, fog, and cloud computing layers. By harmonizing these technologies, the specialist enables the organization to leverage the full spectrum of data processing capabilities, from the edge to the cloud, fostering innovation, agility, and strategic insights in an increasingly data-driven world.

4.4.9 Empowering Operations with AI: The Big Industrial Data and AI Projects Specialist

In the rapidly evolving landscape of OT, the Big Industrial Data and AI Projects Specialist emerges as a key innovator. This specialist harnesses the power of AI and advanced data analytics to revolutionize industrial operations. This specialist is adept at transforming the deluge of data from OT systems into actionable intelligence, leveraging ML, AI, and predictive analytics to foresee maintenance needs, boost operational efficiency, and refine system performance [59].

Strategic Data Analytics and AI Integration: Central to this role is crafting and executing sophisticated data analytics frameworks designed to sift through and make sense of the vast amounts of data generated by OT systems. The specialist deploys ML models and AI algorithms that predict potential equipment failures and optimize maintenance schedules, enhancing operations' efficiency and reliability.

Collaborative Technological Synergy: Integrating AI and ML into the OT infrastructure is a collaborative endeavor that requires close coordination between the OT and IT teams. This synergy ensures that AI-driven capabilities are seamlessly woven into the existing systems, facilitating a unified data analysis and application landscape across the OT spectrum.

Pioneering Predictive Maintenance: By leading predictive maintenance initiatives, the specialist leverages data analytics to preempt equipment failures and avoid costly downtime, thereby extending the lifespan of critical machinery and components. This proactive approach to maintenance is grounded in a deep analysis of data trends and patterns, providing a strategic advantage in operational management.

Innovative Quality Control through Image Processing: The specialist utilizes cutting-edge image processing and computer vision techniques to enhance quality control measures, identify defects, and streamline processes within OT environments. This application of AI in visual inspections and monitoring contributes significantly to maintaining high standards of quality and efficiency.

Continuous Model Optimization: The lifecycle of AI and ML models in industrial settings requires constant evaluation and refinement. The specialist meticulously monitors the performance of these models, making necessary adjustments to ensure they remain aligned with operational dynamics and organizational objectives.

Advancements and Continuous Learning: The technological landscape is evolving, so the specialist remains at the forefront of developments in data analytics, ML, and AI. This commitment to continuous learning ensures that the organization's OT systems benefit from the latest innovations and best practices.

Key Competencies:

- Expertise in data analytics, ML, and AI, particularly their application in industrial and operational contexts.
- Proficiency in developing and managing predictive models, underpinned by a solid foundation in statistical analysis, algorithm design, and data visualization.
- Skilled in image processing and computer vision, applying these techniques to enhance industrial automation, monitoring, and quality control.
- Capable of integrating sophisticated AI and ML solutions into existing OT infrastructures, ensuring harmonious operation and system compatibility.
- Exceptional problem-solving abilities, converting complex data sets into tangible strategies and actionable solutions.
- Collaborative communicator, able to bridge diverse teams and disciplines to realize the potential of AI-driven enhancements in OT environments.

The Big Industrial Data and AI Projects Specialist is a visionary role, pivotal in steering OT systems toward a future where operations are data-informed and AI-enhanced. This specialist propels the organization toward enhanced efficiency, predictive precision, and operational excellence through strategic AI and ML implementation.

4.4.9.1 Software Development Specialists: The Linchpins of IT-OT Convergence

In organizations' vanguard of Digital Transformation, Software Development Specialists are essential pillars, driving the integration and functional sophistication of message queuing telemetry transport (MQTT) and OPC-supported software. These professionals, embedded within the information management systems teams, are at the heart of developing and refining services and packaged software solutions that bridge the gap between diverse systems such as PLM, ERP, manufacturing resource planning (MRP), and MES.

Armed with expertise in Python, Java, and .NET frameworks, Software Development Specialists are pivotal in crafting solutions that meet the technical requirements and align with the organization's operational dynamics. Their responsibilities extend beyond mere development; they are the architects of system integration, ensuring that software solutions communicate seamlessly with handheld terminals, HMI, and SCADA systems.

The contribution of Software Development Specialists transcends the technical realm, impacting the organization's operational agility and efficiency. By fostering a harmonious IT-OT environment, they enable the seamless execution of various operational mandates and initiatives, ensuring the digital infrastructure is robust, responsive, and adaptive to the evolving business landscape.

Responsibilities:

- Spearheaded the development of MQTT and OPC-supported software, significantly enhancing the organization's capabilities in OT software.
- Forge strong collaborations with software teams across information management systems to ensure holistic integration with core systems such as PLM, ERP, MRP, and MES.

- Led the design and implementation of service-oriented and packaged software solutions tailored for comprehensive system integration, emphasizing functionality and system compatibility.
- Harness the potential of Python, Java, and .NET technologies to drive a wide array of development projects, contributing to the organization's digital evolution.
- Oversee the seamless integration and compatibility of new software solutions with existing handheld terminals, HMI, and SCADA systems, optimizing operational workflows and efficiency.

Key Competencies:

- Mastery in Python, Java, and .NET technologies, with a demonstrated ability to apply these skills effectively in a development context.
- In-depth knowledge and practical experience with MQTT and OPC-supported software, underpinning the development of advanced operational technologies.
- Proven track record of successful collaboration within information management systems, showcasing the ability to work synergistically with diverse software teams.
- Comprehensive understanding of PLM, ERP, MRP, and MES systems, with a keen insight into their integration challenges and opportunities.
- Proficiency in ensuring the smooth integration and compatibility of software solutions with handheld terminals, HMI, and SCADA systems, crucial for maintaining operational continuity and efficiency.

Software Development Specialists are, therefore, central to the realization of IT-OT Convergence in Digital Transformation projects. Their technical insight and deep understanding of OT make them invaluable in navigating the complexities of integrating diverse systems and technologies. As organizations continue to evolve in the digital era, the role of Software Development Specialists will undoubtedly expand, becoming more critical to achieving strategic objectives and sustaining competitive advantage in an increasingly interconnected world.

4.4.9.2 Operational Technology Systems Support Specialist: Bridging IT and OT

The OT Systems Support Specialist is a cornerstone in the intricate Digital Transformation landscape, harmonizing the interplay between IT systems and OT frameworks. Positioned under the aegis of the cybersecurity and OT manager, this role embodies the synthesis of technical understanding and collaborative prowess, ensuring the operational integrity and efficiency of technology infrastructures within the OT domain.

The OT Systems Support Specialist's realm extends beyond mere maintenance; it encompasses the proactive establishment and fine-tuning of systems that underpin the organization's production lines and operational workflows. Their contributions are instrumental in fortifying the organization's technological backbone, enabling a seamless fusion of IT and OT systems that underlies its operational cadence [60].

Responsibilities:

- Spearheaded the upkeep and optimization of IT systems within OT environments, guaranteeing their performance and reliability to uphold operational continuity.
- Foster collaborative engagements with counterparts across diverse industrial systems segments to orchestrate system installations, ensuring adherence to established configurations and parameters for optimal system performance.
- Champion the deployment of new systems within production environments, managing their meticulous installation and overseeing the initiation of their operational functionalities.
- Facilitate the integration and operationalization of software solutions developed by Software Development Specialists, ensuring their alignment with system specifications and operational needs.
- Architect and maintain robust connections between automation systems and Level 2 equipment, paving the way for interoperability and efficient communication across the organization's technological ecosystem.

Key Competencies:

- Deep-seated proficiency in managing IT systems within OT landscapes, coupled with a keen ability to perform maintenance tasks with precision and efficiency.
- Demonstrated experience in collaborative projects involving industrial system installations, showcasing an ability to navigate complex setups and configurations.
- Adept at interpreting and implementing predetermined system configurations and parameters, ensuring systems are optimized for their intended operational roles.
- Proficient in overseeing the integration of new systems into production lines, focusing on ensuring seamless initiation and operational efficacy.
- Skilled in software deployment, with a track record of successfully installing and activating software solutions in alignment with organizational objectives.
- In-depth understanding of the nexus between automation systems and Level 2 equipment, with a proven ability to establish and sustain effective connections that enhance system interoperability.

The OT Systems Support Specialist is thus pivotal in knitting together the fabric of IT-OT Convergence, ensuring that the Digital Transformation journey is underpinned by a resilient, efficient, and integrated technology infrastructure. Their role enhances the organization's operational agility and fortifies its capacity to adapt and thrive in an increasingly digitized industrial landscape.

4.4.9.3 Operational Technology Systems Analysis and Governance: A Comprehensive Approach

In the evolving landscape of Digital Transformation within the industrial sector, the role of the System Analysis, Documentation, Training, and OT Governance Specialist becomes increasingly critical. This specialist is entrusted with the meticulous analysis of OT Systems, ensuring that these systems are optimally configured to meet the organization's operational needs and adhere to the stringent governance standards akin to those in IT Governance.

This multifaceted role encompasses various responsibilities, from conducting detailed system analyses, maintaining rigorous documentation, spearheading OT Governance initiatives, and developing comprehensive training programs. Through their efforts, the specialist ensures that the OT Systems are efficient and compliant and that all organizational stakeholders have a pervasive and profound understanding of these systems.

Responsibilities:

- Undertake thorough analyses of OT Systems, evaluating their alignment with operational processes and demands and providing strategic insights for enhancement and optimization.
- Maintain a comprehensive inventory and up-to-date documentation of all OT Systems, ensuring a clear and accurate representation of the organization's OT infrastructure.
- Champion the development and implementation of OT Governance initiatives, drawing on the principles established in IT Governance to foster a structured and compliant OT environment.
- Lead the organization through certification processes, guiding OT Systems to meet and exceed industry standards and regulatory requirements.
- Conduct audits of OT Systems, rigorously assessing their adherence to established governance standards and identifying opportunities for continuous improvement.
- Create and disseminate educational materials tailored to enhance the organizational understanding of OT Systems, their operational significance, and governance frameworks.
- Design and execute targeted training programs to equip industrial line workers and key stakeholders with the knowledge and skills to effectively navigate and leverage OT Systems.

Key Competencies:

- Exceptional analytical prowess, conducting in-depth analyses of complex OT Systems and deriving actionable insights.
- Advanced inventory management and documentation skills, with a keen eye for detail and an unwavering commitment to accuracy.
- Comprehensive understanding of OT Governance frameworks, with proven experience in orchestrating governance initiatives and aligning OT operations with best practices and regulatory standards.

- Demonstrated ability to orchestrate certification processes and audits, with a track record of leading OT Systems to achieve and maintain compliance with industry standards.
- Adept at creating informative and engaging materials that elucidate the nuances of OT Systems and governance principles tailored to a diverse audience.
- Proficient in communication and pedagogy, able to develop and deliver impactful training programs that foster a deep understanding of OT Systems among stakeholders.

The System Analysis, Documentation, Training, and OT Governance Specialist is pivotal in bridging the gap between OT and organizational governance. By ensuring that OT Systems are analyzed, documented, governed, and understood in a manner that mirrors the rigor applied in IT environments, this specialist lays the foundation for a resilient, compliant, and efficient operational framework crucial for thriving in the digital age.

4.4.9.4 The Role of Sustainability Specialist in IT-OT Convergence for Digital Transformation

In the evolving landscape of IT-OT Convergence, the Sustainability Specialist emerges as a pivotal figure dedicated to embedding sustainable practices and principles into the fabric of an organization's IT and OT ecosystems. This specialist is at the forefront of crafting and steering strategies that propel Digital Transformation and underscore a commitment to environmental stewardship, resource optimization, and energy efficiency.

Tasked with aligning IT-OT integration efforts with overarching sustainability objectives, the Sustainability Specialist ensures that technological advancements reduce the organization's carbon footprint, enhance waste management, and foster the adoption of renewable energy solutions. Through a blend of strategic foresight and environmental understanding, this role encapsulates the essence of responsible innovation, ensuring that Digital Transformation journeys are progressive and planet-friendly.

Responsibilities:

- Architect and execute comprehensive sustainability frameworks within IT-OT Convergence endeavors, ensuring alignment with corporate environmental aspirations.
- Undertake rigorous environmental impact evaluations for IT-OT projects, infusing sustainability into the lifecycle of technological systems and initiatives.
- Champion integrating eco-friendly technologies and renewable energy solutions into IT and OT infrastructures, advocating for a greener technological footprint.
- Collaborate with multidisciplinary teams to weave sustainability into IT-OT system design, operation, and ongoing management, promoting eco-conscious practices across all phases.

- Continuously monitor and analyze the environmental performance of IT-OT systems, leveraging insights to drive enhancements in sustainability metrics.
- Spearheaded initiatives to minimize waste and optimize resource utilization within IT-OT operations, setting new benchmarks in sustainable industrial practices.
- Maintain an up-to-date understanding of environmental regulations, sustainability standards, and industry best practices, developing policies and processes that elevate the organization's environmental compliance and leadership.

Key Competencies:

- A robust foundation in environmental sciences and sustainable development, specifically focusing on their application within IT and OT frameworks.
- Expertise in environmental impact assessment methodologies, equipped to integrate and apply sustainability standards and frameworks within technological contexts.
- A deep reservoir of knowledge concerning green technologies, renewable energy solutions, and best practices for sustainable operations in IT and OT settings.
- Demonstrated leadership in driving sustainability initiatives, with a proven track record of implementing eco-conscious changes that deliver tangible environmental improvements.
- Exceptional communication and collaborative prowess, capable of galvanizing diverse stakeholder groups around sustainability goals and initiatives.
- Analytical understanding to critically evaluate the eco-efficiency of IT-OT systems, identifying and actioning opportunities for sustainability enhancement.

The Sustainability Specialist's role in IT-OT Convergence is not merely operational but transformative, setting the stage for an era where technological advancement and environmental responsibility converge. By steering Digital Transformation projects toward sustainability, this specialist ensures that the march toward technological innovation is in lockstep with the principles of ecological stewardship, marking a new chapter in responsible industrial progress.

4.4.9.5 The Role of Net Zero and Carbon Management Specialist in IT-OT Convergence for Sustainable Transformation

Within the ambit of IT-OT Convergence, the Net Zero and Carbon Management Specialist emerges as a crucial architect of sustainable transformation, championing the organization's commitment to achieving net zero emissions. This role is intricately designed to fuse the realms of IT and OT with strategic environmental stewardship, guiding the organization toward minimized carbon footprints and enhanced ecological responsibility.

Tasked with the intricate challenge of melding technological innovation with carbon management imperatives, this specialist spearheads the development and meticulous execution of comprehensive strategies to monitor, report, and ultimately reduce greenhouse gas emissions across the spectrum of IT and OT operations. Through a blend of analytical prowess and strategic foresight, the Net Zero and Carbon Management Specialist ensures that the organization's Digital Transformation journey is environmentally sustainable and aligned with global net zero aspirations.

Responsibilities:

- Architect and steward the organization's carbon management and reduction blueprint, aligning IT-OT Convergence efforts with the goal of net zero emissions.
- Conduct detailed analyses of the carbon footprint inherent within IT and OT systems, pinpointing critical areas where emissions can be curtailed and energy efficiency can be elevated.
- Seamlessly weave carbon management methodologies and tools into IT-OT Convergence initiatives, guaranteeing that operational advancements are inherently sustainable.
- Foster collaborative synergies with sustainability, IT, and OT divisions to roll out energy conservation measures, embrace renewable energy solutions, and champion waste minimization practices.
- Lead the meticulous reporting and verification of carbon emissions data, ensuring its integrity and adherence to established standards and regulatory frameworks.
- Keep pace with evolving climate policies and breakthroughs in carbon reduction technologies, ensuring the organization remains at the forefront of sustainable practices in the IT-OT landscape.
- Cultivate a culture of environmental consciousness and energy responsibility among IT and OT personnel through targeted education and skill development initiatives.

Key Competencies:

- Profound expertise in carbon footprint quantification, greenhouse gas accounting, and formulating emissions reduction strategies.
- An in-depth understanding of IT and OT systems' energy dynamics and environmental implications, coupled with the insight to devise bespoke carbon reduction interventions.
- Mastery over sustainability reporting frameworks, environmental compliance mandates, and carbon verification protocols, ensuring organizational actions are both transparent and accountable.
- Exceptional project management capabilities, enabling the orchestration of multidisciplinary projects that drive the organization toward its net zero ambitions.
- Stellar communication and stakeholder engagement skills, essential for building a widespread organizational ethos of sustainability and environmental stewardship.

- An innovative mindset, adept at harnessing emerging technologies and methodologies to bolster carbon management efforts within the confluence of IT and OT systems.

The Net Zero and Carbon Management Specialist thus plays an instrumental role in redefining the trajectory of IT-OT Convergence, steering it toward a future where technological advancements and environmental sustainability are inextricably linked. Through strategic interventions and collaborative endeavors, this specialist ensures that the Digital Transformation journey is innovative, efficient, and conscientiously aligned with the imperative of global sustainability.

4.4.10 Optimizing Team Composition for IT-OT Convergence: A Guide to Academic and Professional Development

In the evolving Digital Transformation landscape, the confluence of IT and OT necessitates a strategic approach to team formation, underscored by a deliberate selection of academic backgrounds and professional skill sets. The fusion of IT and OT domains heralds a new era of interdisciplinary collaboration, where the distinct expertise of each field is leveraged to foster innovative solutions and seamless system integrations across industries. Below, we delineate the recommended academic foundations and professional qualifications for personnel within these converging spheres.

4.4.10.1 Academic Foundations for IT Personnel

IT professionals, pivotal in driving the technological aspects of IT-OT Convergence, should ideally possess a robust academic foundation in fields that underscore the principles of computing, system architecture, and digital security. Relevant undergraduate and vocational qualifications include:

Bachelor's Degrees in:

- IT
- Computer Science
- Software Engineering
- Cybersecurity
- Management Information Systems (MIS)
- Network Engineering

4.4.10.2 Suggested Coursework for IT Professionals

To equip IT personnel with the necessary competencies for IT-OT integration, the following coursework is recommended, covering a broad spectrum from programming to systems analysis:

- Core Programming Languages: Proficiency in Java, Python, C++, and. NET frameworks, facilitating versatile software development and system integration.

- Data Structures and Algorithms: Fundamental understanding of computational structures and algorithmic efficiency to optimize data processing and storage solutions.
- Database Management: Mastery of relational databases and NoSQL systems is crucial for managing traditional data and the voluminous, varied datasets characteristic of IIoT projects.
- Network Security and Infrastructure: Deep insights into network architectures and security protocols to safeguard information flow within and between IT and OT systems.
- Web Development: Skills in designing and implementing web-based applications and services, enhancing interface usability and system accessibility.
- Operating Systems: Comprehensive knowledge of various operating environments, ensuring system compatibility and performance optimization.
- System Analysis and Design: Aptitude for dissecting complex systems and architecting solutions that align with organizational objectives and operational workflows.
- IT Project Management: Strategies for overseeing IT projects, ensuring timely delivery, budget adherence, and alignment with stakeholder requirements.

4.4.10.3 Professional Certifications for IT Personnel

While formal education lays the groundwork, professional certifications can further validate an IT professional's expertise and commitment to continuous learning:

- **CompTIA A+:** A foundational certification attesting to broad IT operational and troubleshooting skills.
- **Cisco's CCNA:** Certification demonstrating proficiency in network fundamentals, access, connectivity, and security.
- **Microsoft's MCSE:** Credentials affirming advanced skills in Microsoft server technologies and solutions.
- **Certified Information Systems Security Professional (CISSP):** A globally recognized certification in information security.

In synthesizing IT personnel's diverse academic backgrounds and professional qualifications, organizations can cultivate a workforce adept at navigating the complexities of IT-OT Convergence. This strategic team formation approach enhances the IT domain's technical proficiency. It ensures that IT professionals are well-equipped to collaborate effectively with their OT counterparts, driving forward the unified Digital Transformation goals.

4.4.11 Optimizing Team Composition for OT in IT-OT Convergence: Academic and Professional Pathways

As Digital Transformation initiatives increasingly require the integration of OT with IT, the academic and professional training of OT personnel becomes crucial. OT professionals are instrumental in ensuring that physical systems and processes are efficiently and securely integrated with digital technologies. The following delineates the recommended academic pathways and professional qualifications for personnel specializing in OT within the context of IT-OT Convergence.

4.4.11.1 Academic Foundations for OT Personnel

OT professionals, pivotal in managing and optimizing physical systems and processes, should ideally possess a solid academic foundation in engineering and automation. Relevant undergraduate and vocational qualifications include:

Bachelor's Degrees or Vocational School Degrees in:

- Engineering (specializations in Aerospace, Electrical, Mechanical, Industrial, or Systems Engineering)
- Automation Technology or Industrial Automation
- Mechatronics
- Similar fields that blend mechanical, electrical, and computing disciplines

4.4.11.2 Suggested Coursework for OT Professionals

To equip OT personnel with the necessary skills for effective IT-OT integration, the following coursework is recommended, covering essential aspects from automation to network security:

- **Automation and Robotics:** Foundational knowledge in automation technologies and robotic systems is crucial for modernizing and enhancing industrial operations.
- **Industrial Control Systems (ICS):** Understanding the operation, management, and security of ICS, which are fundamental to OT environments.
- **Electrical Engineering Fundamentals:** Basic electrical concepts and applications are essential for OT professionals who work with electrical systems and components.
- **SCADA Systems:** Mastery over SCADA systems, enabling the monitoring and control of industrial processes.
- **PLCs:** Proficiency in using PLCs for automated control over machinery and processes.
- **Embedded Systems:** Insights into integrating computer systems within industrial machines and devices for dedicated functions.
- **Process Control & Instrumentation:** Techniques for managing and optimizing industrial processes through accurate measurement and control systems.
- **Industrial Network Security:** Strategies for securing industrial networks against cyber threats and maintaining the integrity of OT systems.
- **Manufacturing Systems:** Comprehensive understanding of modern manufacturing technologies and systems for optimized production.
- **HMI:** Skills in designing and implementing interfaces that facilitate interaction between users and machines.
- **Industrial Communication Networks:** Knowledge of networking principles tailored to industrial settings, ensuring effective communication between various systems and devices.

4.4.11.3 Professional Certifications for OT Personnel

In addition to formal education, professional certifications can serve to validate further an OT professional's expertise and dedication to their field:

- **Certified Automation Professional (CAP):** A credential recognizing proficiency in automation and control systems.
- **Certified Control Systems Technician (CCST):** This certification demonstrates expertise in the calibration, maintenance, and installation of control systems.
- **Certified Industrial Cybersecurity Professional (CICP):** A designation that signifies specialized knowledge in securing industrial automation and control systems.

By carefully sculpting the academic and professional trajectory of OT personnel, organizations can foster a workforce adept at overseeing the sophisticated interplay between physical operations and digital technologies. This strategic approach to team composition ensures that OT professionals are equipped with a deep understanding of industrial systems and prepared to engage collaboratively with IT counterparts, driving the collective success of IT-OT Convergence initiatives.

4.4.12 Interdisciplinary Expertise for IT-OT Convergence with a Sustainability Focus

In the dynamic realm of IT-OT Convergence, professionals poised at this intersection are increasingly required to integrate sustainability and environmental stewardship into Digital Transformation initiatives. This necessitates a holistic academic and professional framework that amalgamates the technical prowess of IT and OT with a profound understanding of environmental sustainability. Such interdisciplinary expertise drives technologically advanced, environmentally responsible, and sustainable innovations.

4.4.12.1 Academic Foundations for Sustainability-Focused IT-OT Professionals

A multidisciplinary educational background is essential to cultivate a workforce adept at navigating the complexities of IT-OT systems while championing sustainability goals. This blend of technical, environmental, and sustainability education equips professionals to design, implement, and manage IT-OT systems contributing to the organization's sustainability objectives. Relevant degrees include:

Bachelor's or Master's Degrees in:

- **Environmental Engineering:** This field applies engineering principles to solve environmental issues, making it perfect for integrating sustainable practices in IT-OT systems.
- **Sustainable Energy Systems:** Providing insights into renewable energy technologies and their application in industrial systems.

- **Environmental Informatics:** Bridging the gap between environmental sciences and IT to leverage data for sustainability.
- **Industrial Ecology:** Studying material and energy flows through industrial systems to enhance sustainability and efficiency.
- **Environmental Science and Policy:** Offering a broad understanding of environmental issues and policy frameworks crucial for compliance and strategic planning.
- **Systems Engineering Specializing in Sustainable Practices:** Integrating systems thinking with sustainability principles to design and manage complex IT-OT systems.

4.4.12.2 Suggested Coursework for an Integrated Approach

Professionals at the nexus of IT, OT, and sustainability should pursue coursework that spans across these domains, encompassing:

- **Sustainable Practices in IT and OT**: Exploring strategies to enhance sustainability in the design and operation of IT-OT systems.
- **Energy Efficiency in Industrial Processes:** Techniques to reduce energy consumption and improve efficiency in industrial settings.
- **Renewable Energy Integration:** Understanding how to incorporate renewable energy into IT and OT infrastructures.
- **Environmental Impact Analysis:** Assessing the ecological footprint of IT-OT systems and identifying mitigation strategies.
- **Data Analytics for Sustainability:** Leveraging data analytics to drive environmental performance and sustainability insights.
- **Green Supply Chain Management:** Principles of sustainable supply chain practices and their application in technology sectors.
- **Lifecycle Assessment:** Evaluating the environmental impact of products and systems throughout their lifecycle, supporting circular economy models.
- **Environmental Regulations:** Navigating the complex landscape of environmental regulations pertinent to the tech industry.

4.4.12.3 Professional Certifications and Training

To further bolster their credentials and expertise, professionals should consider obtaining certifications that underscore their commitment to sustainability and environmental management:

- **Certified Sustainability Practitioner (CSP):** Recognizing proficiency in implementing and managing sustainability projects.
- **LEED Green Associate or LEED AP:** Validating knowledge of green building practices and principles.
- **ISO 14001 Environmental Management System Auditor:** Demonstrating expertise in auditing environmental management systems.
- **Certified Energy Manager (CEM):** Credentialing energy efficiency and management expertise.

- **Green IT Professional:** Certifying skills in environmentally sustainable IT practices.
- **Certificate in Carbon Footprint Management:** Specializing in measuring and managing carbon footprints for organizations and products.

By embracing this comprehensive academic and professional development framework, individuals working at the intersection of IT and OT with a focus on sustainability are better equipped to lead their organizations toward achieving net zero emissions and minimizing environmental impacts, all the while harnessing the transformative power of digital technologies.

4.4.13 Strategic Approaches and Security Considerations for IT-OT Convergence

4.4.13.1 Implementing IT-OT Convergence with a Focus on Impact Analysis and Sustainability

Establishing specialized units and robust infrastructures plays a crucial role in the intricate process of IT-OT Convergence. This enhances operational efficiency while adhering to sustainability and environmental goals. A dedicated OT Systems Unit comprising experts from diverse technical backgrounds fosters seamless integration and maintains system integrity. This unit should include network hardware specialists, policy officials, software development experts, and technical support personnel working in unison to ensure a cohesive OT environment.

Critical Strategies for IT-OT Integration:

- **Infrastructure Development:** Building an industrial-grade OT network with isolated physical cabling and customizable devices is paramount. Such infrastructure must be designed to withstand the unique demands of OT environments, ensuring reliability and sustainability.
- **Secure Connectivity:** Implementing advanced security measures like firewalls, VPNs, and secure remote access protocols is critical to protecting the IT-OT interface. These mechanisms should be supported by rigorous policies and personnel training to manage and monitor access effectively.
- **Software Management:** Establishing stringent protocols for software updates and security enhancements in OT systems is essential. This includes comprehensive pre-implementation analysis, sandbox testing, and contingency planning to minimize operational disruptions.
- **Access Control:** Rigorous user credentials management and adherence to password policies are vital for maintaining system security. Efforts should be made to update legacy systems and enforce stringent access controls to prevent unauthorized entry.
- **Network Security:** Ensuring the secure transmission of data between IT and OT networks involves allowing only approved applications and data types to traverse network boundaries, supported by physical and automated routing solutions.

- **Monitoring and Analysis:** Continuous monitoring of IT-OT network interactions is crucial for identifying potential threats. Leveraging ML and advanced analytics can enhance threat detection and response mechanisms.

4.4.13.2 Navigating Cybersecurity Risks in OT Systems

The convergence of IT and OT systems introduces complex cybersecurity challenges, necessitating a nuanced understanding of the security landscape and the development of robust defense mechanisms. The historical isolation of OT systems provided security through obscurity, which is no longer viable in interconnected environments. The integration exposes OT systems to a broader range of threats, underscoring the need for tailored security strategies considering the low tolerance for downtime in OT environments.

Addressing Cybersecurity Challenges:

- **Return on Investment (ROI):** Demonstrating the economic viability of IoT and IIoT initiatives is essential for securing executive support and investment in large-scale deployments. Establishing clear metrics for evaluating the cost-effectiveness of these technologies is crucial.
- **Interoperability:** The diverse protocols used by connected devices pose significant interoperability challenges. A standardized communication framework is essential for seamless device integration and data exchange.
- **Legacy System Vulnerabilities:** Many industrial automation systems, designed for simplicity and continuous operation, lack modern security features. Addressing these vulnerabilities requires a concerted effort to update and secure legacy systems against contemporary threats.
- **Operational Practices:** Reinforcing the importance of cybersecurity practices such as robust password policies, system backups, and high-level protection is vital for safeguarding OT systems against malware and other cyber threats.
- **Silent Risks:** Like the Stuxnet worm, stealthy modifications to PLC programs represent a particularly insidious threat. Implementing advanced detection and response systems is critical for identifying and mitigating such risks.

In summary, the successful convergence of IT and OT systems, mainly focusing on sustainability and environmental objectives, demands a comprehensive approach encompassing specialized team formation, secure infrastructure development, and vigilant cybersecurity practices. Addressing these multifaceted challenges is essential for realizing the full potential of IT-OT integration in driving Digital Transformation while ensuring operational resilience and environmental stewardship.

4.4.14 Optimizing Workforce Strategies for IT-OT Integration: A Comprehensive Approach to Shift Management and Working Models

In the intricate ecosystem of IT-OT Convergence, the orchestration of staff shifts and working models plays a pivotal role in maintaining the integrity and uninterrupted functionality of cyber-physical systems. With its indispensable mandate for continuous operation, OT demands a meticulously structured approach to workforce management that ensures around-the-clock oversight of critical infrastructure processes. Contrastingly, IT operations, while vital, generally adhere to conventional working hours, focusing on tasks such as data management, system development, and network maintenance, with provisions for on-call support under challenging circumstances.

4.4.14.1 Strategic Shift Management for Continuous OT Operations

Given OT's indispensable role in overseeing continuous industrial processes, organizations are tasked with implementing a robust shift-based working model that guarantees 24/7 system monitoring and support. This model is essential for ensuring that OT systems, particularly those involved in Level 2 automation and integrated directly into production lines, receive the requisite attention to prevent unscheduled downtimes that could lead to significant operational setbacks.

4.4.14.2 Critical Aspects of Shift-Based OT Working Model

- **Continuous Support:** A shift-based framework ensures that expert support staff are always available to maintain and troubleshoot critical OT applications, thus upholding the seamless operation of production processes.
- **Shift Planning and Scheduling:** Effective shift scheduling is paramount, requiring careful consideration of staff availability, technical expertise, and the physical and mental demands of shift work to ensure comprehensive coverage.
- **Transition Protocols:** Establishing clear protocols for shift transitions is crucial to maintain operational continuity. This includes detailed handovers and communication processes to ensure incoming personnel are fully apprised of ongoing issues and system statuses.
- **Professional Development:** Continuous training programs are essential to equip staff with the skills necessary to address a range of operational and emergency scenarios, thereby enhancing response efficacy and system reliability.

4.4.14.3 Office Working Model for IT and Administrative OT Functions

An office working model is typically employed for roles focused on IT and OT's managerial, developmental, and policy-oriented aspects. This model, characterized by standard daytime working hours, supports OT's planning, administrative, and developmental facets, facilitating a structured and collaborative work environment conducive to strategic planning and innovation.

4.4.14.4 Components of the Office Working Model

- **Structured Work Hours:** Personnel operating under this model adhere to standard business hours, concentrating on managerial and developmental tasks that underpin the strategic direction of OT systems.
- **Task Specialization:** Key responsibilities encompass network management, security policy development, software updates, and other administrative tasks crucial for the optimal functioning of IT and OT systems.

- **Continuous Learning:** Allocating ongoing education and skill development time is vital to staying abreast of evolving industry trends, standards, and cybersecurity practices.
- **Collaborative Planning:** The structured nature of this model facilitates synchronized interactions with various departments, enhancing cross-functional collaboration and strategic alignment across the organization.

In synthesizing these working models, organizations can strike a harmonious balance between operational imperatives and workforce well-being. By tailoring shift arrangements to the unique demands of IT and OT functions and fostering a work environment that prioritizes both productivity and employee satisfaction, companies can navigate the complexities of IT-OT Convergence more effectively, ensuring resilience, efficiency, and sustainability in their Digital Transformation journeys.

4.5 CASE STUDY: TRANSITIONING FROM INTELLIGENT TO SMART MANUFACTURING IN INDUSTRY 4.0 THROUGH ADVANCED AI TECHNOLOGIES

In recent years, big data has become a crucial area of focus within the academic and industrial sectors, particularly as it intersects with AI. Notably, scientists outlined the evolution of business intelligence and analytics (BI&A) through three distinct phases. The initial phase, BI&A 1.0, primarily focused on database management systems (DBMS) and structured data. This was followed by BI&A 2.0, which expanded to include text and web analytics targeting unstructured data from the web. The most recent stage, BI&A 3.0, incorporates analytics derived from mobile sources and sensor data. This progression underscores the rapid growth in data production, fueled by the widespread digitalization and integration of the IoT within various industrial settings, particularly in manufacturing. This sector is recognized as one of the top five areas where big data can have a transformative impact.

Simultaneously, intelligent manufacturing has begun taking shape, garnering significant interest across academic and industrial landscapes. Defined by the Smart Manufacturing Leadership Coalition as the enhanced use of sophisticated intelligence systems for the swift production of new products, adaptive responses to product demand, and real-time optimizations, smart manufacturing represents a paradigm shift in production methodologies. It also raises questions about the relationship and convergence between smart manufacturing (SM), intelligent manufacturing (IM), big data, and AI. What are the intersections and divergences among these concepts? How are they evolving to meet the demands of modern industry?

This case study aims to explore these questions, providing a detailed examination of the current landscape of manufacturing intelligence and innovative manufacturing initiatives. It seeks to pinpoint how next-generation AI technologies are being integrated into these frameworks and identify critical areas primed for future development in IM and SM. By doing so, this study will offer insights into the ongoing evolution and potential future directions of these interconnected fields.

4.5.1 AI Evolution

The concept of AI, established in 1956, has traversed significant fluctuations, enduring two notable periods of stagnation known as the AI winters during 1974–1980 and 1987–1993. AI methodologies are broadly categorized into two types: symbolic AI, which includes logic and rule-based systems, and subsymbolic AI, which encompasses neural networks, fuzzy systems, and evolutionary algorithms.

During the initial wave in the 1960s, symbolic AI achieved remarkable success in handling high-level cognitive tasks within controlled "toy problem" environments. Conversely, during this era, neural networks and cybernetics saw diminished focus, mainly due to the influential critiques by Minsky and Papert in their 1969 work "Perceptrons," which exposed limitations in early neural network models. The subsequent Lighthill Report and reduced funding from DARPA further contributed to AI's first winter from 1974 to 1980.

The 1980s heralded a resurgence of interest in AI with the advent of expert and knowledge-based systems, which integrated domain-specific expert knowledge into structured formats. These systems were particularly effective in complex manufacturing processes requiring simultaneous operations by multiple tools. This period also marked the reemergence of interest in neural networks and other sub-symbolic methods, thanks to pioneering work by researchers like Hopfield and Rumelhart. Unfortunately, this resurgence was short-lived, as the collapse of the Lisp machine market and the decline in expert systems ushered in the second AI winter in the late 1980s and early 1990s.

The 1990s saw the rise of distributed AI (DAI) and multi-agent systems (MAs), transforming traditional centralized AI architectures into decentralized networks of agents that interact, learn from each other, and operate collectively. This transition facilitated a more collaborative and scalable approach to AI.

Entering the 2000s, the explosion of the internet and the advent of Web 2.0 technologies brought forth an unprecedented volume of structured and unstructured data. This era emphasized the need for advanced data mining and processing techniques to handle the growing data deluge, setting the stage for the next wave of AI innovation.

The most recent wave of AI interest began around 2010, fueled by three interlinked factors: the proliferation of Big Data from diverse sources such as e-commerce and social media, significant advancements in ML algorithms, and the increased computational power available to process this data. This period has been marked by a shift from traditional symbolic AI (AI 1.0) to AI 2.0, which favors ML and deep learning approaches. These methods typically handle unstructured data and are characterized by decentralized control structures.

In addition, this era introduced an intermediary phase termed AI 1.5X, encompassing both Web AI (AI 1.5W) and Distributed AI (AI 1.5D). These serve as transitional forms, blending elements of symbolic and sub-symbolic approaches. This hybridization addresses the limitations of earlier AI models and enhances the system's ability to derive intelligence from vast, unstructured datasets. See Figure 4.2.

Parallel to these developments in AI, there has been a significant evolution in computing technologies. From the era of mainframe computers in the 1950s and

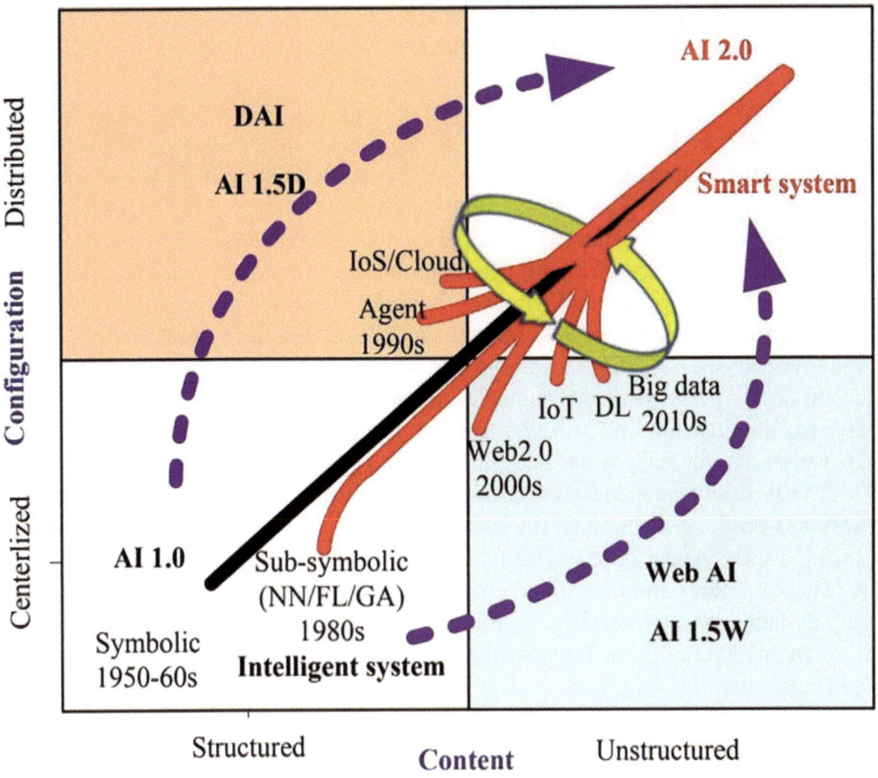

FIGURE 4.2 AI evolution from the perspectives of content and control [61].

1960s to personal computers in the 1980s and 1990s, and now to the modern landscape dominated by IoT, cloud computing, and ubiquitous computing, each phase has significantly impacted AI's capabilities and applications. Today's AI tools must be adept at collecting and analyzing the vast data streams of these contemporary technologies.

In summary, as outlined in Table 4.1, the evolution of AI is deeply intertwined with advancements in computing and data processing technologies, continually expanding the boundaries of what AI systems can achieve.

4.5.2 From Intelligent Manufacturing to Smart Manufacturing Along AI Evolution

4.5.2.1 Intelligent Manufacturing

Intelligent manufacturing (IM) represents the convergence of AI and manufacturing practices, evolving in tandem with advances in AI, as depicted in Figure 4.3. The publication of the seminal book marks the inception of IM, (Manufacturing Intelligence), in 1988. This era saw the integration of AI methodologies into manufacturing, leading to the development of various specialized IM systems in areas such as design, scheduling, production, inspection, diagnosis, modeling, and control, as observed during the second AI wave (refer to Figure 4.3(a)).

Significant scholarly work has focused on AI's role in the manufacturing sector. Notably, Teti and Kumara reviewed AI applications in manufacturing up to 1997, categorizing the technologies into knowledge-based/expert systems (KBSs/ESs), neural networks (NNs), fuzzy logic (FL), multi-agent systems (MAs), and other techniques like evolutionary algorithms and simulated annealing (SA). These applications have facilitated the creation of intelligent components for computer-integrated manufacturing (CIM), including intelligent CAD (CAD), CAP, CAM, and CAQ, as well as intelligent robotics. KBSs/ESs dominated the IM landscape; however, as the field matured, NNs, case-based reasoning, GA, and FL also gained traction. The pinnacle of this early period was the launch of the international Intelligent Manufacturing System initiative in 1995, originally from Japan in 1989, with contributions from industrial nations such as the USA, the EU, and Japan.

The 1990s introduced agent-based systems in IM, which were succeeded by web-services-based systems and the concepts of Enterprise 2.0 and crowdsourcing in the 2000s. Agent-based systems offered a promising paradigm for intelligent CIM components and overall IM structures. These systems are adept at managing the dynamic and uncertain conditions typical in modern software applications, though most remain in the research and prototyping phase within laboratory settings.

4.5.2.2 Smart Manufacturing

The 2010s marked a significant shift from traditional intelligent technologies (Symbolic AI) to a new era of "smart" technologies (referred to as "smart AI" in contrast to Symbolic AI) in manufacturing. This transition is poised to revolutionize the management of manufacturing enterprises throughout the product lifecycle, enhancing customer options, as illustrated in Figure 4.3.

Smart manufacturing employs a broad spectrum of technologies, initially focusing on IoT technologies and expanding to include Internet of Services (IoS), Cyber-Physical Systems (CPS), Big Data, and advanced robotics. These technologies are at the forefront of the second generation of intelligent manufacturing (IM 2.0), also known as smart manufacturing. Integrating IoT/CPS has transformed products into more interconnected and accessible entities, generating vast amounts of data that enable precise targeting and proactive enterprise management through timely and informed decisions. Moreover, the synergy of human intelligence, data, and intelligent algorithms significantly enhances manufacturing efficiency.

In this context, Big Data is primarily associated with data analytics. At the same time, CPS encompasses a broader range of functionalities compared to IoT or IoS, which is becoming increasingly critical in the manufacturing domain. Smart manufacturing is a cyber-physical production system that merges IoT and IoS. In the cyber realm, manufacturing resources are virtualized as cloud services accessible via IoS, which, due to their complexity, often require intelligent optimization algorithms like particle swarm optimization, differential evolution, and bee colony algorithms to solve service composition and selection challenges. These optimized business

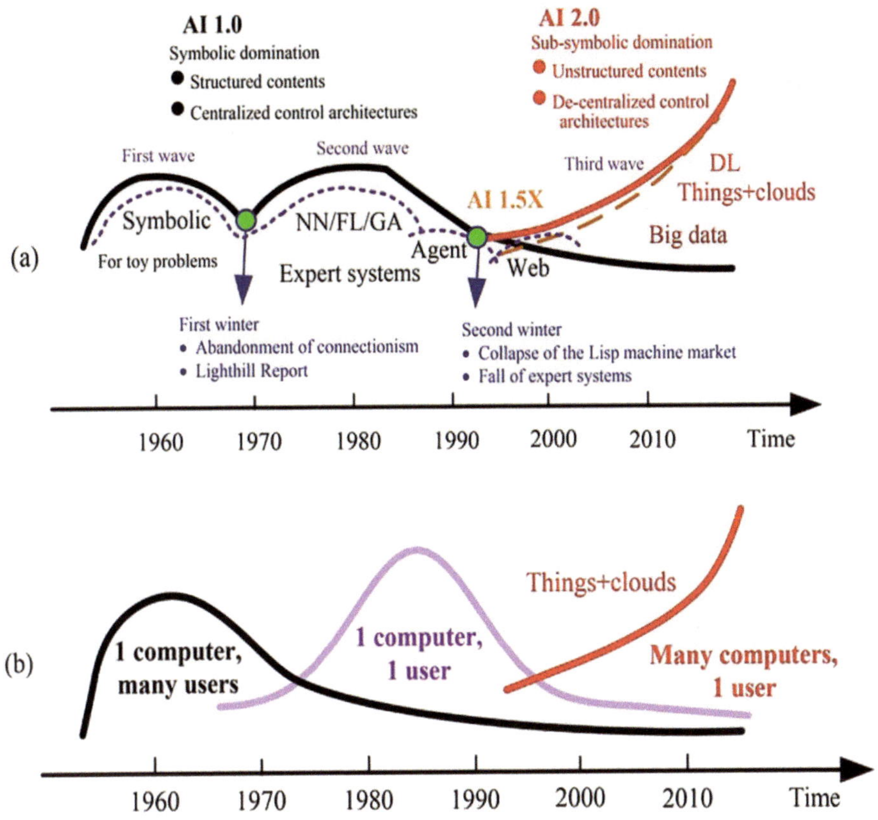

FIGURE 4.3 AI evolution versus computing's: (a) AI evolution; (b) computing evolution [61].

TABLE 4.1
AI Evolution [61]

Age	1950s—1960s	1980s	1990s	2000s	2010s
AI Focus	Symbolic	Expert system & Sub-symbolic	Agent	Web	Smart
Computation	Mainframes	PCs	PCs	Networks	Things+clouds
Processing Content/Focus	DBMS-based structured content/ Knowledge representation	Computational intelligence/soft computing/Data analytics & statistical methods	Distributed computing intelligence	Unstructured user-created content/Web analytics and web intelligence	IoT- based big data/Context-aware analysis/Deep learning
Control Structure	Centralized	Centralized	Distributed	Web-service based	CPS-based distributed

processes are then implemented on the physical shop floor, with continuous feedback from IoT sensors ensuring adherence to the operational plans.

4.5.3 Comparison of IM and SM

Centralized configurations and structured content management, such as databases, knowledge bases, and intelligent CAD systems characterize traditional IM systems. They are typically implemented within specific enterprise departments on a relatively small scale. In contrast, the advent of the internet has propelled manufacturing enterprises toward web-based platforms, introducing unstructured data from social media into the mix. With the rise of IoT and intelligent technologies, the manufacturing sector is increasingly adopting innovative manufacturing practices, confronting the challenges posed by the exponential growth of Big Data. Thus, enterprises are now compelled to leverage Big Data analytics for enhanced prediction, proactive maintenance, and production capabilities, capabilities that traditional, even agent- or web-based manufacturing systems lack due to their limited data acquisition and processing abilities (see Figures 4.4 and 4.5).

As depicted in Figure 4.6, while IM is rooted in knowledge-based approaches, SM evolves into a data-driven and knowledge-enabled paradigm. The advent of Big Data has shifted the focus from knowledge to data within the Data, Information, Knowledge, and Wisdom hierarchy. The processing of large data volumes through ML and intense learning enables the extraction of high-level data representations, facilitating data-driven decision-making over traditional expert systems, which rely on mimicking human-expert rules.

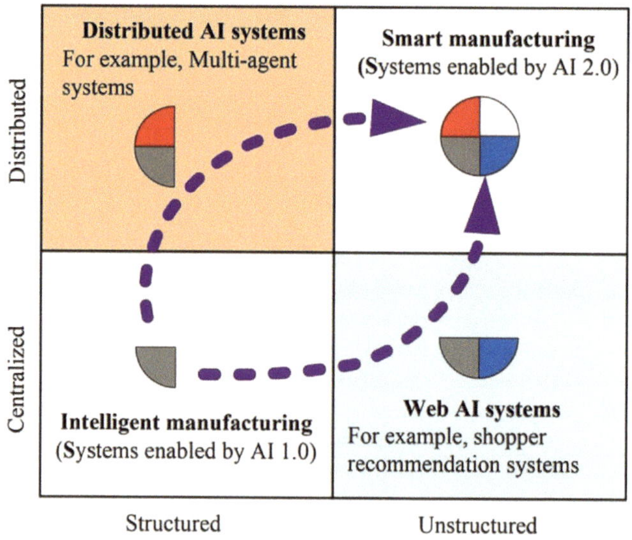

FIGURE 4.4 Intelligent manufacturing evolution, along with AI [61].

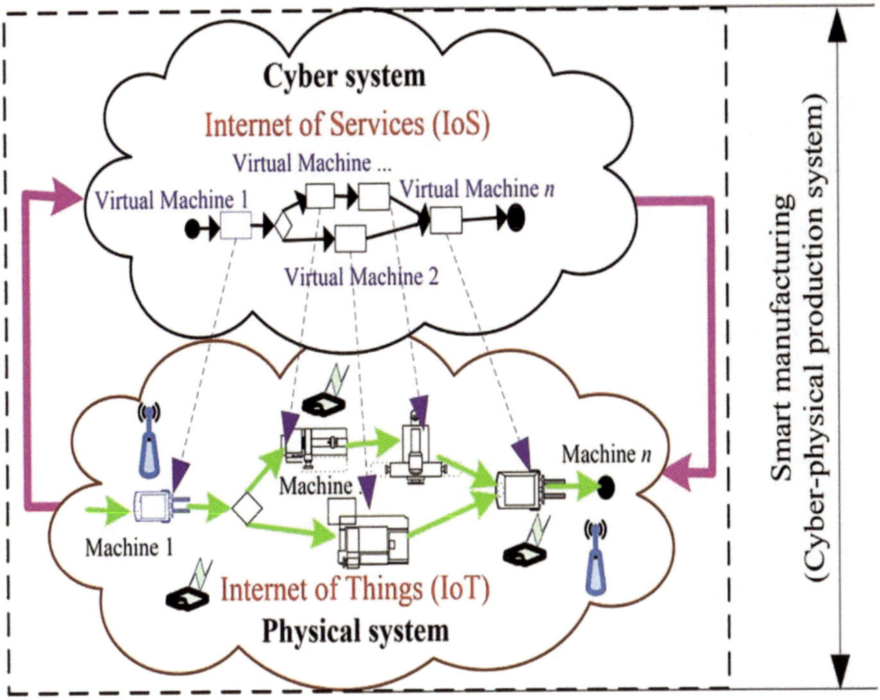

FIGURE 4.5 Smart manufacturing is exemplified as a cyber-physical production system [61].

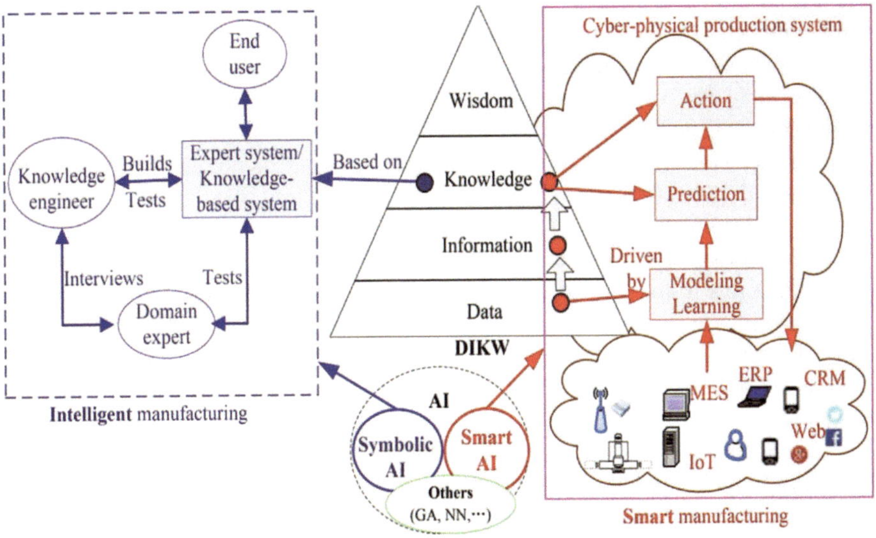

FIGURE 4.6 Intelligent manufacturing versus smart manufacturing [61].

4.5.4 Further Development of Intelligent Manufacturing for Industry 4.0

Industry 4.0, a term originating from a high-tech initiative by the German government, signifies a revolutionary shift toward "smart factories." This modern industrial phase, following the first Industrial Revolution of mechanization, the second of mass production, and the third of automation, integrates cyber-physical systems (CPS), the IoT, and the Internet of Services (IoS) to enhance manufacturing processes. The concepts underpinning intelligent factories and Industry 4.0 often overlap and reinforce each other, frequently illustrated within CPS architectures.

However, unlike traditional manufacturing systems, Industry 4.0 embraces a socio-technical approach, recognizing the integral role of social dynamics in manufacturing. China's "Made in China 2025 Strategy" similarly adopted this perspective, which focuses on integrating intelligent manufacturing techniques. The increasing demand for customized and sustainable manufacturing has also spurred developments in Enterprise 2.0, socialized enterprises, crowdsourcing, social manufacturing, and open innovation. Consequently, the social dimension is increasingly considered essential in intelligent manufacturing and Industry 4.0 frameworks, as depicted in Figure 4.7.

To address these multifaceted needs, the concept of "wisdom manufacturing" or "wise manufacturing" has emerged and is characterized by social cyber-physical systems (SCPS). This new model extends traditional CPS-based manufacturing to include social elements, effectively reviving craft production within a modern context through technologies like 3D printing. Wisdom manufacturing integrates the Internet of Things, Services, Content and Knowledge (IoCK), and People (IoP) within the manufacturing sector, collectively referred to as IoTSKP (Internet of Things, Services, Knowledge, and People). This integration facilitates a comprehensive approach to manufacturing that considers the physical and cyber aspects and the social impacts and interactions.

The rapid rise of IoT, IoS, and IoP technologies has led to an overwhelming influx of data, posing significant challenges and opportunities for manufacturing enterprises. This era of data-intensive computing demands innovative approaches to process and derive value from Big Data, transcending traditional experimental, theoretical, and simulation methodologies. As depicted in Figure 4.8, SCPS-based manufacturing holistically integrates the physical, cyber, and social systems, covering six semiotic levels from physical to social. This integration generates a broad spectrum of data from various sources, including social media networks, Web 2.0 platforms, crowdsourcing communities, mobile technologies, and digital manufacturing tools like NC/CAD/CAM/CAE/CAPP/PDM/ERP, simulation, and virtual manufacturing (see Figure 4.9).

In this context, the blending of traditional "symbolic" AI with modern "smart" AI leads to the development of "wise" AI, or Artificial Wisdom. This evolution represents a significant shift from AI 1.0 (symbolic) through AI 2.0 (smart) to AI 3.0 (wise), merging symbolic AI, smart AI, and other innovative approaches. Similarly, the manufacturing industry is evolving from intelligent to smart and now to wise (wisdom) manufacturing. This new phase integrates not only symbolic AI and intelligent

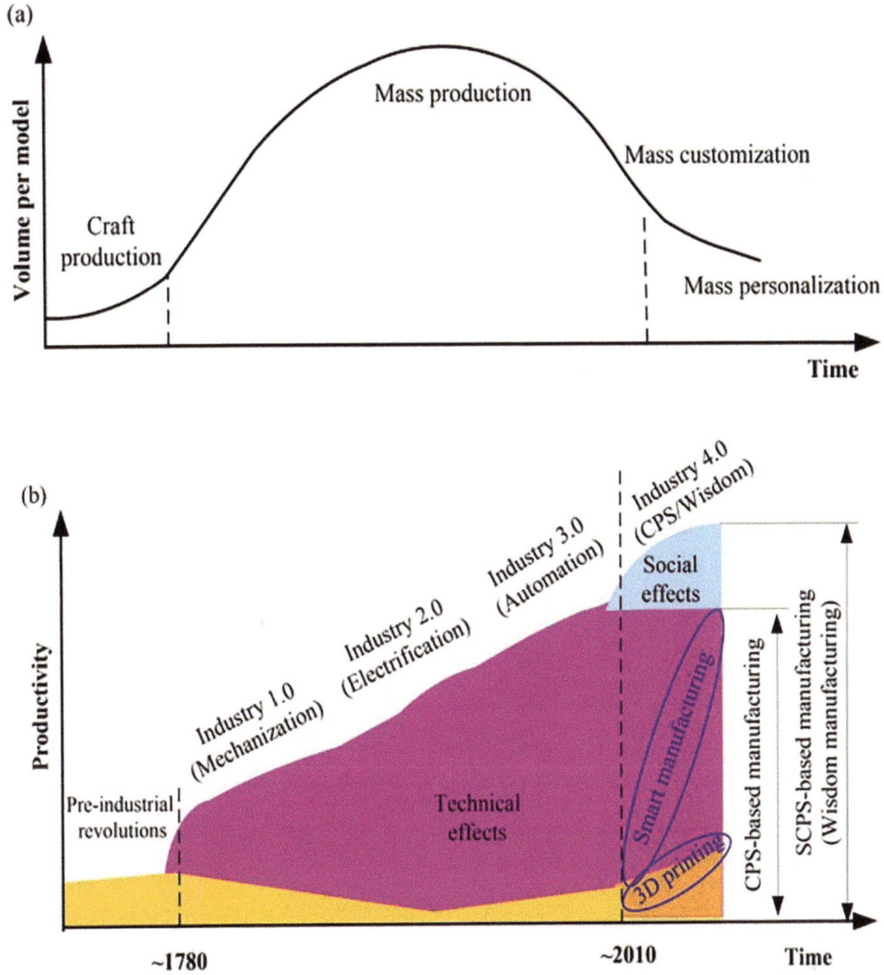

FIGURE 4.7 Industry 4.0 is a social-technical revolution for producing customized/personalized products: (a) manufacturing paradigm shift; (b) industrial revolutions [61].

technologies but also human intelligence and wisdom, creating a holistic manufacturing environment that includes humans, computers, machines, ubiquitous collective intelligence, and human knowledge and experience. This integration supports innovative business models such as "Everything-as-a-Service" and "Pay-per-use" in cloud-based design and manufacturing, enabling on-demand access to "Design-as-a-Service" and "Product-as-a-Service." Therefore, as we enter the next generation of intelligent manufacturing—smart manufacturing—factories are increasingly capable of sensing, understanding, thinking, and responding proactively to our needs. The comparative benefits of IM and SM are summarized in Table 4.2, clearly demonstrating the significant advantages that smart manufacturing offers over traditional intelligent manufacturing approaches.

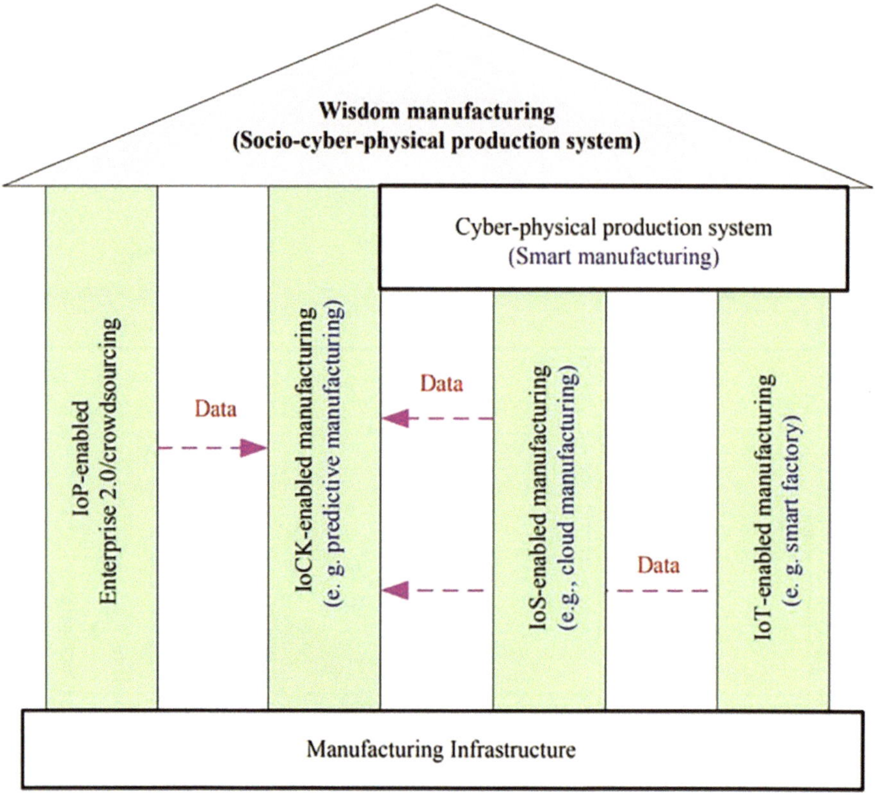

FIGURE 4.8 Wisdom manufacturing versus other emerging manufacturing models with big data in common [61].

4.5.5 Challenges and Opportunities in Smart Manufacturing

As the landscape of intelligent manufacturing continues to evolve within the framework of Industry 4.0, it presents unique challenges and opportunities that must be addressed to realize its full potential.

Challenges:

1. **Integration Complexity:** Integrating diverse technologies such as IoT, IoS, CPS, and AI into existing manufacturing systems presents significant technical and managerial challenges. It requires seamless interoperability between different systems and platforms, which can be technically demanding and costly.
2. **Data Security and Privacy:** With the increasing reliance on data-driven processes, ensuring the security and privacy of data becomes paramount. Vast networks of interconnected devices and systems increase vulnerability to cyberattacks and data breaches.

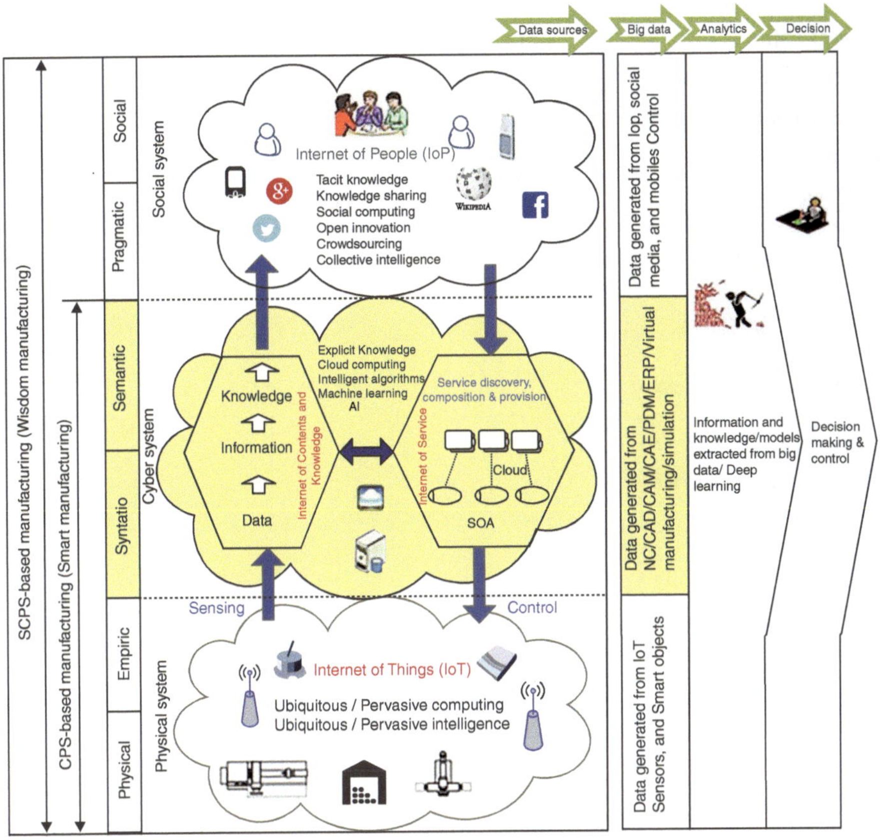

FIGURE 4.9 A framework for SCPS-based manufacturing [61].

3. **Skill Gap:** The shift toward advanced digital technologies necessitates a workforce adept in IT and core manufacturing skills. There is a significant skill gap that needs to be bridged through targeted education and training programs.
4. **Regulatory and Ethical Issues:** As technologies advance, they often outpace the existing regulatory frameworks. There is a need for updated regulations that address the ethical considerations and potential risks associated with automated and data-driven manufacturing processes.
5. **Economic and Cultural Barriers:** Adopting new technologies often requires substantial investment, which can be a barrier for small and medium-sized enterprises (SMEs). In addition, cultural resistance within organizations to changing traditional methods and workflows can exist.

Opportunities:

1. **Increased Efficiency and Productivity:** Smart manufacturing technologies enable higher operational efficiency and productivity through automation

TABLE 4.2
The Comparison of IM and SM [61]

Characteristics	IM	SM
Structure	Centralized	Distributed
Optimal scale	Usually local	Global
Structured content (data)		
Big data (unstructured content)		
IoT/CPS		
IoS/Cloud computing		
Deep learning		
Entire value chain support		
Ubiquitous access		
Virtualization		
Everything-as-a-Service		
Visibility		
Proactivity		
Adaptability		
Self-organization		
Self-predictiveness		
Context-awareness		
System of systems		

and data analytics, which can optimize production processes and reduce downtime.

2. **Customization and Flexibility:** Advanced manufacturing technologies allow for greater customization of products to meet specific customer demands without significant increases in cost or production time.
3. **Sustainable Manufacturing:** Smart manufacturing facilitates more environmentally friendly production processes through improved resource management and waste reduction, aligning with global sustainability goals.
4. **Supply Chain Optimization:** IoT and AI can vastly improve supply chain management by providing real-time data that helps predict demand, optimize inventory, and reduce supply chain disruptions.
5. **New Business Models and Revenue Streams:** Leveraging data to optimize manufacturing processes and create new business models, such as product-as-a-service, can open new revenue streams and change industry competitive dynamics.

4.5.6 Future Outlook and Strategic Directions

As intelligent manufacturing continues developing, it is poised to transform the landscape fundamentally. Enterprises must strategically plan their adoption of these technologies, considering the immediate benefits and long-term implications. Strategic directions could include:

1. **Investing in Talent and Training:** Building a skilled workforce that can manage and operate advanced manufacturing technologies is crucial. Companies should invest in continuous training and development programs to keep their employees at the cutting edge of technology.
2. **Developing Strong Cybersecurity Measures:** As manufacturing becomes more connected, the importance of robust cybersecurity protocols cannot be overstated. Implementing advanced security measures will protect sensitive data and maintain the integrity of manufacturing operations.
3. **Collaboration Between Industry and Academia:** To foster innovation and bridge the skill gap, there should be stronger collaborations between manufacturing companies and academic institutions. These partnerships can drive research and development and provide a steady pipeline of skilled professionals.
4. **Leveraging Government Incentives:** Governments worldwide support the transition to intelligent manufacturing through various incentives and grants. Companies should use these opportunities to defray the costs associated with technology adoption.
5. **Adapting to Regulatory Changes:** Staying informed about and compliant with new regulations regarding smart manufacturing is essential for legal and operational security.

By addressing these challenges and leveraging the opportunities, the future of smart manufacturing promises to enhance operational efficiencies and revolutionize the way products are designed, manufactured, and delivered, setting a new standard in the industrial sector.

4.6 CONCLUSION

As we draw this chapter to a close, it is evident that mastering the convergence of IT and OT is not merely a technical endeavor but a transformative journey that reshapes the entire organizational landscape. Integrating IT and OT paves the way for a robust Digital Transformation, aligning with the principles of Industry 4.0 to create more innovative, more agile enterprises. This chapter has endeavored to demystify the complexities of this convergence, offering a structured framework and actionable strategies to guide organizations through their Digital Transformation journeys. The insights provided here emphasize that successfully integrating IT and OT requires more than advanced technology—it demands a holistic approach to project management, skill development, team coordination, and collaboration. Organizations can adopt sophisticated project management methodologies to ensure that their Digital Transformation initiatives are executed precisely and aligned with their strategic objectives. Similarly, tailored skill enhancement programs equip the workforce with the necessary competencies to effectively navigate and leverage new technologies. Furthermore, fostering a cooperative work ethos and efficient team coordination practices are fundamental in cultivating an environment that supports continuous improvement and innovation. Such an environment is essential for leveraging the collective strengths of diverse teams, enhancing problem-solving capabilities, and driving the organization toward operational excellence. As organizations look to the future, the IT-OT Convergence journey offers challenges and substantial opportunities for growth and competitive advantage. The case studies and real-world examples discussed in this chapter highlight the transformative potential of digital technologies when integrated thoughtfully across organizational processes. These narratives testify to the power of a well-orchestrated digital strategy in overcoming operational hurdles and enhancing business outcomes. In conclusion, this chapter has provided a comprehensive blueprint for navigating the intricate landscape of Digital Transformation and Industry 4.0. It is a call to action for business leaders, executives, and industry practitioners to embrace the challenges and seize the opportunities presented by this new era. Organizations can survive and thrive in the dynamic and ever-evolving digital landscape by fostering an innovative culture, continuously developing skills, promoting teamwork, and leveraging the synergies between IT and OT. Let this chapter serve as both a guide and an inspiration for those ready to lead their organizations into a future where Digital Transformation drives sustainable growth and enduring success.

REFERENCES

[1] Garimella, P.K., IT-OT integration challenges in utilities. In *2018 IEEE 3rd international conference on computing, communication and security (ICCCS)*. 2018: IEEE.

[2] Mantravadi, S., et al., Securing IT/OT links for low power IIoT devices: design considerations for Industry 4.0. *IEEE Access*, 2020. 8: p. 200305–200321.

[3] Alcaraz, C., Secure interconnection of IT-OT networks in Industry 4.0. In *Critical infrastructure security and resilience: theories, methods, tools and technologies*. 2019: Springer International Publishing. p. 201–217.

[4] Ehie, I.C. and M.A. Chilton, Understanding the influence of IT/OT Convergence on the adoption of Internet of Things (IoT) in manufacturing organizations: an empirical investigation. *Computers in Industry*, 2020. 115: p. 103166.

[5] Heeks, R., Do information and communication technologies (ICTs) contribute to development? *Journal of International Development*, 2010. 22(5): p. 625–640.

[6] Aljawarneh, N. and Z. Al-Omari, The role of enterprise resource planning systems ERP in improving customer relationship management CRM: an empirical study of safeway company of Jordan. *International Journal of Business and Management*, 2018. 13(8): p. 86–100.

[7] Ruivo, P., T. Oliveira, and A. Mestre, Enterprise resource planning and customer relationship management value. *Industrial Management & Data Systems*, 2017. 117(8): p. 1612–1631.

[8] Olanipekun, A.O. and M. Sutrisna, Facilitating digital transformation in construction—a systematic review of the current state of the art. *Frontiers in built Environment*, 2021. 7: p. 660758.

[9] Fenech, R., P. Baguant, and D. Ivanov, The changing role of human resource management in an era of digital transformation. *International Journal of Entrepreneurship*, 2019. 22(2): p. 166–175.

[10] Plenert, G., Focusing material requirements planning (MRP) towards performance. *European Journal of Operational Research*, 1999. 119(1): p. 91–99.

[11] Whybark, D.C. and J.G. Williams, Material requirements planning under uncertainty. *Decision Sciences*, 1976. 7(4): p. 595–606.

[12] Stark, J., Product lifecycle management (PLM). In *Product lifecycle management, Vol. 1: 21st Century paradigm for product realisation*. 2022: Springer. p. 1–32.

[13] Kuusk, A. and J. Gao, Factors for successfully integrating operational and information technologies. In *2015 Portland international conference on management of engineering and technology (PICMET)*. 2015: IEEE.

[14] Thomas, M.S., P. Kumar, and V.K. Chandna, Design, development, and commissioning of a supervisory control and data acquisition (SCADA) laboratory for research and training. *IEEE Transactions on Power Systems*, 2004. 19(3): p. 1582–1588.

[15] Upadhyay, D. and S. Sampalli, SCADA (Supervisory Control and Data Acquisition) systems: vulnerability assessment and security recommendations. *Computers & Security*, 2020. 89: p. 101666.

[16] Kamal, S., et al., IT and OT convergence-opportunities and challenges. In *SPE intelligent energy international conference and exhibition*. 2016: SPE.

[17] Behrendt, A., et al., *Leveraging industrial IoT and advanced technologies for digital transformation*. McKinsey & Company, 2021: p. 1–75.

[18] Zimmer, D. and D. Rhodes, Human-machine interfaces. *IEEE Industry Applications Magazine*, 2006. 12(2): p. 29–35.

[19] Kumar, N. and S.C. Lee, Human-machine interface in smart factory: a systematic literature review. *Technological Forecasting and Social Change*, 2022. 174: p. 121284.

[20] Chinotaikul, P. and S. Vinayavekhin, Digital transformation in business and management research: bibliometric and co-word network analysis. In *2020 1st International conference on big data analytics and practices (IBDAP)*. 2020: IEEE.

[21] Dalenogare, L.S., et al., The expected contribution of Industry 4.0 technologies for industrial performance. *International Journal of Production Economics*, 2018. 204: p. 383–394.

[22] Lasi, H., et al., Industry 4.0. *Business & Information Systems Engineering*, 2014. 6: p. 239–242.

[23] Gorecky, D., et al., Human-machine-interaction in the Industry 4.0 era. In *2014 12th IEEE international conference on industrial informatics (INDIN)*. 2014: IEEE.

[24] Remane, G., et al., Digital maturity in traditional industries-an exploratory analysis. In *Proceedings of 25th European conference on information systems*. 2017: ECIS.

[25] Chatterjee, S., R. Chaudhuri, and D. Vrontis, Does data-driven culture impact innovation and performance of a firm? An empirical examination. *Annals of Operations Research*, 2024. 333(2): p. 601–626.

[26] Sawyer, B.D., et al., Cyber vigilance. *American Intelligence Journal*, 2015. 32(2): p. 151–159.

[27] Collins, J., Lifelong learning in the 21st century and beyond. *Radiographics*, 2009. 29(2): p. 613–622.

[28] Gilchrist, A., *Digital success: a holistic approach to digital transformation for enterprises and manufacturers*. 2018: Alasdair Gilchrist.

[29] Bangert, P., *Optimization for industrial problems*. 2012: Springer Science & Business Media.

[30] Montero, J. and M. Finger, Digitalizing infrastructure: active management for smarter networks. In *A modern guide to the digitalization of infrastructure*. 2021: Edward Elgar Publishing. p. 1–42.

[31] Lorincz, J., A. Capone, and D. Begušić, Optimized network management for energy savings of wireless access networks. *Computer Networks*, 2011. 55(3): p. 514–540.

[32] Bhardwaj, A., et al., Smart IoT and machine learning-based framework for water quality assessment and device component monitoring. *Environmental Science and Pollution Research*, 2022. 29(30): p. 46018–46036.

[33] Xu, K., Y. Qu, and K. Yang, A tutorial on the internet of things: from a heterogeneous network integration perspective. *IEEE Network*, 2016. 30(2): p. 102–108.

[34] Alzoubi, H., BIM as a tool to optimize and manage project risk management. *International Journal of Mechanical Engineering*, 2022. 7(1).

[35] Lewis, F.L., et al., *Cooperative control of multi-agent systems: optimal and adaptive design approaches*. 2013: Springer Science & Business Media.
[36] Marciniak, J.J. and D.J. Reifer, *Software acquisition management*. 1990: New York: John Wiley & Sons, Inc.
[37] Farbey, B. and A. Finkelstein, Software acquisition: a business strategy analysis. In *Proceedings fifth IEEE international symposium on requirements engineering*. 2001: IEEE.
[38] Enck, W., et al., Configuration management at massive scale: system design and experience. *IEEE Journal on Selected Areas in Communications*, 2009. 27(3): p. 323–335.
[39] Jordan, E. and B. Tricker, Information strategy: alignment with organization structure. *The Journal of Strategic Information Systems*, 1995. 4(4): p. 357–382.
[40] Ebad, S.A., Lessons learned from offline assessment of security-critical systems: the case of Microsoft's active directory. *International Journal of System Assurance Engineering and Management*, 2022. 13(1): p. 535–545.
[41] Reeve, A., *Managing data in motion: data integration best practice techniques and technologies*. 2013: Newnes.
[42] Doherty, P., et al., A collaborative framework for 3d mapping using unmanned aerial vehicles. In *PRIMA 2016: principles and practice of multi-agent systems: 19th International conference, Phuket, Thailand, August 22–26, 2016, Proceedings 19*. 2016: Springer.
[43] Feder, M., G. Pearson, and L. Katehi, *Engineering in K-12 education: understanding the status and improving the prospects*. 2009: National Academies Press.
[44] Saenz de Ugarte, B., A. Artiba, and R. Pellerin, Manufacturing execution system–a literature review. *Production Planning and Control*, 2009. 20(6): p. 525–539.
[45] Saelens, B.E., J.F. Sallis, and L.D. Frank, Environmental correlates of walking and cycling: findings from the transportation, urban design, and planning literatures. *Annals of Behavioral Medicine*, 2003. 25(2): p. 80–91.
[46] Scaffa, M.E., et al., The role of occupational therapy in disaster preparedness, response, and recovery. *American Journal of Occupational Therapy*, 2011. 65(6): p. S11.
[47] Rahumed, A., et al., A secure cloud backup system with assured deletion and version control. In *2011 40th International conference on parallel processing workshops*. 2011: IEEE.
[48] Lum, B., et al., Streamlining workflows and redesigning job roles in the theatre sterile surgical unit. *BMJ Open Quality*, 2019. 8(3): p. e000583.
[49] Flowerday, S. and R. Von Solms, Real-time information integrity= system integrity+ data integrity+ continuous assurances. *Computers & Security*, 2005. 24(8): p. 604–613.
[50] Glass, C., A. Cook, and A.R. Ingersoll, Do women leaders promote sustainability? Analyzing the effect of corporate governance composition on environmental performance. *Business Strategy and the Environment*, 2016. 25(7): p. 495–511.
[51] Gupta, A.K., K.G. Smith, and C.E. Shalley, The interplay between exploration and exploitation. *Academy of Management Journal*, 2006. 49(4): p. 693–706.
[52] Sapeta, P., et al., Adaptation and continuous learning: integrative review of coping strategies of palliative care professionals. *Palliative Medicine*, 2022. 36(1): p. 15–29.
[53] Jamieson, G.A. and K.J. Vicente, Ecological interface design for petrochemical applications: supporting operator adaptation, continuous learning, and distributed, collaborative work. *Computers & Chemical Engineering*, 2001. 25(7–8): p. 1055–1074.
[54] Hughes, D.E., J. Le Bon, and A. Malshe, The marketing–sales interface at the interface: creating market-based capabilities through organizational synergy. *Journal of Personal Selling & Sales Management*, 2012. 32(1): p. 57–72.
[55] Carrapico, H. and B. Farrand, "Dialogue, partnership and empowerment for network and information security": the changing role of the private sector from objects of regulation to regulation shapers. *Crime, Law and Social Change*, 2017. 67: p. 245–263.
[56] Thomas, J., Individual cyber security: empowering employees to resist spear phishing to prevent identity theft and ransomware attacks. *International Journal of Business Management*, 2018. 12(3): p. 1–23.

[57] Munir, A., P. Kansakar, and S.U. Khan, IFCIoT: integrated fog Cloud IoT: a novel architectural paradigm for the future Internet of Things. *IEEE Consumer Electronics Magazine*, 2017. 6(3): p. 74–82.
[58] Amayreh, K., The role of knowledge management in delivering the organization to the state of performance excellence: mediating role of technological vigilance. *Management Science Letters*, 2021. 11(4): p. 1277–1286.
[59] Jagatheesaperumal, S.K., et al., The duo of artificial intelligence and big data for Industry 4.0: applications, techniques, challenges, and future research directions. *IEEE Internet of Things Journal*, 2021. 9(15): p. 12861–12885.
[60] Hickson, D.J., D.S. Pugh, and D.C. Pheysey, Operations technology and organization structure: an empirical reappraisal. *Administrative Science Quarterly*, 1969: p. 378–397.
[61] Yao, X., et al., From intelligent manufacturing to smart manufacturing for Industry 4.0 driven by next generation artificial intelligence and further on. In *2017 5th International conference on enterprise systems (ES)*. 2017: IEEE.

5 Harnessing Industrial Internet of Things

Enabling Artificial Intelligence and Machine Learning for Optimized Industrial and Manufacturing Processes

Federico Walas Mateo, Andrés Redchuk, and Ali Soofastaei

5.1 INTRODUCTION

This chapter develops concepts associated with gathering data from industrial processes and transforming it into information to ease process management through descriptive and prescriptive analytics. This is the key for industrial companies to evolve toward the Industry 5.0 (I5.0) paradigm.

5.1.1 Early Developments in Industrial Automation

Industrial automation's roots can be traced back to the early 20th century, marked by the advent of assembly lines and mechanized manufacturing processes. The introduction of programmable logic controllers (PLCs) in the 1960s and 1970s revolutionized factory automation, enabling more precise control and monitoring of industrial processes.

5.1.2 The Rise of Digitalization

The 1980s and 1990s saw the rise of digitalization, with the widespread adoption of computer-aided design (CAD) and manufacturing (CAM) systems. These technologies facilitated greater precision and efficiency in production processes, laying the groundwork for more advanced data collection and analysis capabilities.

5.1.3 The Emergence of IoT

The Internet of Things (IoT) concept emerged in the late 1990s and early 2000s, initially focusing on consumer applications. However, as sensor technology advanced

DOI: 10.1201/9781003186823-5

and connectivity improved, the potential for IoT in industrial settings became increasingly apparent. "Industrial Internet of Things" (IIoT) describes integrating IoT technologies into industrial and manufacturing environments.

5.1.4 IIoT and the Advent of Big Data

The proliferation of IIoT devices in the 2010s led to an explosion of data generated from connected machinery and sensors. This surge in data volume, variety, and velocity necessitated the development of advanced data analytics tools and platforms capable of processing and making sense of this information. Big Data technologies emerged as a critical component of IIoT, enabling the storage and analysis of vast datasets.

5.1.5 Integration of AI and ML

As IIoT systems became more sophisticated, integrating artificial intelligence (AI) and machine learning (ML) technologies became a natural progression. AI and ML offered powerful tools for extracting actionable insights from the massive amounts of data generated by IIoT devices. These technologies facilitated predictive maintenance, real-time process optimization, and enhanced decision-making capabilities in industrial settings.

5.1.6 Modern Advancements and Applications

IIoT is a critical enabler for AI and ML applications in industrial and manufacturing processes. Advanced sensors, edge computing, and cloud-based analytics platforms have become integral components of modern IIoT ecosystems. Companies leverage these technologies to achieve unprecedented levels of efficiency, reduce operational costs, and improve product quality. The ongoing development of 5G networks and advancements in AI algorithms continue to drive innovation in this field, opening new possibilities for the future of industrial automation.

5.1.7 Challenges and Future Directions

Despite the significant progress, challenges still need to be addressed in the widespread adoption of IIoT and AI/ML technologies. These include data security, interoperability, and the need for specialized skills. Looking forward, advancements in AI and ML, coupled with the ubiquity of IIoT devices, promise to further transform industrial and manufacturing processes, ushering in a new era of smart manufacturing and Industry 4.0.

This chapter will explore these historical developments, highlighting key milestones and examining how the convergence of IIoT, AI, and ML shapes the future of industrial and manufacturing processes. Through detailed analysis and case studies, we will illustrate the transformative impact of these technologies and provide insights into the strategies for successful implementation.

This chapter's structure includes vital concepts and an analysis of best practices for integrating data from industrial operations and using it to generate value-added information to improve industrial processes. The chapter presents the case study and the results achieved, ending with the corresponding conclusions.

The market penetration of devices in IIoT architectures, equipped with sensing and communication capabilities, has allowed companies to connect devices in manufacturing plants, developing cyber-physical systems (CPS) capable of generating and collecting data throughout the industrial space [1].

Standard topics on operations technology (OT) and information technology (IT) architecture are discussed, including concepts of IIoT, AI, and ML. Technology (OT/IT) convergence identified by Gartner (2023) is among the main areas of investment in the short term [2].

On the other hand, the link between lean management and I5.0 generates interest. According to Lay et al. [3], eliminating waste from business processes improves efficiency and competitiveness. Facilitating the visibility of operation data impacts the possibility of eliminating tasks that do not add value and identifying opportunities for improvement.

5.2 INDUSTRIAL PROCESS DATA GENERATION, INTEGRATED ARCHITECTURE

To start this topic, let us refer to the article by He and Xu, in which the authors highlight the need to consider the systems approach when addressing research on integrating industrial information [3]. The article describes the modeling and integration of information flow to contextualize business information with operational data through the architecture proposed by IIoT.

The International Society for Automation (ISA) [4] and the International Electrotechnical Commission (IEC) [5] approach operational information collection and its integration. The IEC 62264 multilayer standard, based on the ISA-95 (2010) standard, defines an information model exchange framework that facilitates the integration of solutions in both the IT and OT areas. The firms that comply with this standard define interfaces between control and management functions, allowing them to make informed decisions about the data to exchange so that costs and risks are kept low in case of implementation errors when deploying the solutions.

Figure 5.1 shows the architecture level proposed by the ANSI/ISA-95 standard [6]. This International Standard has been generated to address problems arising during the development of automated interfaces in enterprise management and control systems. It guides the vertical integration of firm information.

The ISA 95 standard defines a functional hierarchy model to categorize the functions of industrial companies. This five-layer model is known as the automation pyramid.

Level 0 is where the production processes are carried out. At that level, the operational frame is measured in milliseconds [7] through sensors (pressure, temperature, flow, etc.) and all field devices (actuators, servo-motors, etc.). This tier also appears to identify devices through RFID or Bluetooth technology, among other alternatives to tracing equipment and goods in the operational process.

Level 1 represents the first logical layer, where data received from the previous level is processed. The operational processes at this level are based on constant feedback mechanisms. In this instance, control elements such as PLC and variable speed drives, among others, appear. The time frame is given in seconds.

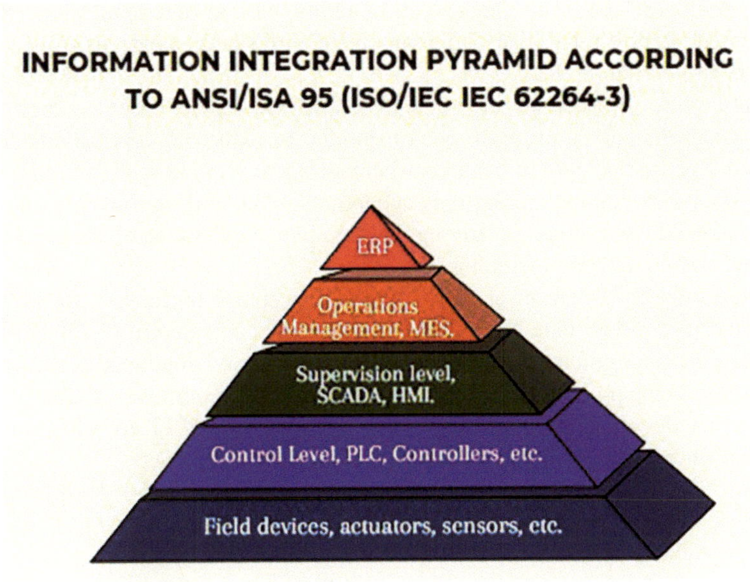

FIGURE 5.1 Information vertical integration model proposed by the ANSI/ISA 95 standard.

Source: Author.

Level 2 represents the automation layer, where control and automation mechanisms are generated. In this instance, the human-machine interfaces and supervision control and data acquisition systems (SCADA) appear. These systems communicate with lower layers, such as PLCs, through protocols and communication standards, such as MODBUS. The time frame at this level is given in minutes.

The next level, 3, is where the contextualization of the manufactured product occurs. This layer defines and maintains the recipes or bill of materials. The system that works in this instance is the manufacturing execution system; operators can enter data using this tool. The time frame is given in hours to days.

The last level, 4, represents the instance of management, planning, and intelligence of operations and business. The critical element is the enterprise resource planning (ERP) system.

To conclude this issue, we must note that although the ISA 95 model and the automation pyramid are still relevant to support innovative production technologies and IIoT, it is necessary to analyze the extension of this model based on the challenges presented by the new variants. Technological. This item is anticipated to be one of the future research lines.

5.3 IIoT ARCHITECTURE. IT AND OT INTEGRATION THROUGH IIOT

OT consists of systems that monitor and control physical processes and manage automated manufacturing processes and associated applications that are typically safety-critical in real time, incorporating additional non-functional properties such as limited latency, reliability, and compliance with safety standards. And industry-specific protection [8, 9].

Until now, IT, such as cloud/edge computing, service-oriented architectures (SOA), and virtualization, have been exploited in industrial applications only in a limited way, that is, only in contexts where stringent requirements were not needed. However, it is becoming increasingly apparent that I5.0 will significantly impact only with full OT/IT convergence, driving the deep joint exploitation of the latest computing and communication technologies.

The article by Patera et al. develops the conceptual framework for IT and OT infrastructure to facilitate the I5.0 model [10]. The convergence of OT/IT is essential for data integration and especially for advancing AI solutions in the industrial decision-making process, providing the foundation for a cognitive-capable plant. The article includes a real case that meets the specific needs of IT and OT, achieving a fast and smooth transfer of large volumes of data to the IT layer.

Lara et al. consider that with the rise of trends such as IoT and cloud manufacturing, which seek the convergence of IT tools in OT networks, IT and OT analysis are highly sought in current industries seeking real-time solutions [11].

5.4 DATA ANALYSIS

To analyze data in the IT and OT domains, it is necessary to use models that focus on describing both domains and show the relationship between them. This article presents a technique that uses operational data produced in an organization from IIoT solutions to model OT and apply analysis methods.

5.5 IIoT AS AN ENABLER OF AI/ML

In this section, we delve into the development of concepts and analyze the current state of the IIoT as a solution that enables and generates opportunities for implementing AI solutions, such as ML.

To initiate this exploration, it is worth referencing the research conducted by Walas Mateo & Redchuk [12]. They conducted a study employing bibliometric analysis to examine the impact of IIoT on the success of AI/ML as a means to optimize processes within Industry 4.0. The study validates the underlying hypothesis, although it underscores the inherent complexity of this type of solution, notes the novelty of the subject, and ultimately points out that it primarily remains within academia, with limited practical application in industry.

Within intelligent production systems, the manufacturing ecosystem comprises diverse devices responsible for collecting data from various industrial processes. Yalcinkaya et al. assert that IIoT represents a new generation of technology enriched by data collection solutions at the plant floor level (e.g., sensors, actuators) with a high degree of precision [13]. Consequently, visibility into operations has advanced to new levels, enabling the acquisition of substantial data volumes and near-instant feedback. This, in turn, facilitates the adoption of AI algorithms geared toward enhancing productivity and process efficiency.

In a study conducted by Silveira et al., the authors examine the Industry 4.0 model within the semiconductor industry. In this sector, high reliability and low operating costs are pivotal to success [14]. This work proposes an Industry 4.0 pilot, summarizing lessons learned while developing a reference design for a semiconductor testing and packaging company. The document delves into cleanroom requirements, sensors, data acquisition boards, and performance details and configurations related to visualization tools and alert notifications from AI tools.

According to Yang et al., smart manufacturing (SM) represents a new paradigm ushering manufacturing into its fourth revolution by harnessing next-generation sensors, communication technologies, and computing capabilities like IIoT [15]. SM aims to enhance manufacturing flexibility and adaptability using high-performance computing and advanced modeling. The authors approach this methodology by reviewing the combined use of knowledge-based and data-based hybrid models (HM) and discussing how these techniques seamlessly integrate into the SM platform. Furthermore, they discuss the new HM paradigms enabled by the SM platform, underscoring their importance in future large-scale SM applications.

It is articulated that the convergence of IIoT, big data, data analysis, and cloud computing is reshaping the landscape of the manufacturing industry. Smart manufacturing and data analytics play a pivotal role in confronting these challenges. In this regard, integrating prescriptive analytics in manufacturing can significantly bolster productivity. The document highlights the prerequisites for production control based on prescriptive analysis, referred to as prescriptive automation, and ultimately outlines the field of activities within this domain.

Finally, Khakifirooz et al. discuss big data analytics as a catalyst for practical manufacturing intelligence in semiconductor manufacturing [16]. This sector is deemed one of the most complex due to its strictly regulated production processes, reentrant process flows, advanced equipment, fluctuating demands, and intricate product mix. The growing adoption of multimode sensors, innovative equipment, and robotics has paved the way for the evolution of IIoT and big data analytics within semiconductor manufacturing. The study introduces a framework founded on Bayesian inference and Gibbs sampling to scrutinize intricate semiconductor manufacturing data for fault detection and powering smart manufacturing. Empirical validation and simulation have demonstrated the practical feasibility of this approach.

5.6 ARTIFICIAL INTELLIGENCE IN INDUSTRIAL PROCESSES

Recent studies on brilliant production employing ML algorithms span various industrial domains, including production planning, energy consumption optimization, machine scheduling, product design, and sustainable machining [17]. Integrating emerging technologies such as IIoT, AI, data analytics, and digital delivery services reshapes innovative manufacturing practices in the Industry 4.0 era [18–20]. Some studies suggest that when applied to sustainable manufacturing, these advanced technologies reduce total energy consumption, decrease labor requirements, and improve condition-based maintenance predictions [21].

Conversely, some authors [19, 22] underscore the challenge of handling vast data volumes. While this data holds the potential to inform decision-making, it requires proper organization and analysis through data modeling tools.

Industries are deploying AI and ML to enhance efficiency, employee safety, and product quality. In manufacturing companies, the ongoing maintenance of production lines and machinery constitutes a significant expense, substantially impacting the bottom line of asset-dependent production operations [23].

ML techniques in manufacturing have gained prominence in the past two decades [23]. In the industrial realm, ML tools find application in various domains, such as problem-solving, control, and optimization [24].

ML techniques, a subset of AI, can learn and adapt to system changes [25]. Priore et al. argue for the necessity of ML techniques in the manufacturing sector, highlighting their capacity to adapt to changing demands and learn from the environment [26].

Referring to the integration of information and its architecture to supply data to ML algorithms, the ISA 95 standard is noteworthy. As discussed earlier, data collection and control activities predominantly occur at lower levels, with complexity increasing as we move up the hierarchy. The top two levels facilitate data sharing and communication for planning and management platforms. Pedone & Mezgar delve into the adaptation of highly heterogeneous systems in the Industry 4.0 context, encompassing cloud models, IIoT, and CPS [27]. They emphasize the pivotal challenges of interoperability and data portability in adopting new technologies within the complex Industry 4.0 ecosystem. The authors also point out that productive systems reap the benefits of adopting cloud computing, big data, AI, and ML in the highest two levels.

5.6.1 A Case in the Food Industry

An adoption case is developed to illustrate the scope and strengths of IIoT architecture. The case occurs in a food-sector industrial company south of the suburbs. The company's production system responds to the continuous process scheme. At the beginning of the project, it had a SCADA architecture for managing industrial processes and an ERP platform for business management.

Based on the initiative of the company's Management, a diagnosis of digital maturity was made, making visible the need to advance in greater data integration to evolve in the I5.0 model.

Within this framework, it was decided to advance in adopting an IIoT architecture. A gateway-type device was incorporated to take the plant operation data found in the OT network to the cloud. For the integration of the process information, the OPC UA server that incorporates the existing SCADA platform in the company is used and enables the data's interoperability so that it can finally be viewed on the IIoT Mindsphere platform developed by the German company Siemens (2023).

The solution Mindsphere is an open, cloud-based IIoT operating system capable of connecting all plant equipment and systems, extracting industrial process data, and converting it into information. This platform has an open action protocol and various functionalities, such as remote access to Amazon Web Service cloud services or the PaaS (Platform-as-a-Service) service.

As an open platform, Mindsphere allows connections with other open platforms, such as NodeRed [28] and Grafana [29].

NodeRed is a programming tool for connecting hardware devices, APIs, and online services. It provides an editor over a web browser, making it easy to develop flows using preconfigured nodes. This tool is event-driven and based on Node.js.

On the other hand, Grafana is an open-source platform for data visualization and monitoring. It allows users to create and share dashboards that display real-time data from various sources, including databases, servers, and cloud services. According to Rani and Chetana, Grafana supports many data sources. It also includes functions such as alerts, annotations, and plugins for data visualization and integration with other tools. It is commonly used in systems, IoT, and network monitoring [28].

Once the operation information from SCADA was integrated, a dashboard system was developed to show different operation KPIs. These can be viewed outside the plant environment without affecting the security conditions required by the OT network. Figures 5.2 and 5.3 show different ways of viewing the KPIs and equipment status generated with operation data.

Lastly, it is worth noting that adopting the new solution was carried out through a co-creation process facilitated by working with an agile methodology to manage change. This way, process experts could get involved early in the solution's adoption and contribute knowledge of the industrial domain to technology providers. Finally, the generated data is stored in a cloud database, easing the evolution of the data architecture toward a prescriptive analytics model.

5.7 CONCLUSIONS

As the first emerging piece of this work, it should be noted that a robust IIoT structure that connects various solutions facilitates data convergence in the manufacturing environment and evolves toward a more mature digital architecture.

One issue that stands out among the results is the possibility of extracting data outside the plant without violating cybersecurity protocols. Achieving the security of

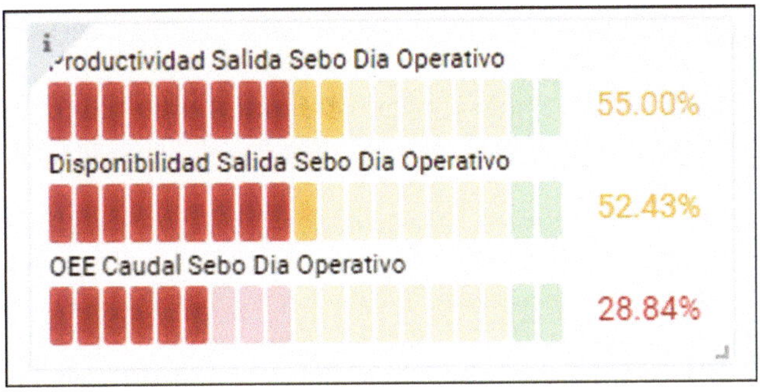

FIGURE 5.2 KPI of the daily operation shown in the dashboard generated on the IIoT platform.

Source: Authors.

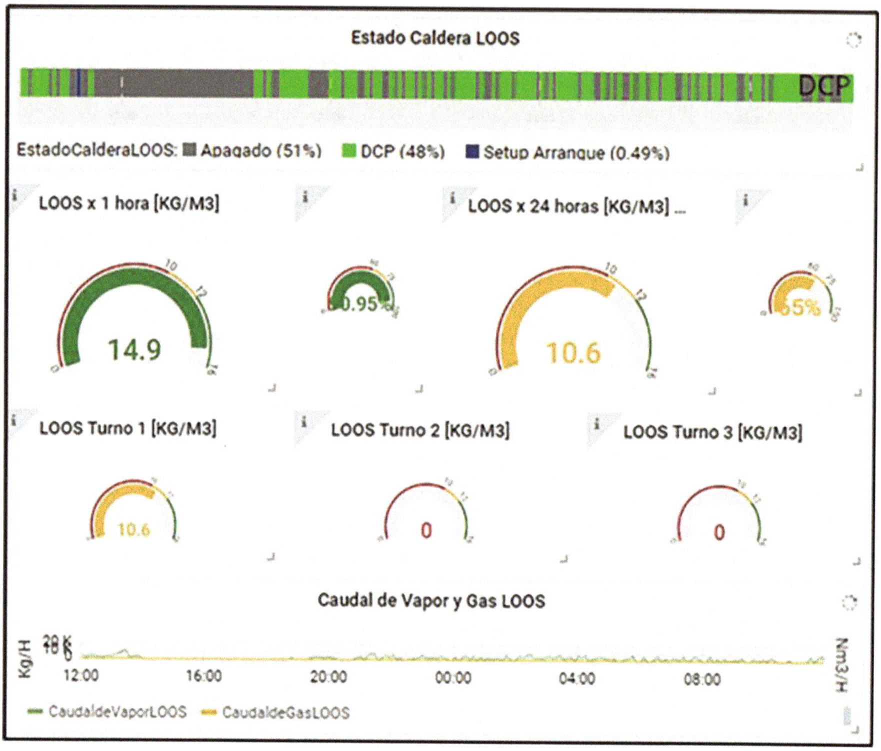

FIGURE 5.3 Dashboard showing the status of a boiler based on data gathered by the IIoT architecture.

Source: Authors.

the SCADA data is an added value that generates the project's value proposition. It is worth mentioning that during emergencies that required access to the SCADA from outside the plant, essential divergences arose with the company's security standards. The most notorious event occurred during the restrictions imposed by the COVID-19 pandemic.

On the other hand, the empowerment of the people involved in the process was achieved by visualizing the data friendlier through dashboards in the manufacturing plant.

An observation that deserves consideration is that the OT infrastructure had a state-of-the-art SCADA platform that incorporated the functionality of the OPC UA server. Without this functionality, the project would have become more complex and consumed more resources.

The IIoT platform generates information that streamlines the continuous improvement of the company's industrial processes. The production state is visualized through digital platforms on monitors in the boiler area and the process control room. This allows for the analysis of the state of the assets and the operational processes to address the waste elimination proposed by the Lean Manufacturing approach. This view, integrated with the I5.0 strategy, leads the company to operate within the Lean 5.0 model.

Finally, it highlights the importance of open platforms to facilitate the dynamic deployment of complex solutions and save development resources. The paragraph mentioning the OPC UA standard in this section provides an example.

REFERENCES

[1] Tapoglou, N., J. Mehnen, and J. Butans, Energy efficient machining through evolutionary real-time optimization of cutting conditions on CNC-milling controllers. In *Experiments and simulations in advanced manufacturing*. 2021: Springer. p. 1–18.

[2] Pettey, C., *When IT and operational technology converge*. 2017. Available from: https://www.gartner.com/smarterwithgartner/when-it-and-operational-technology-converge (cited 2024, 5 April).

[3] He, W. and L. Da Xu, Integration of distributed enterprise applications: a survey. *IEEE Transactions on Industrial Informatics*, 2012. 10(1): p. 35–42.

[4] International Society of Automation (ISA). Available from: https://www.isa.org/ (cited 2024, 5 April).

[5] International Electrotechnical Commission (IEC). Available from: https://www.iec.ch/ (cited 2024).

[6] American National Standards Institute (ANSI), in ISA–95.00.01– 2010, ISA–95.00. 02 – 2010, ISA–95.00. 03 – 2013, ISA–95.00. 04 – 2012, ISA–95.00. 05 – 2013.

[7] Åkerman, M., *Implementing shop floor IT for Industry 4.0*. 2018: Chalmers Tekniska Hogskola (Sweden).

[8] Garbugli, A., A. Bujari, and P. Bellavista, End-to-end QoS management in self-configuring TSN networks. In *2021, the 17th IEEE international conference on factory communication systems (WFCS)*. 2021: IEEE.

[9] Arestova, A., et al., A service-oriented real-time communication scheme for AUTOSAR adaptive using OPC UA and time-sensitive networking. *Sensors*, 2021. 21(7): p. 2337.

[10] Patera, L., et al., A layered middleware for ot/it convergence to empower industry 5.0 applications. *Sensors*, 2021. 22(1): p. 190.

[11] Lara, P., M. Sánchez, and J. Villalobos, Enterprise modeling and operational technologies (OT) application in the oil and gas industry. *Journal of Industrial Information Integration*, 2020. 19: p. 100160.

[12] Mateo, F.W. and A. Redchuk, IIoT/IoT and artificial intelligence/machine learning as a process optimization driver under Industry 4.0 model. *Journal of Computer Science and Technology*, 2021. 21(2): p. 170–176.

[13] Yalcinkaya, E., et al., Empowering ISA95 compliant traditional and intelligent manufacturing systems with the blockchain technology. *Manufacturing Review*, 2021. 8: p. 15.

[14] Silveira, G. d. N., et al., I4. 0 pilot project on a semiconductor industry: implementation and lessons learned. *Sensors*, 2020. 20(20): p. 5752.

[15] Yang, S., et al., Hybrid modeling in the era of intelligent manufacturing. *Computers & Chemical Engineering*, 2020. 140: p. 106874.

[16] Khakifirooz, M., C.F. Chien, and Y.-J. Chen, Bayesian inference for mining semiconductor manufacturing big data for yield enhancement and intelligent production to empower Industry 4.0. *Applied Soft Computing*, 2018. 68: p. 990–999.

[17] Javaid, M., et al., Industry 4.0 technologies and their applications in fighting COVID-19 pandemic. *Diabetes & Metabolic Syndrome: Clinical Research & Reviews*, 2020. 14(4): p. 419–422.

[18] F. Walas Mateo and A. Redchuk, Artificial intelligence as a process optimization driver under Industry 4.0 framework and the role of IIoT, a bibliometric analysis. *Journal of Industrial Integration and Management*, 2022: p. 1–16.
[19] Jamwal, A., et al., Industry 4.0 technologies for manufacturing sustainability: a systematic review and future research directions. *Applied Sciences*, 2021. 11(12): p. 5725.
[20] Machado, C.G., M.P. Winroth, and E.H.D. Ribeiro da Silva, Sustainable manufacturing in Industry 4.0: an emerging research agenda. *International Journal of Production Research*, 2020. 58(5): p. 1462–1484.
[21] Khan, M., et al., Significant data challenges and opportunities in the hype of Industry 4.0. In *2017 IEEE international conference on communications (ICC)*. 2017: IEEE.
[22] Kumar, R., S.P. Singh, and K. Lamba, Sustainable, robust layout using Big Data approach: a key towards Industry 4.0. *Journal of Cleaner Production*, 2018. 204: p. 643–659.
[23] Sharp, M., R. Ak, and T. Hedberg, Jr., A survey of the advancing use and development of machine learning in intelligent manufacturing. *Journal of Manufacturing Systems*, 2018. 48: p. 170–179.
[24] Syafrudin, M., et al., Performance analysis of IoT-based sensor, big data processing, and machine learning model for real-time monitoring system in automotive manufacturing. *Sensors*, 2018. 18(9): p. 2946.
[25] Loyer, J.-L., et al., Comparison of machine learning methods applied to the estimation of manufacturing cost of jet engine components. *International Journal of Production Economics*, 2016. 178: p. 109–119.
[26] Priore, P., D. de la Fuente, and R. Pino, Learning-based scheduling of flexible manufacturing systems using case-based reasoning. *Applied Artificial Intelligence*, 2001. 15(10): p. 949–963.
[27] Pedone, G. and I. Mezgár, Model similarity evidence and interoperability affinity in cloud-ready Industry 4.0 technologies. *Computers in Industry*, 2018. 100: p. 278–286.
[28] Rani, E.G. and D. Chetana, Using GitHub and Grafana tools: data visualization (DATA VIZ) in big data. In *Computer vision and robotics: proceedings of CVR 2022*. 2023: Springer. p. 477–491.

6 Digitizing the Palate
Exploring Opportunities for Digital Transformation in the Food Industry

Guillermo Garcia-Garcia, Hana Trollman, Carlos Parra-López, Carmen Carmona-Torres, Sandeep Jagtap, Yang Luo, and Ali Soofastaei

6.1 INTRODUCTION

The food industry is essential in globalized food supply chains because it provides the population with adequate quantities and food quality. The food industry will need to continue to grow in size and improve the efficiency of its operations to meet a growing demand for food, with estimates suggesting that total global food demand will increase by 35% to 56% between 2010 and 2050 [1]. In addition, the food industry is under increasing pressure to keep prices low so that food is affordable for all, and there is a need to comply with government regulations and meet consumer demands for healthier foods. On top of that, the food sector generates very significant environmental impacts. For example, 71% of global freshwater use is for agriculture alone [2], food production generates 26% of global greenhouse gas emissions, and 78% of global ocean and freshwater eutrophication is caused by agriculture [3]. The food industry and its supply chain must be modernized to provide more food while reducing costs and environmental impacts.

Digitalization is the application of digital technologies to improve business operations. Such technologies include artificial intelligence (AI), simulation, big data analytics, blockchain, and the Internet of Things (IoT). They are often included in the concept of Industry 4.0. These technologies enable collecting and analyzing large amounts of data and provide solutions to optimize industrial performance. The global Industry 4.0 market is estimated to grow from $109.7 billion in 2022 to $430.1 billion in 2030 at a compound annual growth rate of 18.6% [4]. The food industry has already begun to embrace the opportunities that digitalization technologies offer to improve its performance, such as real-time resource-efficient production, resilient and productive food supply chains, and digital technologies to enhance consumer engagement [5, 6].

This chapter introduces and describes the leading digitalization technologies, analyzes their advantages and weaknesses, discusses their application in the food

DOI: 10.1201/9781003186823-6

industry, and shows how they can support advanced analytics in an industrial environment. The technologies examined include AI, big data analytics, blockchain, the IoT, and simulation and modeling.

6.2 ARTIFICIAL INTELLIGENCE

Artificial intelligence (AI) analytics is a subset of business intelligence (BI) that uses machine learning (ML) to uncover patterns and relationships in data to unveil valuable business insights. BI can be differentiated from business analytics (BA) as BI is predominantly concerned with investigating historical data, whereas BA focuses on future outlooks and how things could be improved [7]. This section follows Gartner's Analytics Ascendancy Model [8], as shown in Figure 6.1, to investigate the use of AI in the food industry.

6.2.1 Descriptive Analytics

Descriptive analytics examine past or current data to account for what happened. They can aid in identifying a business's strengths and weaknesses and significantly affect decision-making in formulating sustainable business strategies [9, 10].

6.2.1.1 Real-Time Data Visualization

Data visualization uses charts, plots, graphs, etc., to support understanding patterns, associations, and trends in data. In the context of AI, the types of data being visualized may be big data (large numbers of data points) or high-dimensional data (many dimensions/variables/features/columns). For example, IBM Watson is a tool that includes edge analytics for pattern detection and ML [11].

Dimensionality reduction is commonly used to visualize high-dimensional data in lower dimensions. High-dimensional data often requires effective processing

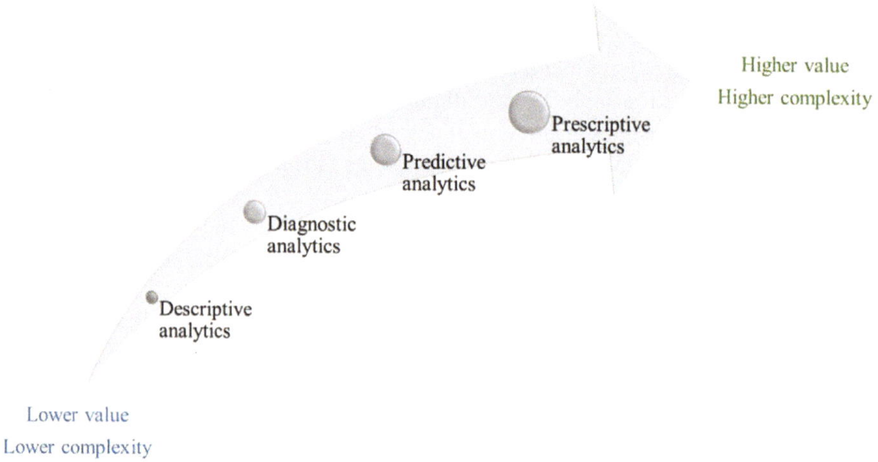

FIGURE 6.1 Gartner's Analytics Ascendancy Model [8].

methods that use ML algorithms. Two basic ML algorithms are density-based spatial clustering of applications with noise and support vector machine [12]. Dynamic visualization of high-dimensional data for whole food supply networks to identify contamination has been proposed.

6.2.1.2 Descriptive Statistics of Processes and Detection of Anomalies

The Industrial Internet of Things may be an innovative approach for monitoring machine components and their related processes in food production. Manufacturers can monitor system conditions and identify device failure, which may be visualized remotely in real time using tools such as the PTC ThingWorx IoT platform [11, 13]. PTC ThingWorx leverages advanced AI and ML techniques with easy-to-use, easy-to-understand visualizations and tools to facilitate advanced edge data analytics [11].

Real-time monitoring of food temperature in a cold chain is critical to the integrity of the distribution system [14], and ML has been shown to improve cold chain management [15]. Kibana is an example of a browser-based visualization tool that applies ML methods in anomaly detection. It has been used to optically record food data better to satisfy growing consumer demand for highly individualized food products [16].

6.2.2 Diagnostic Analytics

Diagnostic analytics are more insight-driven than descriptive analytics, seeking to identify the hidden factors associated with a typically negative outcome. Diagnostic analytics may locate key performance indicators to determine the priority of predictors for outcome improvement, including sustainable value analysis [17]. For example, random forest, a class of ensemble methods combining decision trees, may be used as a "wrapper" algorithm [18]. Feature selection schemes using the wrapper method are based on the learning algorithm used to train the model instead of "filter" methods, which are independent of the learning algorithm.

6.2.2.1 Digital Twins

A digital twin is a virtual representation of an object or system spanning its life cycle, and it is updated using data in real time. Digital twins use simulations of multiple processes, ML, and reasoning (combining information, alternatives, and rules) to aid decision-making. Different types of digital twins exist, including component or parts twins, asset twins (when two or more components work together), system or unit twins (for the interaction of assets), and process twins (the macro level used to reveal, e.g., the workings of an entire production facility). The potential benefits of using digital twins are more effective research and design of products and production, improved efficiency in the whole manufacturing process, and management of by-products of production and end-of-life products such as food packaging disposal by the consumer. A digital twin framework may be used to detect, diagnose, and improve the use of critical resources using diagnostic analytics [19] and unsupervised ML [20]. A digital twin concept has been proposed to model interactions among consumer demand, plant-level constraints, unit operations, and consumer sensory preferences with applications in the food industry (e.g., conceptual case studies of cream cheese fermentation and meat freezing) [21]. Digital twins of food supply chains may improve resilience, reduce food waste, and improve sustainability [22, 23]. Dell Statistica is edge-computing software that provides data analysis capabilities combined with ML and visualization to identify outliers.

6.2.2.2 Audits

The food industry is subject to frequent audits for consumer safety reasons. Audit companies may use AI in their assurance and advisory practices [24], combining company and exogenous data, e.g., using weather indicators (e.g., temperature, humidity) for a multi-location retail company [25]. Auditors, accounting firms, and their clients may be at various stages of digital transformation. Organizations that attempt to implement the most advanced analytics without the necessary foundation provided by less complex analytics (descriptive, diagnostic) will have a reduced likelihood of successful implementation [26]. Auditors can use diagnostic analytics to identify further and evaluate patterns in the data to understand why events have occurred, but where large numbers of anomalies have been identified, auditors often have to choose between population testing and sample testing [25].

6.2.3 Predictive Analytics

Predictive analytics are used for forecasting, using statistical methods and ML. Numerous ML algorithms, such as the stacked long short-term memory (LSTM) model and the term frequency–inverse document frequency (TF-IDF) vectorizer, have been utilized for different types of prediction [27].

6.2.3.1 Prediction of Anomalies and Alerts

Predicting shelf life is essential for ensuring food safety. However, determining shelf life in a laboratory is time-consuming and expensive. Feedforward backpropagation AI and cascade-forward AI models have been used to predict the shelf life of instant coffee-flavored sterilized drinks as a valuable solution for coffee shop owners and food researchers [28]. When combined with sensors, ML may predict the shelf life of, e.g., fresh pizza, fruit, vegetable spray drying [29], and beef cuts [30].

Packaging may become compromised anywhere along the supply chain. ML methods can inform the selection of packaging type as part of the new product development process [31]. Machine vision and ML have been combined to swiftly identify printing defects on packaging, decreasing the cost associated with human sorting and increasing production effectiveness [32].

Predictive maintenance supported by AI can yield higher output, better reliability, and technical equipment availability by determining the time and cost of repair [33].

6.2.3.2 Demand Estimation

Accurate demand forecasting is critical in the food industry, where many products have a short shelf life. Poor stock management can lead to large amounts of business waste. Deep learning models such as random forest regressor, gradient boosting regressor, light gradient boosting machine regressor (LightGBM), extreme gradient boosting regressor (XGBoost), cat boost regressor, LSTM, and bidirectional LSTM

(BiLSTM) show potential for demand forecasting with LSTM demonstrating superiority over the other algorithms [34].

6.2.3.3 Forecasting Process Outcomes Based on the Values of Variables

Consumer demand for quality has shifted the focus of many food-processing industries from low cost to nutritional value and sensory characteristics [33]. AI and ML are promising approaches for adapting to varying consumer demands [35]. ML has been used to predict the quality of foods such as sliced Korean cabbage kimchi [36] and mayonnaise [37]. Electronic noses are particularly useful for quality control [38, 39], and several are commercially available on the market [40].

6.2.4 Prescriptive Analytics

Prescriptive analytics employ stochastic simulation and quantitative optimization to generate responsive action plans. Prescriptive analytics can be applied to various real-world situations, such as pricing, inventory management, maintenance management, logistics, or multi-shift staffing [41]. However, prescriptive analytics is still in a nascent stage of development [42]. The frameworks typically apply various ML techniques using not only historical data but also auxiliary data. However, challenges include the identification of significant predictors, capturing dependencies among predictors for stochastic simulation, and incorporating the impacts of leading predictors on lagging outcomes. Bayesian belief networks have been proposed for such prescriptive analytics challenges [18]. ThingOptimizer of ThingWorx Analytics provides prescriptive scoring and optimization capabilities [11].

6.2.4.1 Generation of Scenarios to Recommend Actions

Scenarios are helpful simulations of future events that allow us to proactively test "what-if" situations [45]. Panic buying at a retailer is one example of how various rationing policies can be tested. This helps store managers decide on the right policy during food shocks to improve access to essential products and mitigate prolonged stockouts that can damage reputations [43].

6.2.4.2 Analysis of the Evolution and Search for Maximum and Minimum Fundamental Values

Perishable food products may be subject to markdowns, but determining the optimal price for inventory control and revenue maximization is difficult. A semi-parametric model that connects black-box ML and the economic model has been proposed to achieve counterfactual prediction for the best discount to maximize overall profit [44].

6.2.4.3 Optimal Autonomous Logistics Solutions with Proactive Updating

Logistics operations face ubiquitous uncertainty in the form of travel time due to weather and traffic conditions, fluctuating prices due to varying supply-demand relationships, and varying transportation demands related to economic and societal changes [7, 45]. Uncertainty increases running costs, decreased resource usage, and reduced customer satisfaction [7, 46]. Regarding optimization problems, uncertainty may exist in the objective function and constraints. State-of-the-art prescriptive analytics include the predict-then-optimize framework, the innovative predict-then-optimize framework, the weighted sample average approximation framework, the empirical risk minimization framework, and the kernel optimization framework [47]. Future research in prescriptive analytics for logistics should focus on tailored learning algorithms, new methodologies, and tools for ML and optimization and validation for real-world industrial problems [47].

6.2.4.4 Prescriptive Maintenance

In an operational context, various actions must occur regularly and repeatedly, offering an opportunity to make impactful data-driven decisions. Prescriptive analytics approaches use data to determine the best decision [48]. Using AI to assess maintenance decisions proactively can increase asset availability by up to 15% and reduce maintenance costs by up to 25% [49]. These can be significant savings as production-intense businesses may have maintenance budgets of 40–50% of the operational budget [50]. Deep reinforcement learning offers a framework for addressing maintenance management with multiple machines and resource constraints, with a cost-saving potential of up to 70% compared to current best practices [51]. Hybrid systems combining structured data with unstructured data, such as text and images, and human input from experts are expected to be the focus of future research.

6.3 BIG DATA

Big data (BD) refers to large and complex data collections that traditional approaches cannot manage [52]. Big data analytics (BDA) uses sophisticated analytical tools and methods to deal with these enormous data volumes [53]. Therefore, BDA involves applying advanced analytics to BD. The food industry can significantly benefit from applying BD and BDA in its operations, from descriptive to predictive and prescriptive analytics and optimization, as shown in the following subsections.

6.3.1 Descriptive Analytics

6.3.1.1 Food Safety and Quality

BD has emerged as a crucial tool for enhancing food safety and quality and highlighted its role in risk monitoring, food tracking and classification, and understanding consumer preferences. Applying BD, AI, and blockchain technologies has become crucial in addressing food safety issues. Furthermore, the role of BD, ML, and AI in mitigating food safety concerns such as food fraud and authenticity and tracking foodborne illnesses has been discussed. Regarding food quality, the concept of Food Quality 4.0, a term introduced by [33], encapsulates the use of Industry 4.0 technologies in food analysis for optimized assessments of food quality.

6.3.1.2 Advanced Techniques and Technologies in Food Analysis

BD is associated with several advanced techniques and technologies used in food analysis. High throughput sequencing (HTS) offers unprecedented resolution and BD potential in the food industry, particularly in identifying and differentiating pathogens in the food supply chain [54]. Electronic nose (e-nose) technology has been highlighted for its potential in food classification, monitoring storage conditions, contamination detection, and volatile compound identification [55]. In addition, spectroscopic techniques have proven efficient in authenticating spices by identifying external adulterants, geographical origin, and material composition [56].

6.3.1.3 Food Production and Supply Chain Management

The application of BD extends to food production and supply chain management, where it helps to address challenges such as climate change, food waste, food security, and sustainability [57]. IoT and BD technologies are used in agriculture to monitor farm conditions, intelligent farm machinery, and drone-based crop imaging [58]. Kamble and others discuss using descriptive, predictive, and prescriptive analytics to achieve sustainable objectives in the agri-food supply chain [6]. Furthermore, using BDA to identify food supply chain and logistics limitations through social media data analysis has been recognized [59].

6.3.1.4 Descriptive Analytics in the Food Business and Consumer Insights

BD has also been instrumental in gaining consumer insights and driving business strategies in the food industry. Advanced text mining techniques are widely used to characterize dietary patterns, provide insights into user preferences, and design food formulations [60]. Crowdsourcing initiatives can be used to obtain real-time monitoring of crowd data, supporting traditional surveillance and restaurant inspection systems [61]. In addition, AI and BDA have been used to analyze various parameters in the food industry, such as quality, appearance, texture, and overall consumer acceptance [62].

6.3.2 Predictive Analytics

6.3.2.1 Food Safety and Quality Control

BD applications transform the food industry by predicting food insecurity, ensuring food safety through risk alert systems, and providing innovative tools through genomic sequencing technologies [63]. ML models are being used to enhance food safety, such as mango grading, wine quality analysis, and dried vegetable quality detection [64]. BD and AI also optimize batch mixing, support quality prediction in processed foods [33], and maintain product quality through real-time data monitoring systems [6]. Spectroscopic techniques coupled with data fusion and BD also contribute to food safety, as demonstrated by their application in spice authentication [56].

6.3.2.2 Precision Agriculture and Food Production

In precision agriculture, BD, ML, and AI are instrumental in yield improvement, optimal harvest time, and prediction of agricultural challenges [65]. HTS offers metagenomic applications that have predictive value in understanding the microbial ecology of factories/ingredients and investigating quality and spoilage incidents [54]. BD is also used to tap consumption markets, enable quantitative production, and plan agriculture and livestock based on weather forecasts [66]. Despite the potential benefits, adopting digital technologies in the agri-food industry faces challenges such as data complexity [67].

6.3.2.3 Supply Chain Management

Predictive analytics are incorporated into supply chain performance measurements, facilitating predictive maintenance and reducing downtime [59, 68]. IoT devices and BDA are modernizing supply chains by enabling real-time monitoring of products in transit and quality control inspections [58]. They also support dynamic pricing of perishable products based on their current quality characteristics [69].

6.3.2.4 Market Analysis and Consumer Behavior

BDA is used to develop effective marketing strategies based on purchasing patterns and demographics [63]. It is used to predict consumer behavior, forecast sales, and understand consumer needs [62, 65]. Online search data can predict market trends in the food industry due to its high correlation with actual market data [70]. Predictive analytics can also inform customers of delivery times for takeaway services, improving customer service [66].

6.3.2.5 Sustainability and Environmental Impact

BDA facilitates decision-making, improves forecasting, and reduces uncertainty in investment decisions, making businesses more predictive, cost-effective, profitable, and sustainable [71]. They are also being applied to support waste management [75] and to guide food-sourcing decisions by modeling weather uncertainty and predicting the impact of climate change on food sourcing [72].

6.3.3 Prescriptive Analytics

6.3.3.1 Quality Control in the Food Industry

Advanced analytics, such as convolutional neural networks, stacked autoencoders, and other deep learning algorithms, have proven instrumental in improving food safety and quality control [64]. These algorithms have been used to qualitatively detect vegetables, fruits, and meat, helping to identify defects, recognize varieties, and assess the quality of meat products [64]. In addition, BDA and AI have been incorporated into intelligent refrigeration systems, providing suggestions on stored products' age and shelf life [62].

6.3.3.2 New Product Development and Operations Optimization

In the field of new product development, BD has demonstrated significant applications. For example, a case study of a beverage company showed that factors such as production performance and cost facilitated the selection of a recipe that was in line with the company's production facilities and strategy [73]. BD has also been effective in improving real-time operational efficiency and developing shorter supply chains in the agri-food industry [52]. Multi-criteria decision-making techniques have also been used to optimize supplier selection and assess supply chain sustainability performance [6].

6.3.3.3 Sustainability in the Food Industry

BD has been used to create more sustainable approaches to food production [57]. It assists in developing standards, training, and testing procedures, thereby ensuring safer food production methods [57]. The application of BD has also been instrumental in helping decision-makers choose actions that reduce harmful emissions during production stages or choose a new mix of raw materials [71]. Furthermore, environmental data guides food-sourcing decisions, considering the impact of climate change on food sourcing [72].

6.3.3.4 Innovative Applications in the food Industry

Innovative applications of BD in the food industry include DNA traceability for the authentication of olive products, an example of intelligent agriculture [52]. In addition, crowdsourcing has been highlighted as a valuable tool within BD that can improve food safety and business practices [61]. Other application areas include the development of intelligent fruit marketing models in e-commerce and the generation of healthy food recommendations in nutrition-based vegetable systems [74]. Furthermore, the application of BD in precision nutrition and health management has led to personalized recipe recommendations and diet therapies [66].

6.3.4 Optimization

6.3.4.1 Food Production and Processing

BD and other Industry 4.0 technologies are revolutionizing food production and processing by increasing operational efficiency, reducing waste, and improving environmental impact. These technologies are helping to create "smart factories" and optimize various food processing techniques [33, 63, 75]. For example, BDA has been used to speed up the new product development process, resulting in significant cost reductions and shorter development times [73]. In addition, innovations such as high-throughput sequencing provide unprecedented resolution in food safety management systems [54]. In parallel, using AI and BD in food chemistry has created new recipes and flavor combinations [66].

6.3.4.2 Authenticating Food Products and Enhancing Food Safety

BD is expanding into areas such as food authentication and food safety. For spice authentication, spectroscopic techniques enhanced by data fusion are used to optimize the performance of individual spectroscopic methods [56]. In addition, critical assessment of industry needs and high-impact areas, supported by BD tools, can maximize food safety and quality [76]. Notably, the management of chilled food supply chains has been improved through sensor data-driven dynamic pricing models, illustrating the potential of BD in strategic supply chain innovation [69].

6.3.4.3 Supply Chain and Inventory Management

BD has become vital in optimizing the food supply chain and inventory management. Digital text data analytics can help minimize food loss in the supply chain [60]. Techniques such as advanced time-temperature indicators are used through crowdsourcing to transform inventory management and promote public trust in science [61]. Furthermore, integrated with Industry 4.0 technologies, BDA facilitates network optimization, supplier collaboration, and inventory management [68]. Although integrating IoT, BD, and AI into business processes is still in its early stages, it is expected to reduce inefficiencies, costs, emissions, and social impacts [58].

6.3.4.4 Sustainability and Environmental Considerations in the Food Industry

BD can also help promote sustainability and address environmental challenges in the food industry. Integrating BD concepts and sustainability assessments can improve the valorization of agricultural waste, as demonstrated in the pretreatment of lignocellulosic biomass in the rice supply chain [77]. In addition, using environmental data in sourcing decisions helps to understand the impact of climate change on food production [72]. Integrating BD technologies with production, sales, and logistics helps in achieving a sustainable and profitable production unit [71].

6.4 BLOCKCHAIN

Blockchain is a digital technology that transparently and securely records transaction and operations data. It enables the storage and transmission of information [54] among network members [80]. Blockchain's characteristics of transparency, immutability, and decentralization have favored its use in the food industry to address various challenges and optimize operations [78]. This section explores blockchain's applications in the food industry, considering the four main areas of advanced analytics.

6.4.1 Descriptive Analytics

6.4.1.1 Improving Traceability with Blockchain

The food industry is witnessing a paradigm shift with the application of blockchain, primarily in traceability and transparency [79–83]. Blockchain's ability to provide a granular, real-time description of each food production and supply step is critical to enhancing food integrity and facilitating anomaly detection and benchmarking [83]. Singh and Sharma further highlighted the importance of blockchain in enabling real-time tracking of products and advanced data visualization [81], which is also supported by the work on blockchain-enabled traceability [84]. Case studies such as the detailed tracking of Greek table olive production using blockchain [85] and the use of non-fungible tokens (NFTs) to track premium food products [86] illustrate its broad potential.

6.4.1.2 Blockchain Integration in Supply Chain Management

The scope of blockchain extends well beyond traceability, making significant inroads into supply chain management by facilitating the collection and descriptive analysis of food supply chain data [58, 87]. And demonstrate how blockchain can improve procurement, logistics, warehousing, inventory management processes, and the safety and quality of food supply chains [88, 89]. Its integration with other Industry

4.0 technologies, such as IoT, AI, and BD, is another transformative factor. Combining blockchain with these technologies enhances the descriptive data layer of supply chains, thus facilitating sophisticated data analytics capabilities [58, 87, 90] and improving supply chain visibility.

6.4.1.3 Blockchain for Improved food Safety

The potential of blockchain is also being used to affect food safety significantly. Real-time recording of production data is a feature that has proven to be instrumental in providing information about the manufacturing process [91]. The Hierarchical Multi-Domain Blockchain network is designed for food safety monitoring, automatically detecting substandard food within the industrial chain and triggering alerts about it. It is a notable application of the technology's ability to improve quality monitoring systems [66, 92]. In addition, the blockchain's immutable record of transactions ensures the authenticity of food products, which is essential to mitigating food fraud and guaranteeing geographical and biological origin [93, 94].

6.4.1.4 Challenges and Prospects of Blockchain in the Food Industry

Despite its promising potential, blockchain implementation in the food industry presents challenges. High costs, scalability issues, low stakeholder awareness, privacy concerns, and the need for standardization and data governance mechanisms are some of the most relevant obstacles [79–81, 95, 96]. Despite these obstacles, the future of blockchain descriptive analytics in the food industry is foreseen to be crucial due to the growing trend toward digital transformation [65, 74, 97]. Blockchain is poised to play a vital role in Industry 4.0, driving sustainable solutions for public health, the environment, and economic development, thus contributing to the holistic optimization of the food industry.

6.4.2 Predictive Analytics

6.4.2.1 Overview of Blockchain Applications and Predictive Analytics

Blockchain technology has shown immense potential to redefine predictive analytics capabilities in the food industry through its transparency, security, and decentralization [59, 98–100]. It significantly improves food traceability and increases consumer trust, providing an opportunity to develop predictive models that estimate consumer behavior [84]. The technology's ability to track and trace food products in real time helps to predict potential problems such as food fraud and product recalls, thereby enabling preventive action. This further highlights the role of blockchain in predicting consumer attitudes toward the organic food sector [97].

6.4.2.2 Blockchain Implementation and Predictive Analytics in Supply Chain Management

The role of blockchain in the food industry extends to supply chain management, contributing to improvements in safety and quality [89]. It facilitates real-time data interaction, enhances the credibility of information, and predicts potential problems in the supply chain through its proof-of-work mechanism and traceability capabilities [88,

90]. Notably, Walmart and Carrefour have leveraged these benefits to reduce tracking time and provide detailed food information to consumers, respectively [99]. In the agri-food sector, blockchain intelligent contracts enhance data traceability and monitoring efficiency, contributing to predictive analytics [101].

6.4.2.3 Case Studies and Blockchain Predictive Analytics

Case studies have demonstrated how blockchain, combined with IoT, Edge Computing, and AI, leads to optimized processes and improved decision-making [102]. In a dairy farm, this combination led to significant improvements. At the same time, its application in grain storage and delivery resulted in increased efficiency in data retrieval, reduced storage costs, and improved overall reliability [103]. In addition, predictive analytics have been used to examine the likelihood of blockchain adoption based on perceived benefits, compatibility, complexity, and level of support from senior management, particularly in SMEs [104].

6.4.2.4 Challenges and Future Directions in Blockchain Implementation

Blockchain technology poses resource constraints, data privacy issues, scalability, and high implementation costs [59, 81]. Nevertheless, these challenges can be mitigated with strong government support, enhanced information and communications technology infrastructure, and integration of AI [87, 105]. Future research directions will likely focus on improving security, reducing complexity in blockchain systems [106], and implementing machine-learning techniques to predict potential health issues in livestock [102]. The growing demand for transparency, traceability, and sustainability in food supply chains further highlights the integration of blockchain technology [98].

6.4.3 Prescriptive Analytics

6.4.3.1 Blockchain for Improved Traceability and Supply Chain Management

Blockchain technology, which has been extensively explored for its potential to enhance transparency and consumer trust in the food industry, offers notable benefits for supply chain management [81, 89, 90, 107]. Various studies have highlighted how blockchain can solve problems of information opacity, improve traceability, and meet legislative requirements while aiding decision-making in emergencies [62, 84, 89]. Despite implementation challenges such as resource constraints and privacy issues, benefits such as cost reduction, time saving, and improved traceability have been demonstrated, mainly when guided by prescriptive analytics approaches [6, 81, 88].

6.4.3.2 Organizational Factors in Blockchain Adoption and Role in Quality Control

The adoption of blockchain in the food industry requires consideration of various organizational factors and can play a crucial role in enhancing security and quality control in supply chains [81, 108–110]. The adoption capacity of organizations, especially small and medium-sized enterprises (SMEs), is influenced by perceived benefits, ease of use, compatibility, complexity, and top management support. Here, prescriptive analytics can serve as a guiding tool for organizations' planning adoption [104, 108]. In the context of quality control, integrated blockchain systems and other technologies, such as TinyML, can prescribe actions against data tampering, thereby ensuring better data security [109].

6.4.3.3 Integrating Blockchain with Innovative Technologies

Advanced technologies such as IoT, BD, AI, and blockchain have been identified as critical to the advancement of the agri-food sector, with a particular focus on system productivity, commercial market optimization, and sustainability [52, 58, 87]. Despite the challenges of internet connectivity issues in rural areas and high technical skill requirements, these technologies can benefit significantly from blockchain integration. As well as building trust between parties, it can reduce transaction costs, automate immutable contracts, and establish tamper-proof voting systems [52, 102]. Prescriptive analytics can guide the integration of these technologies, focusing on data security and improving real-time transaction processes [52, 58].

6.4.3.4 Potential Areas for Further Research and Future Innovation

Several articles suggest areas for further research and improvement in implementing blockchain in the food industry. These areas include the study of the implementation process, scalability issues, the creation of regulations, consumer attitudes toward blockchain [89, 97], and integrating blockchain with other technical trends for better interoperability and protection against attacks [83]. The combination of blockchain with other technical trends, such as IoT, radio-frequency identification, sensor devices, cloud computing, and machine learning, can effectively address these issues and illustrate the authoritarian role of blockchain in enhancing food industry processes, increasing operational efficiency, and improving security [97].

6.4.4 Optimization

6.4.4.1 Improving Supply Chain Efficiency and Transparency

Blockchain technology can significantly optimize the food supply chain's transparency, trust, and efficiency by creating a shared, immutable, and highly secure database [78, 81, 89, 100, 101, 105]. Blockchain enhances traceability and transparency and can ensure regulatory compliance, increase transaction speed, and digitize assets for better trade [94, 100, 101, 105, 111]. Furthermore, the potential of blockchain to improve sustainability by providing complete data on product shelf life and reducing food waste has been well documented [85, 89, 99, 100, 112, 113].

6.4.4.2 Reducing Costs, Risk and Enhancing Security

The role of blockchain in reducing costs and risk in the food industry is a recurring theme in the literature [86, 88, 100, 112, 113]. Blockchain-based systems, such as the one proposed by [86], using NFTs and decentralized InterPlanetary File System storage, can optimize the supply chain by reducing the risks associated with data loss, tampering, and manipulation. Similarly, the works by [88] to [99] pointed out that blockchain can reduce costs, save time, and optimize inventory, logistics, and warehouse management processes. Furthermore, blockchain technology can improve food supply chains' safety, security, and quality despite potential barriers such as standardization, scalability, privacy, and data storage issues [90, 114].

6.4.4.3 Integrating Blockchain with Other Technologies for Optimized Outcomes

Integrating blockchain with other innovative technologies, such as IoT, BD, and AI, has been highlighted as a path to enhanced optimization in the food industry [52, 65, 66, 87, 91, 102, 103, 107]. Such integration can make the food supply chain safer, more efficient, and more sustainable and facilitate intelligent data collection, making the process more efficient, safer, and more thoughtful [52, 74, 107]. The integration of IoT, edge computing, AI, and blockchain in the dairy industry led to a reduction in data traffic, improvement in the reliability of communications, shorter response times, higher quality of service, and enhanced security [102].

6.4.4.4 Addressing Challenges and Future Directions

While the potential benefits of blockchain in the food industry are significant, it is equally important to address the challenges and future directions for its implementation [81, 88, 95, 108, 114]. These challenges include the need for standardization, interoperability issues, scalability concerns, privacy, and legal and regulatory compliance [95, 114]. In addition, lack of awareness, resistance to change, and the need for collaboration and partnership among food industry stakeholders are also seen as potential barriers [81, 108]. Therefore, future research and development should focus on addressing these challenges, and appropriate strategies and policies should be designed to facilitate blockchain adoption and successful implementation in the food industry. Developing education and training programs to raise awareness of blockchain's benefits and practical applications could also be beneficial [88].

6.5 INTERNET OF THINGS

The IoT is a system that connects devices to the Internet, allowing them to collect and share data in real time. This way, it can support descriptive, diagnostic, predictive, and prescriptive analytics to analyze collected data and provide valuable information to stakeholders. One industry that is starting to see the benefits of IoT is the food industry. With the help of IoT, food companies can improve their efficiency, reduce waste, and provide better-quality products to their customers [5, 115]. A typical IoT architecture is shown in Figure 6.2.

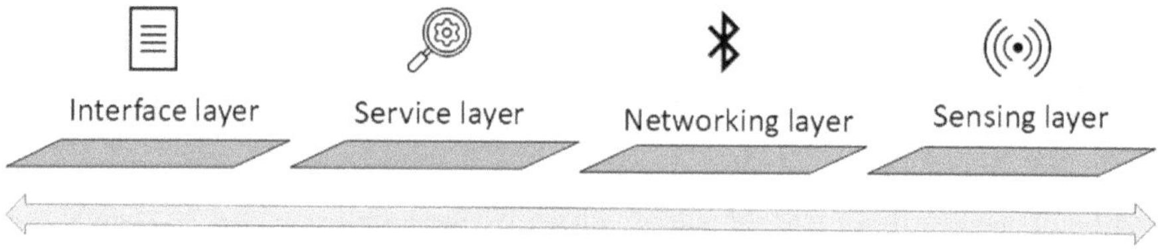

FIGURE 6.2 An example of IoT architecture [65].

6.5.1 IoT Applications in the Food Industry

This section explores the various applications of IoT in the food industry and how it is changing food production, distribution, and consumption.

6.5.1.1 Smart Agriculture

Smart agriculture is one of the primary applications of IoT in the food industry. IoT technology can be used in intelligent farming to optimize crop yields and reduce waste by monitoring the environmental factors affecting crop growth and identifying issues in real time [27]. IoT sensors can be placed in fields to monitor moisture levels, soil acidity, and other environmental factors affecting crop growth [116, 117]. By collecting this data and using predictive and prescriptive analytics, farmers can optimize their irrigation and fertilizer schedules, resulting in better crop yields and less waste [27]. This data can also be analyzed in real-time to identify any issues and take corrective action. This technology can also help reduce the use of fertilizers and other inputs by providing real-time information about the soil conditions and other factors affecting crop growth. In addition to environmental monitoring, IoT can also be used to monitor livestock health. Sensors can be attached to animals to track their activity levels, heart rate, and other vital signs. This information can help farmers identify sick animals early, preventing the spread of disease and reducing the need for antibiotics [115, 118]. Intelligent irrigation systems are another example of the use of IoT in agriculture. These systems use sensors to monitor soil moisture levels and weather conditions. The data collected can be used to adjust irrigation schedules, ensuring that crops receive the right amount of water at the right time. This helps farmers to save water, reduce costs, and improve crop yields [115].

6.5.1.2 Smart Logistics

IoT can also improve the logistics of food distribution to make sure that food is safe and secure throughout the distribution process. Sensors can be placed on shipping containers to track their location, temperature, and humidity. This information can be used to optimize the shipping routes by using predictive and prescriptive analytics, reducing the time it takes to transport food and ensuring that it arrives at its destination in optimal condition [46]. In addition, IoT can be used to improve inventory management in warehouses and stores. Sensors can be placed on products to track their location.

6.5.1.3 Smart Manufacturing

IoT can also improve the manufacturing process of food products. Sensors can be placed on machinery to track their performance and identify potential problems before they cause downtime. This can help reduce the maintenance cost and improve the production line's efficiency [73, 119]. In addition, IoT can be used to monitor the quality of food products as they are being manufactured. Sensors can be placed on the production line to track the temperature, humidity, and other environmental factors that affect the quality of the product. This information can be processed with predictive analytics to identify and address issues before they cause a product recall [33].

6.5.1.4 Smart Packaging

The IoT technology can create innovative packaging that can provide consumers with valuable information. Smart packaging may include sensors that detect the freshness of the product and provide real-time updates to the consumer. Sensors can be placed in packaging to monitor the temperature, humidity, and other environmental factors that affect the quality of the product. This information can be used to ensure that the product remains fresh and safe to consume [115, 116]. This technology can also track the product's location and provide information about the product's origin [33, 120]. Innovative packaging can help consumers make informed decisions about their products by providing real-time information about their freshness and origin. The IoT technology can also help reduce waste by providing information about the expiration date and suggesting ways to use the product before it expires [118].

6.5.1.5 Smart Retail

IoT can also improve the retail experience for customers. Sensors can be placed in stores to track customer traffic and monitor product inventory levels. This information can be used to optimize store layouts, ensure that products are restocked before they run out, and reduce the time it takes for customers to find what they are looking for [115, 121]. IoT sensors can be placed on shelves, refrigerators, and freezers to monitor the stock levels and notify the staff when a product needs to be restocked. Inventory management is critical in retail to ensure that the products are available when needed and that expired products cause no waste. The IoT technology can help reduce waste and optimize inventory levels by monitoring the stock levels and notifying the staff when a product needs to be restocked or is approaching its expiration date [46]. In addition, IoT can be used to personalize customers' shopping experience. By collecting customer preferences and behavior data, retailers can offer personalized recommendations and promotions to their customers.

6.5.2 Benefits of IoT in the Food Industry

The benefits of IoT in the food industry are numerous. This section describes some of them.

6.5.2.1 Improved Efficiency

IoT can help improve the efficiency of food production, distribution, and retail by optimizing processes and reducing waste [122]. For example, farmers can use intelligent irrigation systems to reduce water use, minimize costs, and improve crop yields. The food industry has many machines and equipment that must be maintained regularly. By monitoring the performance of machinery and using predictive analytics, manufacturers can identify potential problems before they cause downtime, reducing the cost of maintenance and improving the efficiency of the production line [123–125]. IoT can help ensure the equipment is maintained regularly and any issues are identified and addressed early [115]. By tracking inventory levels and customer traffic, retailers can optimize store layouts, ensuring that products are restocked before they run out and reducing the time customers need to find what they are looking for [45, 126–128]. Predictive maintenance can help reduce downtime and increase efficiency in the food industry by identifying potential issues before failure occurs.

6.5.2.2 Increased Transparency

Through descriptive analytics, IoT can bring greater transparency to the food supply chain, allowing consumers to track the origin and quality of the food they consume. This helps build trust between consumers and food companies and promotes more sustainable and ethical food production practices [117, 129]. The food industry has complex supply chains that are often difficult to manage [46]. IoT can track products and raw materials moving through the supply chain [129]. IoT-enabled sensors can be placed on trucks and shipping containers to monitor the temperature, humidity, and other environmental factors that may affect the quality of the products [46]. This data can be transmitted to a central database to be analyzed in real time to ensure the products are transported under the correct conditions [121]. Supply chain tracking is essential to ensure the products reach their destination in optimal condition. The IoT technology can help reduce the risk of food spoilage by monitoring factors that can affect the quality of the products. This data can help ensure that the products are transported under the correct conditions and that any issues are identified and addressed in real time [126].

6.5.2.3 Improved Safety

IoT can improve the safety of food products by monitoring the temperature, humidity, and other environmental factors that affect the quality of the product [5]. This can help prevent contamination and reduce the risk of foodborne illness. By collecting data on the performance of equipment and machinery, companies can identify potential safety issues before they become significant problems, making proactive changes to equipment and processes to prevent accidents and injuries [116, 130].

6.5.2.4 Reduced Waste

IoT can help reduce waste in the food industry by optimizing processes and ensuring that products are consumed or at least sold before they expire. By tracking inventory levels and expiration dates, companies can reduce the amount of food that goes to waste, reducing their environmental footprint and improving their bottom line [117, 128]. IoT sensors can be placed in garbage bins and recycling containers to monitor the level of waste and notify the staff when it is time to empty the bins. This technology can also be used to track the amount of waste generated by the facility and identify areas where waste can be reduced by applying prescriptive analytics [116, 124, 128].

6.5.2.5 Improved Quality

IoT can help improve the quality of food products by monitoring the environmental factors that affect their quality and addressing any issues before they cause a product recall [33, 131]. By monitoring the temperature, humidity, and other environmental factors that affect product quality, companies can ensure that their products are of the highest quality, reducing the risk of customer complaints and improving their reputation [33, 132].

6.5.2.6 Personalized Experiences

IoT can help retailers provide personalized shopping experiences to their customers by collecting data on their preferences and behavior. By offering personalized recommendations and promotions, retailers can improve customer satisfaction and loyalty and increase their revenue and market share [120, 133].

6.5.3 Challenges of IoT in the Food Industry

While the benefits of IoT in the food industry are clear, some challenges need to be addressed. This section presents some of these critical challenges.

6.5.3.1 Data Privacy

With the collection and storage of data, there is a risk of data breaches and privacy concerns. Organizations must ensure robust data security measures, including encryption and secure data storage [132].

6.5.3.2 Implementation Costs

Implementing IoT technology can be costly, especially for smaller companies. Before investing in IoT technology, it is essential to consider the investment return [132].

6.5.3.3 Technical Challenges

IoT technology can be complex and requires specialized skills and knowledge to implement and maintain. Companies may need to invest in training or hire specialized personnel to manage their IoT systems [132].

6.5.3.4 Regulatory Challenges

Regulations govern the production and distribution of food products. Companies must ensure that their IoT systems comply with these regulations, including food safety and labeling [132].

6.6 MODELING AND SIMULATION

Data modeling and simulation methodologies present advantageous instruments for examining and enhancing the food supply chain. Using mathematical models and algorithms facilitates the simulation of diverse scenarios, assessment of the influence of multiple factors, and identification of optimal strategies to enhance the food supply chain's performance, thereby empowering decision-makers. Simulation models can depict the dynamics of perishable goods, leading to improved inventory management and decreased product waste. Optimization algorithms can enhance delivery efficiency by optimizing transportation routes and schedules and minimizing costs.

Moreover, these methodologies facilitate the incorporation of sustainability goals into decision-making procedures. The food industry is pressured to implement sustainable practices that mitigate environmental impact and advance responsible resource stewardship. Integrating sustainability indicators and environmental factors into simulation and optimization models enables decision-makers to assess the trade-offs among economic efficiency, customer service, and sustainability objectives.

This integration facilitates the identification of strategies that enhance operational performance while also conforming to broader environmental objectives.

This subsection examines three discrete viewpoints: the operational, system, and strategic perspectives. Various viewpoints emphasize the significance of utilizing data modeling, simulation, and optimization methodologies to tackle distinct obstacles and facilitate advancements within the food sector. The digital optimization timescale, as it pertains to various management levels within the food industry, is depicted in Figure 6.3.

Using simulation and modeling techniques is paramount in improving the efficiency of food industry production processes at an operational level. The methods above offer significant perspectives on the performance of processes, utilization of resources, and strategies for maintenance, ultimately resulting in heightened efficiency and increased productivity [134]. Several optimization techniques exist. Evolutionary algorithms were employed to simulate and optimize processes within the food processing industry. Various methodologies, including genetic algorithms, differential evolution, artificial neural networks, and fuzzy logic, were utilized to tackle optimization problems that could have been more constrained or unconstrained [135]. The utilization of computational fluid dynamics (CFD) is a prevalent approach for modeling food-drying procedures [136]. This facilitates the comprehension of the underlying mechanisms of drying processes, optimizing energy consumption, and improving food quality. CFD enables effective drying systems and process modifications by anticipating fluid flow, heat transfer, and mass transfer during the drying process.

Digital optimization could be done on the system level, where some key points are discussed mainly on resource planning and management, such as optimizing the supplies of energy, water, and raw materials. The measured consumption and

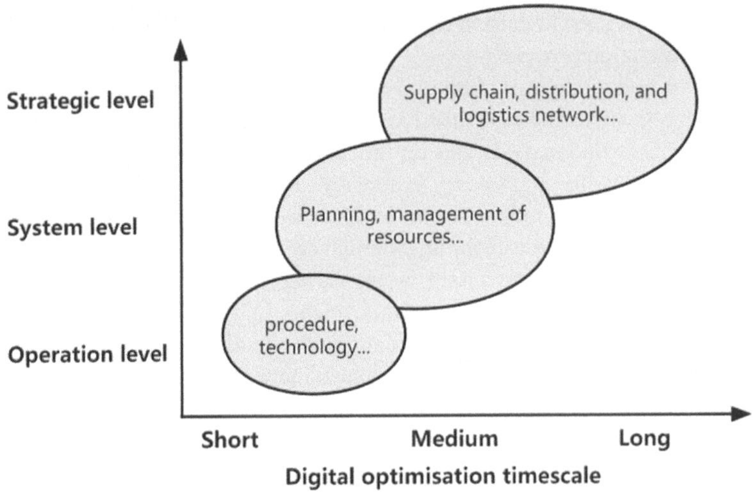

FIGURE 6.3 Digital optimization timescale at different operational levels in the food industry.

energy requirements made it easier to identify energy-saving opportunities [124]. These algorithms optimized various processes such as thermal processing, food quality, process design, drying, fermentation, and hydrogenation. Integrating IoT and simulation-based optimization approaches is highlighted to enhance logistics and refilling strategies in the food supply chain [137]. This approach utilizes intelligent sensor devices to collect real-time data from farm silos and combines biased randomization techniques with simulation-based optimization to improve inventory routing decisions. This approach enables more efficient and cost-effective supply chain operations by leveraging data-driven models and optimization algorithms.

At the strategic level, simulation and modeling techniques contribute to improving the overall performance and sustainability of the food supply chain. Research in this category focuses on systems, implementations, and future research directions in food supply chain management. Data-driven systems are explored to enhance food supply chain management decision-making, potentially integrating with other domains such as databases, enterprise resource planning systems, and management information systems.

Some studies emphasize the significance of data-driven approaches, such as artificial neural networks, adaptive neuro-fuzzy inference systems, and genetic algorithms, in modeling and optimizing energy flows in the food supply chain [138]. Computational intelligent-based systems effectively capture the complexity of energy management and optimize energy flows [139], leading to improved sustainability and efficiency. These techniques enable better decision-making by utilizing large data volumes and simultaneously considering multiple objectives. Fuzzy multi-objective optimization models were proposed for sustainable closed-loop supply chain network design in the food industry [140]. These models used fuzzy theory to address uncertain conditions and maximize supply chain profit and customer satisfaction. The methodologies included mathematical modeling, fuzzy theory, multi-objective optimization, and case studies in the dairy industry. The models incorporated environmental aspects such as carbon footprint, showing the integration of sustainability into the optimization process.

Simulation and modeling techniques have shown great potential in optimizing the food industry from various perspectives. At the process and operation level, these techniques enable the analysis and optimization of complex production processes, improving food quality and energy efficiency. At the system level, simulation and modeling help in energy management, logistics optimization, and decision support. At the strategic level, these techniques enhance the overall performance and sustainability of the food supply chain by providing data-driven insights and optimization approaches. With continued advancements in simulation and modeling, the food industry can achieve higher efficiency and effectiveness, leading to improved productivity, reduced costs, and enhanced sustainability.

6.7 CONCLUSIONS

The food industry is paramount in fulfilling the nutritional requirements of the worldwide population. However, as the demand for food rises due to population growth, urbanization, and changing dietary preferences, the industry faces various challenges in guaranteeing efficient and sustainable food production, distribution, and supply chain management. Simultaneously, the progressions in digitalization technologies present noteworthy prospects for tackling these obstacles and stimulating enhancements in the food sector.

The current status of the food industry is distinguished by a multifaceted and interrelated web of participants encompassing producers, manufacturers, intermediaries, sellers, and buyers. The network covers diverse geographical regions and entails multiple stages, from agricultural cultivation and harvesting to processing, packaging, transportation, and consumption. The food supply chain is characterized by distinct challenges at each stage, such as quality control, perishability, traceability, sustainability, and fluctuations in customer demand.

Efficient data and information management across the supply chain is a significant challenge the food industry faces. The proliferation of digital technologies has resulted in a substantial surge in data generation throughout every phase of the food supply chain. The dataset above details product quality, inventory quantities, transportation circumstances, consumer inclinations, and market patterns. Proficiently utilizing this data to make well-informed decisions is paramount in augmenting operational efficacy, minimizing extra expenditure, and enhancing customer contentment.

Technologies and approaches such as AI, big data analytics, blockchain, the IoT, and simulation and modeling can support the collection and advanced analysis of such data in the food industry. In particular, AI and big data analytics can deal with the vast volumes of data generated by food supply chains and uncover patterns and relationships in data to obtain valuable business insights. Blockchain can be used to store this data transparently and securely. The IoT can connect devices used in the food industry and share data with stakeholders. Finally, modeling and simulation make it easier to simulate different scenarios, assess the impact of multiple factors, and identify the best strategies to improve the performance of food supply chains. Although the food industry has already started to apply these technologies and approaches in its operations, there are still many opportunities to optimize its industrial performance using these digitalization solutions.

REFERENCES

[1] Van Dijk, M., et al., A meta-analysis of projected global food demand and population at risk of hunger for 2010 –2050. *Nature Food*, 2021. 2(7): p. 494–501.

[2] Bank, T.W., *Annual freshwater withdrawals, agriculture (% of total freshwater withdrawal)*. FAO, AQUASTAT Data, 2021. Available from: https://data.worldbank.org/indicator/er.h2o.fwag.zs.

[3] Poore, J. and T. Nemecek, Reducing food's environmental impacts through producers and consumers. *Science*, 2018. 360(6392): p. 987–992.

[4] Markets, R. a., *Industry 4.0: global strategic business report*. 2023. Available from: https://www.researchandmarkets.com/reports/4805798/industry-4–0-global-strategic-business-report.

[5] Rahimifard, S., S. Brewer, G. García-García, and S. Jagtap, Digitalising food manufacturing. *Food Science and Technology*, 2022. 36(3): p. 28–31.

[6] Kamble, S.S., A. Gunasekaran, and S.A. Gawankar, Achieving sustainable performance in a data-driven agriculture supply chain: a review for research and applications. *International Journal of Production Economics*, 2020. 219: p. 179–194.

[7] Wang, W. and Y. Wu, Is uncertainty always wrong for the performance of transportation systems? *Communications in Transportation Research*, 2021. 1: p. 100021.
[8] Eriksson, T., A. Bigi, and M. Bonera, Think with me, or think for me? On the future role of artificial intelligence in marketing strategy formulation. *The TQM Journal*, 2020. 32(4): p. 795–814.
[9] Trollman, H., A novel approach to assessing manufacturer progress toward sustainability. *Procedia CIRP*, 2018. 78: p. 370–375.
[10] Trollman, H. and J. Colwill, A transformational change framework for developing ecologically embedded manufacturing. *Global Journal of Flexible Systems Management*, 2020. 21(4): p. 341–368.
[11] Jayashree, N. and B.S. Babu, Edge data analytics technologies and tools. In *Advances in computers*. 2022: Elsevier. p. 209–236.
[12] Anh, D.T.P., High dimensional data analysis base on machine learning methods. *International Journal of Research in Engineering and Science*, 2022. 10: p. 409–413.
[13] Mohanraj, R., et al., IoT-enabled condition monitoring and intelligent maintenance system for machine. In *Futuristic manufacturing*. 2023: CRC Press. p. 163–177.
[14] Rivas Tucto, J., N. Castillo Talexio, and P. Shiguihara Juárez, Data visualization techniques for monitoring real-time information of cold chain. In *Brazilian technology symposium*. 2019: Springer.
[15] Khanuja, G.S., et al., *Cold chain management uses model-based design, machine learning algorithms and data analytics*. 2018, SAE Technical Paper. SAE International.
[16] Jarschel, T., et al., Practical classification and evaluation of optically recorded Food data using various big-data analysis technologies. *Machines*, 2020. 8(2): p. 34.
[17] Trollman, H. and F. Trollman, A sustainability assessment of smart innovations for mass production, mass customisation and direct digital manufacturing. *Mass Production Processes*, 2019. IntechOpen: p. 1–21.
[18] Wang, C.-H., H.-Y. Cheng, and Y.-T. Deng, Using Bayesian belief network and time-series model to conduct prescriptive and predictive analytics for computer industries. *Computers & Industrial Engineering*, 2018. 115: p. 486–494.
[19] Kumbhar, M., A.H. Ng, and S. Bandaru, A digital twin based framework for detection, diagnosis, and improvement of throughput bottlenecks. *Journal of Manufacturing Systems*, 2023. 66: p. 92–106.
[20] Subramaniyan, M., et al., A generic hierarchical clustering approach for detecting bottlenecks in manufacturing. *Journal of Manufacturing Systems*, 2020. 55: p. 143–158.
[21] Isuru, A., W. Kelton, and C. Bayer, Digital twins in food processing: a conceptual approach to developing multi-layer digital models. *Digital Chemical Engineering*, 2023. 7: p. 100087.
[22] Singh, G., et al., Resilience and sustainability enhancements in food supply chains using Digital Twin technology: a grey causal modelling (GCM) approach. *Computers & Industrial Engineering*, 2023. 179: p. 109172.
[23] Vilas-Boas, J.L., J.J. Rodrigues, and A.M. Alberti, Convergence of Distributed Ledger Technologies with Digital Twins, IoT, and AI for fresh food logistics: challenges and opportunities. *Journal of Industrial Information Integration*, 2023. 31: p. 100393.
[24] Munoko, I., H.L. Brown-Liburd, and M. Vasarhelyi, The ethical implications of using artificial intelligence in auditing. *Journal of Business Ethics*, 2020. 167(2): p. 209–234.
[25] Barr-Pulliam, D., H.L. Brown-Liburd, and I. Munoko, The effects of person-specific, task, and environmental factors on digital transformation and innovation in auditing: a review of the literature. *Journal of International Financial Management & Accounting*, 2022. 33(2): p. 337–374.
[26] Kalsbeek, R., Where to start with the 4 types of analytics. 2020.
[27] Prathibha, S., A. Hongal, and M. Jyothi. IoT based monitoring system in smart agriculture. In *2017 International conference on recent advances in electronics and communication technology (ICRAECT)*. 2017: IEEE.

[28] Badde, D.S., A.K. Gupta, and V.K. Patki, Cascade and feed forward back propagation artificial neural network models for prediction of compressive strength of ready mix concrete. *IOSR Journal of Mechanical and Civil Engineering*, 2013. 3(1): p. 1–6.
[29] Przybył, K. and K. Koszela, Applications MLP and other methods in artificial intelligence of fruit and vegetable in convective and spray drying. *Applied Sciences*, 2023. 13(5): p. 2965.
[30] Feyzioglu, A. and Y.S. Taspinar, Beef quality classification with reduced E-nose data features according to beef cut types. *Sensors*, 2023. 23(4): p. 2222.
[31] Piotrowski, N., Machine learning approach to packaging compatibility testing in the new product development process. *Journal of Intelligent Manufacturing*, 2023: p. 1–13.
[32] Wu, Y. and Y. Lu, An intelligent machine vision system for detecting surface defects on packing boxes based on support vector machine. *Measurement and Control*, 2019. 52(7–8): p. 1102–1110.
[33] Hassoun, A., et al., Food quality 4.0: from traditional approaches to digitalized automated analysis. *Journal of Food Engineering*, 2023. 337: p. 111216.
[34] Panda, S.K. and S.N. Mohanty, Time series forecasting and modelling of food demand supply chain based on regressors analysis. *IEEE Access*, 2023. 11: p. 42679–42700.
[35] Addanki, M., P. Patra, and P. Kandra, Recent advances and applications of artificial intelligence and related technologies in the food industry. *Applied Food Research*, 2022. 2(2): p. 100126.
[36] Park, S.Y., et al., Changes and machine learning-based prediction in quality characteristics of sliced Korean cabbage (Brassica rapa L. pekinensis) kimchi: combined effect of nano-foamed structure film packaging and subcooled storage. *Lwt*, 2022. 171: p. 114122.
[37] Metri-Ojeda, J., et al., Rapid screening of mayonnaise quality using computer vision and machine learning. *Journal of Food Measurement and Characterization*, 2023. 17(3): p. 2792–2804.
[38] Fu, J., et al., Discrimination of geographical indication of Chinese green teas using an electronic nose combined with quantum neural networks: a portable strategy. *Sensors and Actuators B: Chemical*, 2023. 375: p. 132946.
[39] Tian, H., et al., Rapid identification and quantification of vegetable oil adulteration in raw milk using a flash gas chromatography electronic nose combined with machine learning. *Food Control*, 2023. 150: p. 109758.
[40] Loutfi, A., et al., Electronic noses for food quality: a review. *Journal of Food Engineering*, 2015. 144: p. 103–111.
[41] Notz, P.M., P.K. Wolf, and R. Pibernik, Prescriptive analytics for a multi-shift staffing problem. *European Journal of Operational Research*, 2023. 305(2): p. 887–901.
[42] Bhatt Mishra, D., et al., Prescriptive analytics applications in sustainable operations research: conceptual framework and future research challenges. *Annals of Operations Research*, 2023: p. 1–40.
[43] Adulyasak, Y., et al., Using AI to detect panic buying and improve products distribution amid pandemic. *AI & Society*, 2023: p. 1–30.
[44] Hua, J., et al., Markdowns in e-commerce fresh retail: a counterfactual prediction and multi-period optimization approach. In *Proceedings of the 27th ACM SIGKDD conference on knowledge discovery & data mining*, 2021: Association for Computing Machinery. p. 4078–4086.
[45] Jagtap, S., et al., An internet of things approach for water efficiency: a case study of the beverage factory. *Sustainability*, 2021. 13(6): p. 3343.
[46] Jagtap, S., et al., Food logistics 4.0: opportunities and challenges. *Logistics*, 2020. 5(1): p. 2.
[47] Tian, X., et al., Tutorial on prescriptive analytics for logistics: what to predict and how to predict. *Electronic Research Archive*, 2023. 31(4): p. 2265–2285.
[48] Bertsimas, D. and N. Kallus, From predictive to prescriptive analytics. *Management Science*, 2020. 66(3): p. 1025–1044.

[49] Bradbury, S., et al., *Digitally enabled reliability: beyond predictive maintenance*. 2018: McKinsey and Company.

[50] Mather, D., *The future of CMMS*. 2015: Plant Maintenance Resource Center.

[51] Goby, N., T. Brandt, and D. Neumann, Deep reinforcement learning with combinatorial actions spaces: an application to prescriptive maintenance. *Computers & Industrial Engineering*, 2023. 179: p. 109165.

[52] Ayed, R.B., et al., Integration of innovative technologies in the agri-food sector: the fundamentals and practical case of DNA-based traceability of olives from fruit to oil. *Plants*, 2022. 11(9).

[53] Ciccullo, F., et al., Exploring the potential of business models for sustainability and big data for food waste reduction. *Journal of Cleaner Production*, 2022. 340: p. 130673.

[54] Imanian, B., et al., The power, potential, benefits, and challenges of implementing high-throughput sequencing in food safety systems. *NPJ Science of Food*, 2022. 6(1): p. 35.

[55] Al-Dayyeni, W.S., et al., A review on electronic nose: coherent taxonomy, classification, motivations, challenges, recommendations and datasets. *IEEE Access*, 2021. 9: p. 88535–88551.

[56] Oliveira, M.M., J. Cruz-Tirado, and D.F. Barbin, Nontargeted analytical methods as a powerful tool for the authentication of spices and herbs: a review. *Comprehensive Reviews in Food Science and Food Safety*, 2019. 18(3): p. 670–689.

[57] Chapman, J., et al., Challenges and opportunities of the fourth revolution: a brief insight into the future of food. *Critical Reviews in Food Science and Nutrition*, 2022. 62(10): p. 2845–2853.

[58] Misra, N., et al., IoT, big data, and artificial intelligence in agriculture and food industry. *IEEE Internet of Things Journal*, 2020. 9(9): p. 6305–6324.

[59] Dadi, V., et al., Agri-food 4.0 and innovations: revamping the supply chain operations. *Production Engineering Archives*, 2021. 27(2): p. 75–89.

[60] Tao, D., P. Yang, and H. Feng, Utilization of text mining as a big data analysis tool for food science and nutrition. *Comprehensive Reviews in Food Science and Food Safety*, 2020. 19(2): p. 875–894.

[61] Soon, J.M. and I.S. Saguy, Crowdsourcing: a new conceptual view for food safety and quality. *Trends in Food Science & Technology*, 2017. 66: p. 63–72.

[62] Sharma, S., et al., Sustainable innovations in the food industry through artificial intelligence and big data analytics. *Logistics*, 2021. 5(4): p. 66.

[63] Chakraborty, D., et al., Big data in food: systematic literature review and future directions. *Journal of Computer Information Systems*, 2023. 63(5): p. 1243–1263.

[64] Zhou, Q., H. Zhang, and S. Wang, Artificial intelligence, big data, and blockchain in food safety. *International Journal of Food Engineering*, 2022. 18(1): p. 1–14.

[65] Hassoun, A., et al., The fourth industrial revolution in the food industry—Part I: Industry 4.0 technologies. *Critical Reviews in Food Science and Nutrition*, 2023. 63(23): p. 6547–6563.

[66] Tao, Q., et al., Application research: big data in food industry. *Foods*, 2021. 10(9): p. 2203.

[67] Carmela Annosi, M., et al., Digitalization in the agri-food industry: the relationship between technology and sustainable development. *Management Decision*, 2020. 58(8): p. 1737–1757.

[68] Govindan, K., et al., Supply chain 4.0 performance measurement: a systematic literature review, framework development, and empirical evidence. *Transportation Research Part E: Logistics and Transportation Review*, 2022. 164: p. 102725.

[69] Li, D. and X. Wang, Dynamic supply chain decisions based on networked sensor data: an application in the chilled food retail chain. *International Journal of Production Research*, 2017. 55(17): p. 5127–5141.

[70] Kim, S.-S., S. Lim, and S. Kim, Real-time analysis and predictability of the health functional food market using big data. *Food Science and Biotechnology*, 2021. 30(13): p. 1667–1674.
[71] Hämäläinen, E. and T. Inkinen, How to generate economic and sustainability reports from Big Data? Qualifications of process industry. *Processes*, 2017. 5(4): p. 64.
[72] Srinivasan, R., et al., Modelling food sourcing decisions under climate change: a data-driven approach. *Computers & Industrial Engineering*, 2019. 128: p. 911–919.
[73] Jagtap, S. and L.N.K. Duong, Improving the new product development using big data: a case study of a food company. *British Food Journal*, 2019. 121(11): p. 2835–2848.
[74] Hassoun, A., et al., Emerging trends in the agri-food sector: digitalisation and shift to plant-based diets. *Current Research in Food Science*, 2022. 5: p. 2261–2269.
[75] Režek Jambrak, A., et al., Internet of nonthermal food processing technologies (Iontp): food Industry 4.0 and sustainability. *Applied Sciences*, 2021. 11(2): p. 686.
[76] King, T., et al., Food safety for food security: relationship between global megatrends and developments in food safety. *Trends in Food Science & Technology*, 2017. 68: p. 160–175.
[77] Belaud, J.-P., et al., Big data for agri-food 4.0: application to sustainability management for by-products supply chain. *Computers in Industry*, 2019. 111: p. 41–50.
[78] Rogerson, M. and G.C. Parry, Blockchain: case studies in food supply chain visibility. *Supply Chain Management: An International Journal*, 2020. 25(5): p. 601–614.
[79] Kampan, K., T.W. Tsusaka, and A.K. Anal, Adoption of blockchain technology for enhanced traceability of livestock-based products. *Sustainability*, 2022. 14(20): p. 13148.
[80] Kayikci, Y., et al., Food supply chain in the era of Industry 4.0: blockchain technology implementation opportunities and impediments from the perspective of people, process, performance, and technology. *Production Planning & Control*, 2022. 33(2–3): p. 301–321.
[81] Singh, V. and S.K. Sharma, Application of blockchain technology in shaping the future of food industry based on transparency and consumer trust. *Journal of Food Science and Technology*, 2023. 60(4): p. 1237–1254.
[82] Hemker, A.K., et al., Effects of pressure-assisted enzymatic hydrolysis on functional and bioactive properties of tilapia (Oreochromis niloticus) by-product protein hydrolysates. *Lwt*, 2020. 122: p. 109003.
[83] Suresh Kumar, K., et al., A comparative analysis of blockchain in enhancing the drug traceability in edible foods using multiple regression analysis. *Journal of Food Quality*, 2022. Article ID 1689913: p. 6.
[84] Shahzad, K., et al., The role of blockchain-enabled traceability, task technology fit, and user self-efficacy in mobile food delivery applications. *Journal of Retailing and Consumer Services*, 2023. 73: p. 103331.
[85] Kechagias, E.P., et al., An ethereum-based distributed application for enhancing food supply chain traceability. *Foods*, 2023. 12(6): p. 1220.
[86] Hawashin, D., et al., Using composable NFTs for trading and managing expensive packaged products in the food industry. *IEEE Access*, 2023. 11: p. 10587–10603.
[87] Vyas, S., et al., Integration of artificial intelligence and blockchain technology in healthcare and agriculture. *Journal of Food Quality*, 2022. Article ID 4228448: p. 11.
[88] David, A., C.G. Kumar, and P.V. Paul, Blockchain technology in the food supply chain: empirical analysis. *International Journal of Information Systems and Supply Chain Management (IJISSCM)*, 2022. 15(3): p. 1–12.
[89] Vu, N., A. Ghadge, and M. Bourlakis, Blockchain adoption in food supply chains: a review and implementation framework. *Production Planning & Control*, 2023. 34(6): p. 506–523.
[90] Peng, X., et al., Research on the cross-chain model of rice supply chain supervision based on parallel blockchain and smart contracts. *Foods*, 2022. 11(9): p. 1269.

[91] Yu, B., et al., Food quality monitoring system based on smart contracts and evaluation models. *IEEE Access*, 2020. 8: p. 12479–12490.
[92] Su, J.-P., et al., Applying multi-criteria decision-making to improve the waste reduction policy in Taiwan. *Waste Management & Research*, 2010. 28(1): p. 20–28.
[93] Galvez, J.F., J.C. Mejuto, and J. Simal-Gandara, Future challenges on the use of blockchain for food traceability analysis. *TrAC Trends in Analytical Chemistry*, 2018. 107: p. 222–232.
[94] Kamilaris, A., A. Fonts, and F.X. Prenafeta-Boldύ, The rise of blockchain technology in agriculture and food supply chains. *Trends in Food Science & Technology*, 2019. 91: p. 640–652.
[95] Damoska Sekuloska, J. and A. Erceg, Blockchain technology toward creating a smart local food supply chain. *Computers*, 2022. 11(6): p. 95.
[96] Barbosa, M.W., Uncovering research streams on agri-food supply chain management: a bibliometric study. *Global Food Security*, 2021. 28: p. 100517.
[97] Georgescu, B., et al., Consumer attitude towards the use of blockchain technology. Study on the implementation of the "green deal" strategy for organic foods. *Amfiteatru Economic*, 2022. 24(60): p. 379–394.
[98] Wünsche, J.F. and F. Fernqvist, The potential of blockchain technology in the transition towards sustainable food systems. *Sustainability*, 2022. 14(13): p. 7739.
[99] Li, K., J.-Y. Lee, and A. Gharehgozli, Blockchain in food supply chains: a literature review and synthesis analysis of platforms, benefits and challenges. *International Journal of Production Research*, 2023. 61(11): p. 3527–3546.
[100] Rejeb, A., et al., Blockchain technology in the food industry: a review of potentials, challenges and future research directions. *Logistics*, 2020. 4(4): p. 27.
[101] Peng, X., et al., A review on blockchain smart contracts in the agri-food industry: current state, application challenges and future trends. *Computers and Electronics in Agriculture*, 2023. 208: p. 107776.
[102] Alonso, R.S., et al., An intelligent Edge-IoT platform for monitoring livestock and crops in a dairy farming scenario. *Ad Hoc Networks*, 2020. 98: p. 102047.
[103] Hong, Y., New model of food supply chain finance based on the internet of things and blockchain. *Mobile Information Systems*, 2021. 2021: p. 1–8.
[104] Hamdan, I.K., et al., Forecasting blockchain adoption in supply chains based on machine learning: evidence from Palestinian food SMEs. *British Food Journal*, 2022. 124(12): p. 4592–4609.
[105] Panghal, A., et al., Benchmarking the interactions among challenges for blockchain technology adoption: a circular economy perspective. *International Journal of Mathematical, Engineering and Management Sciences*, 2022. 7(6): p. 859–872.
[106] Tao, Q., et al., Food safety supervision system based on hierarchical multi-domain blockchain network. *IEEE Access*, 2019. 7: p. 51817–51826.
[107] Feng, T., H. Li, and Z. Sun, Application of blockchain technology in fresh food supply chain under COVID-19 environment in China. *International Journal of e-Collaboration (IJeC)*, 2022. 18(3): p. 1–11.
[108] Dehghani, M., A. Popova, and S. Gheitanchi, Factors impacting digital transformations of the food industry by adoption of blockchain technology. *Journal of Business & Industrial Marketing*, 2022. 37(9): p. 1818–1834.
[109] Tsoukas, V., et al., Enhancing food supply chain security through the use of blockchain and TinyML. *Information*, 2022. 13(5): p. 213.
[110] Majdalawieh, M., et al., Blockchain-based solution for secure and transparent food supply chain network. *Peer-to-Peer Networking and Applications*, 2021. 14: p. 3831–3850.
[111] Iftekhar, A., et al., Application of blockchain and Internet of Things to ensure tamper-proof data availability for food safety. *Journal of Food Quality*, 2020. 2020: p. 1–14.

[112] Demestichas, K., et al., Blockchain in agriculture traceability systems: a review. *Applied Sciences*, 2020. 10(12): p. 4113.
[113] Motta, G.A., B. Tekinerdogan, and I.N. Athanasiadis, Blockchain applications in the agri-food domain: the first wave. *Frontiers in Blockchain*, 2020. 3: p. 6.
[114] Tanwar, S., et al., Blockchain adoption to secure the food industry: opportunities and challenges. *Sustainability*, 2022. 14(12): p. 7036.
[115] Martindale, W., L. Duong, and S. Jagtop, *Food Industry 4.0: unlocking advancement opportunities in the food manufacturing sector.* 2022: CABI.
[116] Jagtap, S., et al., IoT technologies in the food supply chain. In *Food technology disruptions*. 2021: Elsevier. p. 175–211.
[117] Jagtap, S. and S. Rahimifard, The digitisation of food manufacturing to reduce waste–Case study of a ready meal factory. *Waste Management*, 2019. 87: p. 387–397.
[118] Gupta, S., et al., Development of a structural model for the adoption of Industry 4.0 enabled sustainable operations for operational excellence. *Sustainability*, 2022. 14(17): p. 11103.
[119] Garcia-Garcia, G., Y. Singh, and S. Jagtap, Optimising changeover through lean-manufacturing principles: a case study in a food factory. *Sustainability*, 2022. 14(14): p. 8279.
[120] Echegaray, N., et al., Meat 4.0: principles and applications of Industry 4.0 technologies in the meat industry. *Applied Sciences*, 2022. 12(14): p. 6986.
[121] Jagtap, S., P. Saxena, and K. Salonitis, Food 4.0: implementation of the augmented reality systems in the food industry. *Procedia CIRP*, 2021. 104: p. 1137–1142.
[122] Jagtap, S., G. Garcia-Garcia, and S. Rahimifard, Optimisation of the resource efficiency of food manufacturing via the Internet of Things. *Computers in Industry*, 2021. 127: p. 103397.
[123] Duong, L.N., et al., A review of robotics and autonomous systems in the food industry: from the supply chains perspective. *Trends in Food Science & Technology*, 2020. 106: p. 355–364.
[124] Luo, Y., et al., A framework for recovering waste heat energy from food processing effluent. *Water*, 2022. 15(1): p. 12.
[125] Jagtap, S., et al., The Russia-Ukraine conflict: its implications for the global food supply chains. *Foods*, 2022. 11(14): p. 2098.
[126] Jagtap, S., S. Rahimifard, and L.N. Duong, Real-time data collection to improve energy efficiency: a case study of food manufacturer. *Journal of Food Processing and Preservation*, 2022. 46(8): p. e14338.
[127] Pathare, P.B., A.P. Roskilly, and S. Jagtap, Energy efficiency in meat processing. In *Novel technologies and systems for food preservation*. 2019: IGI Global. p. 78–107.
[128] Jagtap, S., et al., Monitoring potato waste in food manufacturing using image processing and internet of things approach. *Sustainability*, 2019. 11(11): p. 3173.
[129] Hassoun, A., et al., Implementation of relevant fourth industrial revolution innovations across the supply chain of fruits and vegetables: a short update on Traceability 4.0. *Food Chemistry*, 2023. 409: p. 135303.
[130] Bader, F. and S. Jagtap, Internet of things-linked wearable devices for managing food safety in the healthcare sector. In *Wearable and implantable medical devices*. 2020: Elsevier. p. 229–253.
[131] Garcia-Garcia, G., et al., Business process re-engineering to digitalise quality control checks for reducing physical waste and resource use in a food company. *Sustainability*, 2021. 13(22): p. 12341.
[132] Jagtap, S., Utilising the Internet of Things concepts to improve the resource efficiency of food manufacturing (Ph.D. thesis), 2019. Loughborough University.
[133] Hassoun, A., et al., Food processing 4.0: current and future developments spurred by the fourth industrial revolution. *Food Control*, 2023. 145: p. 109507.

[134] Nikkhah, A. and S. Van Haute, Energy flow modeling and optimization trends in food supply chain: a mini review. *Current Opinion in Environmental Science & Health*, 2020. 13: p. 16–22.
[135] Zhong, R., X. Xu, and L. Wang, Food supply chain management: systems, implementations, and future research. *Industrial Management & Data Systems*, 2017. 117(9): p. 2085–2114.
[136] Trystram, G., Modelling of food and food processes. *Journal of Food Engineering*, 2012. 110(2): p. 269–277.
[137] Raba, D., et al., Combining the internet of things with simulation-based optimization to enhance logistics in an agri-food supply chain. In *2019 Winter simulation conference (WSC)*. 2019: IEEE.
[138] Enitan, A.M. and J. Adeyemo, Food processing optimization using evolutionary algorithms. *African Journal of Biotechnology*, 2011. 10(72): p. 16120–16127.
[139] Brandenburg, M., et al., Quantitative models for sustainable supply chain management: developments and directions. *European Journal of Operational Research*, 2014. 233(2): p. 299–312.
[140] Alinezhad, M., et al., A fuzzy multi-objective optimization model for sustainable closed-loop supply chain network design in food industries. *Environment, Development and Sustainability*, 2022: p. 1–28.

7 Constructing Tomorrow

Exploring the Future of Construction in the Era of Industry 4.0

Ali Soofastaei

7.1 INTRODUCTION TO CONSTRUCTION IN INDUSTRY 4.0

At the dawn of Industry 4.0, also known as the Fourth Industrial Revolution, the construction sector is poised for unprecedented transformation. This section introduces the concept of Industry 4.0 within the construction industry context, defines its implications, and explores the pivotal role of advanced analytics in reshaping traditional construction practices.

7.1.1 Defining Industry 4.0 and Its Impact on Various Sectors

Industry 4.0 represents a paradigm shift in manufacturing and production, driven by integrating digital technologies, automation, and data-driven processes. At its core, Industry 4.0 harnesses the power of connectivity, artificial intelligence (AI), and real-time data analytics to enable intelligent, efficient, and agile operations across diverse industries.

Industry 4.0 heralds a new era of innovation, productivity, and sustainability in the construction sector. By leveraging advanced technologies such as building information modeling (BIM), the Internet of Things (IoT), robotics, and augmented reality (AR), construction companies can streamline project workflows, optimize resource utilization, and enhance collaboration throughout the project lifecycle. From design and planning to construction and maintenance, Industry 4.0 promises to revolutionize every facet of the construction industry, driving efficiency gains, cost savings, and improved project outcomes.

7.1.2 The Role of Advanced Analytics in Transforming the Construction Industry

At the heart of Industry 4.0 lies the transformative power of advanced analytics. In the construction industry, advanced analytics encompasses various techniques and tools, including predictive modeling, data visualization, and machine learning algorithms, that enable stakeholders to extract actionable insights from vast amounts of project data.

DOI: 10.1201/9781003186823-7

Advanced analytics can revolutionize construction project management by giving stakeholders real-time visibility into project performance, identifying potential risks and opportunities, and facilitating data-driven decision-making. By analyzing historical project data, predicting future outcomes, and optimizing project schedules and resource allocation, advanced analytics empowers construction companies to enhance productivity, minimize delays, and mitigate project risks.

Moreover, advanced analytics enables construction companies to adopt a proactive approach to quality management and safety by identifying potential hazards and deviations from project specifications in real time. By leveraging sensor data from IoT devices and drones, construction companies can monitor job site conditions, detect safety violations, and implement corrective actions to ensure a safe and compliant work environment.

In conclusion, introducing Industry 4.0 presents a transformative opportunity for the construction industry, ushering in an era of innovation, efficiency, and sustainability. By embracing advanced analytics and leveraging digital technologies, construction companies can unlock new levels of productivity, agility, and competitiveness and position themselves for success in the digital age [1].

7.2 EVOLUTION OF CONSTRUCTION PRACTICES: FROM TRADITIONAL TO MODERN METHODS

The evolution of construction practices reflects the gradual integration of innovation and technology into the industry. This section explores the historical progression from traditional construction techniques to modern methods, culminating in the emergence of digital technologies reshaping the construction landscape.

7.2.1 Historical Overview of Construction Techniques

Construction techniques have evolved throughout history due to materials, tools, and advancements in engineering knowledge. Ancient civilizations developed ingenious methods for building structures using locally available materials, such as stone, mud, and timber. From the monumental pyramids of Egypt to the intricate temples of Greece and Rome, early builders employed manual labor and simple tools to erect impressive architectural marvels that testify their ingenuity and craftsmanship.

The Middle Ages witnessed the rise of Gothic architecture, characterized by soaring cathedrals with intricate stone vaults and flying buttresses. Medieval builders refined stone masonry and carpentry techniques, incorporating mathematical principles and geometric proportions. The Renaissance renewed interest in classical architecture, inspiring architects such as Leonardo da Vinci and Andrea Palladio to explore new methods for constructing domes, arches, and columns.

The Industrial Revolution marked a turning point in construction history, as innovations such as steam power, mechanized manufacturing, and iron and steel production revolutionized building techniques. The advent of mass production and standardized building materials enabled the construction of taller buildings, longer bridges, and more complex structures. The introduction of reinforced concrete and steel-frame construction techniques further expanded the possibilities for architectural expression, leading to the rise of skyscrapers and modern urban landscapes [2].

7.2.2 Introduction to Modern Construction Methods

In the 20th century, the construction industry witnessed the emergence of modern construction methods that prioritized efficiency, speed, and safety. Prefabrication and modular construction techniques revolutionized building assembly, allowing components to be manufactured off-site and assembled on-site, reducing construction time and labor costs. Innovations such as the curtain wall system and precast concrete panels enabled architects to experiment with new forms and aesthetics, leading to the proliferation of modernist architecture and the international style.

The late 20th and early 21st centuries saw the adoption of computer-aided design and BIM software, transforming how buildings are designed, planned, and constructed. BIM enables architects, engineers, and contractors to collaborate in a virtual environment, integrating three-dimensional (3D) models, construction schedules, and cost estimates to streamline project workflows and minimize errors. Advanced materials, such as carbon fiber composites and engineered timber, have expanded the possibilities for sustainable construction and innovative design [3].

7.2.3 The Emergence of Digital Technologies in Construction

The 21st century has witnessed the rapid emergence of digital technologies reshaping the construction industry. From drones and 3D printing to robotics and augmented reality, these technologies are revolutionizing every aspect of the construction process, from site surveying and excavation to building assembly and maintenance.

Drones are used to survey construction sites, monitor progress, and inspect structures, providing real-time data and aerial imagery that inform decision-making and improve safety. 3D printing technology enables architects and engineers to fabricate complex building components precisely and efficiently, reducing waste and labor costs. Robotics are being deployed for bricklaying, welding, and concrete pouring, augmenting human labor and increasing productivity. AR tools overlay digital information onto the physical environment, enabling stakeholders to visualize construction projects and detect clashes or discrepancies before they occur.

In conclusion, the evolution of construction practices reflects a continuous journey of innovation and adaptation to changing technological and societal needs. From ancient civilizations to the digital age, builders have embraced new materials, techniques, and technologies to push the boundaries of what is possible in construction. As we look to the future, digital technologies such as BIM, drones, and robotics promise to revolutionize how we design, build, and inhabit the built environment, ushering in a new era of intelligent, sustainable, and resilient construction practices [4].

7.3 FUNDAMENTALS OF INDUSTRY 4.0 IN CONSTRUCTION

Industry 4.0 represents a paradigm shift in the construction sector, leveraging digital technologies to transform traditional construction practices. This section delves into the fundamentals of Industry 4.0 in construction, including its principles, the integration of critical technologies, and the benefits and challenges associated with its implementation.

7.3.1 Understanding Industry 4.0 Principles

Industry 4.0 principles revolve around interconnectedness, automation, and data-driven decision-making. At its core, Industry 4.0 seeks to create intelligent, interconnected systems that can communicate, analyze data, and adapt in real time. The fundamental principles of Industry 4.0 in construction include [5]:

- **Interoperability:** Different devices, systems, and software can seamlessly communicate and exchange data. In construction, interoperability enables the integration of various technologies and tools, such as BIM software and IoT devices, to streamline project workflows and improve collaboration.
- **Data Transparency:** This enables the accessibility and visibility of data across the entire construction ecosystem, from project stakeholders to subcontractors and suppliers. By providing real-time access to project data, construction companies can make informed decisions, identify potential issues, and optimize project performance.
- **Decentralized Decision-Making:** This empowers frontline workers and project teams to make autonomous decisions based on real-time data and insights. This further enables agile project management, reduces response times, and enhances project flexibility and adaptability.
- **Predictive Maintenance:** This technique utilizes data analytics and predictive algorithms to anticipate equipment failures and maintenance needs before they occur. Predictive maintenance helps construction companies optimize equipment uptime, reduce downtime, and minimize repair costs, improving overall project efficiency and productivity.

7.3.2 Integration of IoT, AI, Big Data, and Automation in Construction

The integration of IoT, AI, big data, and automation is driving the transformation of the construction industry, enabling new levels of efficiency, productivity, and innovation [6]:

- **Internet of Things:** IoT devices such as sensors, cameras, and wearables are being deployed on construction sites to collect real-time data on equipment performance, worker productivity, and job site conditions. IoT enables remote monitoring, predictive maintenance, and automated workflows, enhancing project visibility and control.
- **Artificial Intelligence:** AI-powered algorithms revolutionize construction planning, scheduling, and decision-making processes. AI can analyze vast amounts of project data, identify patterns and trends, and generate actionable insights to optimize project schedules, resource allocation, and risk management strategies.

- **Big Data Analytics:** Big data analytic tools enable construction companies to extract valuable insights from large datasets, such as historical project data, weather patterns, and supply chain information. By analyzing this data, construction companies can identify opportunities for process optimization, cost reduction, and performance improvement.
- **Automation:** Robotics and automation technologies automate repetitive tasks and streamline construction processes, from bricklaying and welding to site grading and material handling. Automation increases productivity, reduces labor costs, and improves safety by minimizing human error and exposure to hazardous conditions.

7.3.3 Benefits and Challenges of Implementing Industry 4.0 in Construction

The implementation of Industry 4.0 in construction offers numerous benefits, including [7]:

- Increased Efficiency: Industry 4.0 technologies streamline project workflows, optimize resource utilization, and reduce project timelines, improving efficiency and productivity.
- Enhanced Safety: IoT devices and AI-powered analytics enable real-time monitoring of job site conditions and worker behavior, helping to identify potential safety hazards and prevent accidents before they occur.
- Cost Savings: Automation and predictive maintenance technologies reduce labor costs, equipment downtime, and project delays, resulting in overall cost savings for construction companies.

However, the implementation of Industry 4.0 in construction also presents several challenges, including [8]:

- **Technological Barriers:** Adopting Industry 4.0 technologies requires significant investment in infrastructure, training, and software integration, which can be prohibitive for smaller construction companies.
- **Data Security and Privacy Concerns:** Collecting and sharing project data raises concerns about data security, privacy, and intellectual property rights, mainly when collaborating with multiple stakeholders and third-party vendors.
- **Cultural Resistance to Change:** Embracing Industry 4.0 requires a cultural shift within construction companies, from top-down leadership to frontline workers, which may encounter resistance from employees accustomed to traditional construction practices.

In conclusion, the fundamentals of Industry 4.0 in construction represent a transformative opportunity for the industry to embrace digital technologies, improve project outcomes, and drive innovation. By understanding the principles of Industry 4.0, integrating key technologies, and addressing associated benefits and challenges, construction companies can position themselves for success in the digital age.

7.4 DIGITAL TWIN TECHNOLOGY IN CONSTRUCTION

Digital twin technology is revolutionizing the construction industry by providing a digital replica of physical assets, processes, and systems. This section explores the fundamentals of digital twin technology, its applications in construction project management, and real-world examples of its implementation in construction projects.

7.4.1 Introduction to Digital Twin Technology

Digital twin technology involves creating a virtual representation, or digital twin, of a physical asset or system. This virtual model is continuously updated with real-time data from sensors, IoT devices, and other sources, enabling stakeholders to monitor, analyze, and simulate the performance of the physical asset or system in a digital environment.

Digital twins create virtual models of buildings, infrastructure projects, and construction processes in the construction industry. These digital twins enable stakeholders to visualize project progress, simulate construction scenarios, and optimize project workflows before implementation in the physical world [9].

7.4.2 Applications of Digital Twins in Construction Project Management

Digital twins offer a wide range of applications in construction project management, including [10]:

- **Design and Planning:** Digital twins enable architects, engineers, and contractors to collaborate on construction projects in a virtual environment. By visualizing project components and simulating construction scenarios, stakeholders can identify potential issues, optimize building layouts, and streamline project schedules.
- **Construction Simulation:** Digital twins facilitate the simulation of construction processes and workflows, allowing stakeholders to identify bottlenecks, optimize resource allocation, and improve construction sequencing. By simulating construction activities in a virtual environment, stakeholders can minimize project delays, reduce costs, and improve overall project efficiency.
- **Monitoring and Control:** Digital twins enable real-time monitoring of construction progress, equipment performance, and job site conditions. By integrating IoT sensors and data analytics into the digital twin model, stakeholders can monitor key project metrics, detect deviations from project plans, and implement corrective actions to ensure project success.
- **Maintenance and Operations:** After construction is complete, digital twins can be used for facility management, maintenance, and operations. Digital twins enable stakeholders to optimize building operations, reduce energy costs, and enhance occupant comfort and safety by capturing data on building performance, energy consumption, and occupancy patterns.

7.4.3 Real-World Examples of Digital Twin Implementation in Construction Projects

Several real-world examples illustrate the successful implementation of digital twin technology in construction projects [11]:

- **The Shard, London:** This iconic skyscraper in London used digital twin technology to optimize construction sequencing, monitor project progress, and simulate building performance. Project stakeholders identified potential issues by creating a digital twin of the building, streamlined construction workflows, and ensured timely project delivery.
- **Marina Bay Sands, Singapore:** Marina Bay Sands, a landmark integrated resort in Singapore, utilized digital twin technology to monitor building systems, optimize energy usage, and enhance guest experiences. By creating a digital twin of the resort, the facility managers could monitor heating, ventilation, and air conditioning systems, lighting controls, and occupancy patterns in real time, improving operational efficiency and guest satisfaction.
- **Crossrail, London:** Crossrail, a major railway project in London, implemented digital twin technology to simulate construction scenarios, optimize tunneling operations, and monitor tunnel stability. By creating a digital twin of the tunneling process, project stakeholders could predict ground settlement, mitigate risks, and ensure the project's safety and success.

In conclusion, digital twin technology offers significant opportunities for improving construction project management, optimizing construction processes, and enhancing project outcomes. By leveraging digital twins to visualize, simulate, and analyze construction projects, stakeholders can minimize risks, reduce costs, and deliver projects more efficiently and effectively.

7.5 BIM IN CONSTRUCTION

BIM has emerged as a transformative technology in the construction industry, revolutionizing how buildings are designed, planned, and constructed. This section explores the fundamentals of BIM, its benefits in construction design and planning, and its applications for collaboration, visualization, and project coordination.

7.5.1 Overview of BIM

BIM is a digital representation of the physical and functional characteristics of a building or infrastructure project. Unlike traditional two-dimensional drawings, BIM models are intelligent, parametric models that contain rich data about the project, including geometry, spatial relationships, materials, and quantities.

BIM enables stakeholders to collaborate, visualize, and simulate construction projects in a virtual environment, facilitating better decision-making, improved coordination, and enhanced project outcomes. By creating a digital twin of the building, BIM allows architects, engineers, contractors, and owners to work together seamlessly throughout the project lifecycle, from conceptual design to facility management [12].

7.5.2 Benefits of BIM in Construction Design and Planning

BIM offers numerous benefits in construction design and planning, including [13]:

- **Enhanced Visualization:** BIM enables stakeholders to visualize construction projects in 3D, providing a clear understanding of the building's spatial layout, form, and function. By visualizing the project in a virtual environment, stakeholders can identify design conflicts, optimize building layouts, and improve overall project clarity and comprehension.
- **Improved Coordination:** BIM facilitates better coordination among project stakeholders by centralizing project information and enabling real-time collaboration. By sharing a unified model, architects, engineers, and contractors can coordinate design changes, resolve conflicts, and avoid costly errors before construction begins.
- **Cost and Time Savings:** BIM enables stakeholders to identify and address potential issues early in the design process, minimizing the need for costly rework and delays during construction. By simulating construction sequences, optimizing material quantities, and improving project scheduling, BIM helps reduce project costs and shorten project timelines.
- **Enhanced Sustainability:** BIM enables stakeholders to evaluate the environmental impact of construction projects and identify opportunities for sustainable design and construction practices. By analyzing energy consumption, carbon emissions, and building performance metrics, BIM helps optimize building efficiency, reduce environmental footprint, and achieve green building certifications.

7.5.3 BIM Applications for Collaboration, Visualization, and Project Coordination

BIM offers a wide range of applications for collaboration, visualization, and project coordination, including [14]:

- **Collaborative Design:** BIM enables architects, engineers, and contractors to collaborate on design concepts, share project information, and coordinate design changes in real time. By working together in a virtual environment, stakeholders can streamline communication, improve decision-making, and enhance project outcomes.
- **Clash Detection:** BIM allows stakeholders to detect clashes and conflicts between different building systems, such as structural, mechanical, and electrical systems, before construction begins. By simulating construction sequences and analyzing spatial relationships, BIM helps identify potential clashes and resolve them early in the design process.

- **Construction Sequencing:** BIM enables stakeholders to simulate construction sequences and visualize the construction process from start to finish. By sequencing construction activities, optimizing material deliveries, and identifying critical path activities, BIM helps streamline project scheduling and improve overall project efficiency.
- **Facility Management:** BIM models can be used for facility management and operations, enabling owners and operators to access building information, track maintenance activities, and plan future renovations. By capturing data on building systems, equipment, and maintenance schedules, BIM helps optimize building performance, reduce operating costs, and enhance occupant comfort and safety.

In conclusion, BIM is revolutionizing the construction industry by providing a digital platform for collaboration, visualization, and project coordination. By embracing BIM technology, stakeholders can streamline construction processes, improve project outcomes, and deliver better buildings that meet the needs of clients, occupants, and communities.

7.6 ROBOTICS AND AUTOMATION IN CONSTRUCTION

Robotics and automation are reshaping the construction industry, offering opportunities to increase efficiency, improve safety, and reduce labor costs. This section explores the automation trends, the use of robotics for specific construction tasks, and the emergence of autonomous construction equipment and vehicles.

7.6.1 Automation Trends in the Construction Industry

Automation is increasingly being adopted in the construction industry to streamline processes, reduce manual labor, and improve productivity. Critical trends in automation include [15]:

- **Prefabrication and Modular Construction:** Prefabrication and modular construction techniques involve manufacturing components off-site in controlled environments before transporting them to the construction site for assembly. Prefabrication enables faster construction timelines, higher quality control, and reduced labor costs, driving the adoption of automation in construction.
- **Robotic Process Automation (RPA):** RPA involves using software robots to automate repetitive tasks and workflows in construction project management. RPA tools can automate data entry, document processing, and project scheduling, freeing up human resources for more value-added activities.
- **Internet of Things and Smart Construction:** IoT devices such as sensors, cameras, and wearables are being deployed on construction sites to collect real-time data on equipment performance, worker productivity, and job site conditions. IoT enables remote monitoring, predictive maintenance, and automated workflows, enhancing project visibility and control.

7.62 Robots for Construction Tasks such as Bricklaying and Welding

Robots are being used to automate specific construction tasks that are repetitive, labor-intensive, and prone to human error. Examples include [16]:

- **Bricklaying:** Robotic bricklaying systems use robotic arms equipped with specialized tools to lay bricks quickly and accurately. These systems can lay thousands of bricks daily, significantly increasing construction productivity and reducing labor costs.
- **Welding:** Robotic welding systems automate the welding process, improving weld quality, consistency, and efficiency. These systems can precisely weld steel beams, columns, and other structural components, reducing the risk of defects and rework.
- **3D Printing:** 3D printing technology automates the fabrication of building components such as walls, columns, and facades. 3D printers can produce complex geometries and custom designs with minimal material waste, offering opportunities for innovative construction techniques and sustainable building practices.

7.6.3 Autonomous Construction Equipment and Vehicles

The emergence of autonomous construction equipment and vehicles is revolutionizing how construction projects are executed. Examples include [17]:

- **Autonomous Excavators:** Autonomous excavators use global positioning system and sensor technology to navigate job sites, excavate trenches, and grade terrain without human intervention. These machines can work around the clock, improving project efficiency and reducing labor costs.
- **Self-Driving Trucks:** Self-driving trucks and vehicles are being used to transport materials and equipment on construction sites, improving logistics, reducing fuel consumption, and enhancing safety. These vehicles can navigate complex job site environments and communicate with other machines to coordinate tasks.
- **Drones:** Drones are used for aerial surveying, mapping, and monitoring of construction sites, providing real-time data and imagery for project planning and management. Drones enable stakeholders to track project progress, identify potential issues, and make informed decisions to keep projects on schedule and within budget.

In conclusion, robots and automation are transforming the construction industry, offering opportunities to increase productivity, improve safety, and reduce costs. By embracing automation trends, leveraging robots for specific construction tasks, and adopting autonomous construction equipment and vehicles, construction companies can enhance project outcomes and remain competitive in a rapidly evolving industry landscape.

7.7 ADVANCED MATERIALS AND SUSTAINABLE CONSTRUCTION PRACTICES

In recent years, there has been a growing emphasis on incorporating advanced materials and sustainable construction practices into the building industry. This section explores the introduction of advanced construction materials, sustainable construction practices, and innovations in materials recycling and waste reduction.

7.7.1 Introduction to Advanced Construction Materials

Advanced construction materials offer superior performance, durability, and sustainability compared to traditional building materials. These materials are engineered to meet the demands of modern construction projects while minimizing environmental impact. Examples of advanced construction materials include [18]:

- **High-Performance Concrete (HPC):** This concrete is engineered to be stronger, more durable, and more workable than conventional concrete. HPC incorporates silica fume, fly ash, and superplasticizers to improve performance and reduce environmental footprint.
- **Engineered Timber:** Engineered timber products such as cross-laminated timber and laminated veneer lumber offer an environmentally friendly alternative to traditional building materials such as steel and concrete. Engineered timber is renewable, lightweight, and easy to work with, making it ideal for sustainable construction projects.
- **Recycled Materials:** Recycled materials such as recycled concrete aggregate, recycled glass, and recycled plastic are being used to replace conventional building materials in construction projects. By incorporating recycled materials, builders can reduce demand for virgin resources, minimize waste, and lower carbon emissions.

7.7.2 Sustainable Construction Practices and Green Building Certifications

Sustainable construction practices aim to minimize the environmental impact of construction projects while maximizing resource efficiency and occupant comfort. Critical sustainable construction practices include [19]:

- **Energy Efficiency:** Buildings are designed to minimize energy consumption through passive design strategies, efficient HVAC systems, and renewable energy technologies such as solar panels and geothermal heating.
- **Water Conservation:** Implementing water-efficient fixtures, rainwater harvesting systems, and water recycling technologies to reduce water consumption and minimize strain on municipal water supplies.
- **Waste Reduction:** Adopting construction waste management plans, recycling construction debris, and using prefabricated and modular construction techniques to minimize waste generation and landfill disposal.

- **Green Building Certifications:** Green building certifications such as LEED (Leadership in Energy and Environmental Design) and BREEAM (Building Research Establishment Environmental Assessment Method) provide standards and guidelines for sustainable building design, construction, and operation. These certifications encourage builders to incorporate sustainable practices and materials into their projects, leading to healthier, more environmentally friendly buildings.

7.7.3 Innovations in Materials Recycling and Waste Reduction in Construction

Innovations in materials recycling and waste reduction are driving the construction industry's transition toward a circular economy. Examples of innovations include [20]:

- **Prefabrication and Modular Construction:** Prefabrication and modular construction techniques involve manufacturing components off-site in controlled environments before transporting them to the construction site for assembly. Compared to traditional construction methods, prefabrication reduces waste, minimizes construction time, and improves quality control.
- **Construction Waste Recycling:** Construction waste recycling facilities separate, process, and recycle construction debris such as concrete, wood, and metal for reuse in new construction projects. Recycling construction waste reduces landfill disposal, conserves resources, and lowers carbon emissions associated with material production.
- **3D Printing:** This technology enables the fabrication of building components using recycled materials such as plastic, glass, and concrete. It printing reduces material waste, allows for complex geometries, and enables customization, making it an attractive option for sustainable construction projects.

In conclusion, the adoption of advanced materials and sustainable construction practices is essential for creating buildings that are environmentally friendly, resource-efficient, and resilient to climate change. By embracing innovative materials, incorporating sustainable practices, and reducing waste generation, the construction industry can contribute to a more sustainable built environment for future generations.

7.8 DATA ANALYTICS FOR CONSTRUCTION PROJECT MANAGEMENT

Data analytics is increasingly becoming integral to construction project management, offering insights that enable better decision-making, enhanced efficiency, and improved project outcomes. This section delves into the importance of data analytics in construction project management, collecting, processing, and analyzing construction project data and applying predictive analytics for project scheduling, cost estimation, and risk management.

7.8.1 Importance of Data Analytics in Construction Project Management

Data analytics is crucial in construction project management by providing stakeholders with actionable insights derived from project data. Key areas of application include [21]:

- **Decision-Making:** Data analytics enables informed decision-making by giving stakeholders real-time visibility into project performance, progress, and potential risks. By analyzing project data, stakeholders can identify trends, patterns, and areas for improvement, leading to more effective decision-making and better project outcomes.
- **Efficiency:** Data analytics streamlines project workflows, optimizes resource allocation, and minimizes project delays by identifying bottlenecks and inefficiencies in construction processes. By leveraging data analytics tools, project managers can identify opportunities for process optimization, reduce project timelines, and improve overall project efficiency.
- **Risk Management:** Data analytics helps to identify and mitigate project risks by analyzing historical project data, thereby predicting potential issues and implementing proactive risk management strategies. By identifying potential risks early in the project lifecycle, stakeholders can implement mitigation measures to minimize their impact on project cost, schedule, and quality.

7.8.2 Collection, Processing, and Analysis of Construction Project Data

The process of collecting, processing, and analyzing construction project data involves several steps [22]:

- **Data Collection:** Construction project data is collected from various sources, including project management software, IoT devices, sensors, and manual data entry. Data may include project schedules, budget estimates, material quantities, equipment usage, and labor productivity.
- **Data Processing:** Once collected, construction project data is processed to ensure accuracy, consistency, and completeness. Data processing involves cleaning, organizing, and standardizing the data for analysis, removing duplicates, errors, and inconsistencies.
- **Data Analysis:** Data analysis involves applying statistical techniques, machine learning algorithms, and data visualization tools to extract insights from construction project data. It may include identifying data trends, patterns, correlations, and outliers to inform decision-making and improve project performance.

7.8.3 Predictive Analytics for Project Scheduling, Cost Estimation, and Risk Management

Predictive analytics utilizes historical project data to forecast future project outcomes, enabling stakeholders to address potential issues and opportunities proactively.

Critical applications of predictive analytics in construction project management include [23]:

- **Project Scheduling:** Predictive analytics enables stakeholders to forecast project schedules, identify critical path activities, and anticipate potential delays. By analyzing historical project data and simulating different scenarios, stakeholders can optimize project schedules, allocate resources efficiently, and minimize project timelines.
- **Cost Estimation:** Predictive analytics helps stakeholders forecast project costs, identify cost drivers, and anticipate budget overruns. By analyzing historical cost data and considering material prices, labor rates, and market conditions, stakeholders can develop accurate cost estimates and mitigate financial risks.
- **Risk Management:** Predictive analytics enables stakeholders to identify and assess project risks, prioritize risk mitigation strategies, and develop contingency plans. By analyzing historical project data and identifying risk factors, stakeholders can predict potential risks, assess their likelihood and impact, and implement proactive risk management measures to minimize their impact on project outcomes.

In conclusion, data analytics transforms construction project management by providing stakeholders with actionable insights from project data. By leveraging data analytics tools and techniques, stakeholders can optimize project performance, improve decision-making, and deliver successful construction projects on time and within budget.

7.9 AR AND VIRTUAL REALITY (VR) IN CONSTRUCTION

AR and VR transform the construction industry by offering immersive, interactive experiences that enhance design visualization, project coordination, and construction training. This section provides an overview of AR and VR technologies, explores their applications in construction design, visualization, and training, and discusses future trends and developments in AR and VR for construction [24].

7.9.1 Overview of AR and VR Technologies

- AR overlays digital information, such as 3D models, annotations, and instructions, onto the real-world environment, enhancing the user's perception of reality. AR applications are typically accessed through smartphones, tablets, or wearable devices, enabling users to interact with digital content in real time.
- VR immerses users in a computer-generated environment, simulating a realistic, 3D experience. VR applications are typically accessed through head-mounted displays or VR goggles, providing a fully immersive experience that transports users to virtual construction sites, buildings, or training scenarios.

7.9.2 Applications of AR and VR in Construction

- **Construction Design and Visualization:** AR and VR enable stakeholders to visualize construction projects in 3D, explore design options, and make informed decisions before construction begins. By immersing users in virtual environments, AR and VR allow architects, engineers, and clients to experience buildings and spaces at full scale, improving design communication and reducing the risk of design errors [25].
- **Project Coordination:** AR and VR facilitate project coordination by enabling stakeholders to visualize project plans, identify clashes, and resolve issues before construction begins. By overlaying digital models onto the real-world environment, AR enables on-site workers to visualize hidden infrastructure, such as pipes and cables, reducing the risk of clashes and rework during construction.
- **Construction Training:** AR and VR provide immersive training experiences for construction workers, enabling them to practice construction tasks, simulate hazardous scenarios, and improve safety awareness in a controlled environment. By replicating real-world construction scenarios, AR and VR training programs help reduce accidents, improve worker proficiency, and enhance overall job performance [26].

7.9.3 Future Trends and Developments in AR and VR for Construction

- **Enhanced Collaboration:** Future AR and VR technology developments will enhance project stakeholders' collaboration. This will allow architects, engineers, contractors, and clients to interact with virtual construction models in real time, regardless of their location.
- **Advanced Visualization Tools:** Future AR and VR applications will incorporate advanced visualization tools, such as real-time rendering, photorealistic graphics, and interactive animations, to provide users with a more immersive and realistic experience.
- **Integration with BIM:** AR and VR will increasingly be integrated with BIM software, enabling seamless visualization and interaction with BIM models in augmented and virtual environments.
- **Wearable Devices:** The development of lightweight AR and VR devices will make immersive experiences more accessible and practical for construction workers, enabling hands-free interaction with digital site content.

In conclusion, AR and VR technologies are revolutionizing the construction industry by offering immersive, interactive experiences that enhance design visualization, project coordination, and construction training. As these technologies continue to evolve and become more accessible, they will play an increasingly important role in shaping the future of construction [27].

7.10 CHALLENGES AND FUTURE DIRECTIONS

As the construction industry continues to embrace Industry 4.0 technologies, it faces various challenges and opportunities. This section explores the key challenges in adopting Industry 4.0 practices, ethical considerations in data-driven construction, and future trends and opportunities shaping the construction industry.

7.10.1 Addressing Technological and Cultural Barriers to Industry 4.0 Adoption

- **Technological Barriers:** One of the main challenges in adopting Industry 4.0 technologies in construction is the industry's complexity and fragmentation. Integrating diverse systems, software, and hardware platforms can be challenging, requiring interoperability standards and investment in technology infrastructure.
- **Cultural Barriers:** Resistance to change and a lack of digital literacy among construction workers and management can hinder the adoption of Industry 4.0 practices. Overcoming cultural barriers requires effective change management strategies, training programs, and leadership commitment to embracing innovation.

7.10.2 Ethical Considerations in Data-Driven Construction Practices

- **Data Privacy:** Collecting, storing, and analyzing construction project data raises concerns about privacy and data security. Stakeholders must implement robust data protection measures and comply with privacy regulations to safeguard sensitive information and prevent unauthorized access or misuse.
- **Bias and Fairness:** Data-driven decision-making in construction may be subject to bias and discrimination, leading to inequitable outcomes for marginalized communities. Stakeholders must address data collection and analysis bias to ensure fair and equitable project outcomes for all stakeholders.
- **Transparency and Accountability:** Transparency and accountability are essential principles in data-driven construction practices. Stakeholders must be transparent about the data they collect, how it is used, and the decisions made based on it. They must also establish mechanisms for accountability and recourse in case of data misuse or unethical practices [28].

7.10.3 Future Trends and Opportunities in the Construction Industry

- **Digital Twin Technology:** Digital twin technology will continue to evolve, enabling stakeholders to create more accurate and detailed digital replicas of construction projects. Advanced digital twins will facilitate real-time monitoring, simulation, and optimization of construction processes, improving project outcomes and reducing risk.

- **Sustainable Construction:** The growing emphasis on sustainability will drive innovation in construction materials, techniques, and practices. Sustainable construction practices, such as green building certifications, renewable energy integration, and circular economy principles, will become increasingly mainstream, leading to more environmentally friendly and resilient buildings.
- **Robots and Automation:** Robots and automation will significantly automate repetitive and hazardous tasks in construction, improving productivity, safety, and efficiency. Advances in robots, such as autonomous construction equipment and drones, will enable greater automation of construction processes and accelerate project delivery.

In conclusion, the construction industry faces challenges and opportunities in embracing Industry 4.0 practices. By addressing technological and cultural barriers, upholding ethical principles, and embracing future trends and opportunities, the industry can harness Industry 4.0's transformative potential to build a more sustainable, efficient, and resilient environment for future generations [29, 30].

7.11 CONCLUSION: THE FUTURE LANDSCAPE OF CONSTRUCTION

The rapid advancement of Industry 4.0 technologies is reshaping the construction industry, presenting stakeholders with challenges and opportunities. This concluding section reflects on key insights and findings, explores Industry 4.0's transformative potential in construction, and calls for embracing innovation and sustainability in construction practices.

7.11.1 Summary of Key Insights and Findings

Throughout this chapter, we have explored the various facets of Industry 4.0 in construction, from adopting advanced analytics and digital technologies to integrating robots, automation, and augmented reality. We have witnessed how these technologies revolutionize construction project management, design visualization, and workforce training, increasing construction practices' efficiency, productivity, and safety.

Key insights and findings include the importance of data analytics for informed decision-making, the role of advanced materials and sustainable practices in reducing environmental impact, and the potential of digital twin technology to optimize construction processes. We have also examined the challenges of adopting Industry 4.0 practices, such as addressing technological and cultural barriers and navigating ethical considerations in data-driven construction.

7.11.2 Reflections on the Transformative Potential of Industry 4.0 in Construction

The transformative potential of Industry 4.0 in construction cannot be overstated. By embracing digital technologies, automation, and innovation, the construction industry can revolutionize how buildings are designed, constructed, and operated. Industry 4.0 enables stakeholders to streamline project workflows, improve project outcomes, and deliver sustainable, resilient buildings that meet the needs of the future.

By integrating digital twin technology, construction stakeholders can visualize, simulate, and optimize construction projects in a virtual environment, improving project coordination, reducing risk, and enhancing efficiency. Robots and automation offer opportunities to automate repetitive tasks, improve worker safety, and accelerate project delivery. At the same time, AR and VR provide immersive, interactive experiences that enhance design visualization and workforce training.

7.11.3 Call to Action for Embracing Innovation and Sustainability in Construction Practices

As we look to the future, the construction industry must embrace innovation and sustainability in its practices. Stakeholders must collaborate, innovate, and invest in technology infrastructure to harness Industry 4.0's full potential and address the industry's pressing challenges, from climate change and resource depletion to labor shortages and productivity gaps.

A call to action is issued for construction stakeholders to prioritize innovation, sustainability, and ethical considerations in their practices. This includes investing in training and education programs to upskill the workforce, adopting sustainable construction practices and materials, and embracing digital technologies to improve project outcomes and deliver environmentally friendly, resilient, and future-proof buildings.

In conclusion, the future landscape of construction is bright and full of potential, thanks to the transformative power of Industry 4.0 technologies. By embracing innovation, sustainability, and collaboration, the construction industry can build a better future for all, one sustainable, efficient, and resilient building at a time.

REFERENCES

[1] Abounia Omran, B., *Application of data mining and big data analytics in the construction industry*. 2016: The Ohio State University.

[2] AbouRizk, S., Simulation-based analytics: advancing decision support in construction. In *Proceedings of the 2nd European and Mediterranean structural engineering and construction conference*. 2018: Curran Associates, Inc.

[3] Amer, A., *Automated segmentation and reconstruction of structural elements for indoor multi-level room environment*. 2020: Concordia University.

[4] Arayici, Y., et al., Technology adoption in the BIM implementation for lean architectural practice. *Automation in Construction*, 2011. 20(2): p. 189–195.

[5] Azhar, S., M. Khalfan, and T. Maqsood, Building information modeling (BIM): now and beyond. *Australasian Journal of Construction Economics and Building*, 2012. 12(4): p. 15–28.

[6] Barlish, K. and K. Sullivan, How to measure the benefits of BIM—a case study approach. *Automation in Construction*, 2012. 24: p. 149–159.

[7] Bhattacharyya, A., et al., Purdue index for construction analytics: prediction and forecasting model development. *Journal of Management in Engineering*, 2021. 37(5): p. 04021052.

[8] Bilal, M., et al., Big data in the construction industry: a review of present status, opportunities, and future trends. *Advanced Engineering Informatics*, 2016. 30(3): p. 500–521.
[9] Boles, D.B., *Cognitive evolution: from single cells to the human mind.* 2022: Routledge.
[10] Cao, Y., S.N. Kamaruzzaman, and N.M. Aziz, Green building construction: a systematic review of BIM utilization. *Buildings*, 2022. 12(8): p. 1205.
[11] Casini, M., *Construction 4.0: advanced technology, tools and materials for the digital transformation of the construction industry.* 2021: Woodhead Publishing.
[12] Chataut, R., A. Phoummalayvane, and R. Akl, Unleashing the power of IoT: a comprehensive review of IoT applications and prospects in healthcare, agriculture, smart homes, smart cities, and Industry 4.0. *Sensors*, 2023. 23(16): p. 7194.
[13] Chen, J., Z. Kira, and Y.K. Cho, Deep learning approach to point cloud scene understanding for automated scan to 3D reconstruction. *Journal of Computing in Civil Engineering*, 2019. 33(4): p. 04019027.
[14] Chen, X., et al., Implementation of technologies in the construction industry: a systematic review. *Engineering, Construction and Architectural Management*, 2022. 29(8): p. 3181–3209.
[15] Durdyev, S., et al., Strategies for implementation of green roofs in developing countries. *Engineering, Construction and Architectural Management*, 2023. 30(6): p. 2481–2502.
[16] Gu, N. and K. London, Understanding and facilitating BIM adoption in the AEC industry. *Automation in Construction*, 2010. 19(8): p. 988–999.
[17] Jin, R., et al., Scientometric analysis of BIM-based research in construction engineering and management. *Engineering, Construction and Architectural Management*, 2019. 26(8): p. 1750–1776.
[18] Koc, K. and A.P. Gurgun, Stakeholder-associated life cycle risks in the construction supply chain. *Journal of Management in Engineering*, 2021. 37(1): p. 04020107.
[19] Lu, W., et al., *Big data in construction waste management: prospects and challenges.* Detritus: 2018.
[20] Marinagi, C., et al., The impact of Industry 4.0 technologies on key performance indicators for a resilient supply chain 4.0. *Sustainability*, 2023. 15(6): p. 5185.
[21] Naderpajouh, N., J. Choi, and M. Hastak, Exploratory framework for application of analytics in the construction industry. *Journal of Management in Engineering*, 2016. 32(2): p. 04015047.
[22] Ngo, J., B.-G. Hwang, and C. Zhang, Factor-based big data and predictive analytics capability assessment tool for the construction industry. *Automation in Construction*, 2020. 110: p. 103042.
[23] Osunsanmi, T.O., et al., Construction supply chain management practice in developed countries. In *Construction supply chain management in the fourth industrial revolution era.* 2022: Emerald Publishing Limited. p. 77–126.
[24] Pan, Y. and L. Zhang, Integrating BIM and AI for smart construction management: current status and future directions. *Archives of Computational Methods in Engineering*, 2023. 30(2): p. 1081–1110.
[25] Rasheed, A., O. San, and T. Kvamsdal, Digital twin: values, challenges and enablers. *arXiv* preprint arXiv:1910.01719, 2019.
[26] Rasheed, A., O. San, and T. Kvamsdal, Digital twin: values, challenges and enablers from a modeling perspective. *IEEE Access*, 2020. 8: p. 21980–22012.
[27] Sacks, R., et al., Interaction of lean and building information modeling in construction. *Journal of Construction Engineering and Management*, 2010. 136(9): p. 968–980.
[28] Sawhney, A., M. Riley, and J. Irizarry (Eds.). *Construction 4.0: An Innovation Platform for the Built Environment.* 2020: Routledge.

[29] Stoyanova, M., An approach to big data analytics in construction industry. *Economics and Computer Science*, 2022(2): p. 6–18.
[30] Sun, Y., C. Liu, and C. Zhang, Mobile technology and studies on transport behavior: a literature analysis, integrated research model, and future research agenda. *Mobile Information Systems*, 2021. 2021: p. 1–24.

8 Leveraging Advanced Analytics for Transforming Logistics
The Road to Logistics 4.0

Ali Soofastaei

8.1 INTRODUCTION

8.1.1 Industry 4.0: The Dawn of a New Era in Manufacturing

The term "Industry 4.0" was first introduced in 2011 as part of a German government initiative to promote the digitization of manufacturing processes. Referred to as the "Fourth Industrial Revolution," Industry 4.0 follows three preceding industrial revolutions, each marked by significant technological advancements that transformed manufacturing and production.

The First Industrial Revolution, which began in the late 18th century, harnessed steam power to enhance the productivity of the iron and textile industries. Just before World War I, the Second Industrial Revolution leveraged electric power to enable mass production, significantly reducing manufacturing costs. The Third Industrial Revolution emerged in the 1980s with personal computers and the internet, dramatically altering the economic landscape [1].

Revived in 2011 by the German economic development agency, Industry 4.0 builds upon these historical transformations by integrating advanced technologies such as additive manufacturing, advanced robotics, artificial intelligence (AI), autonomous vehicles, blockchain, drones, and the Internet of Things (IoT). Unlike its predecessors, the Fourth Industrial Revolution, as Schwab [2] argued, is fundamentally different due to its focus on connectivity and communication among billions of devices. These technologies, combined with vast amounts of real-time data, are set to revolutionize manufacturing and service operations along global supply chains, altering the dynamics between humans (consumers and supply chain partners) and machines.

Many companies are exploring ways to capitalize on these Industry 4.0 technologies to create value. According to the McKinsey Global Institute, optimizing operations and equipment in factory settings alone could generate up to $3.7 trillion in value by 2025 [2]. More recently, Frank A. examined various economic and technological drivers that compel companies to adopt Industry 4.0 technologies [3]. Their survey of 92 manufacturing companies revealed that intelligent manufacturing is a critical driver in this technological shift.

DOI: 10.1201/9781003186823-8

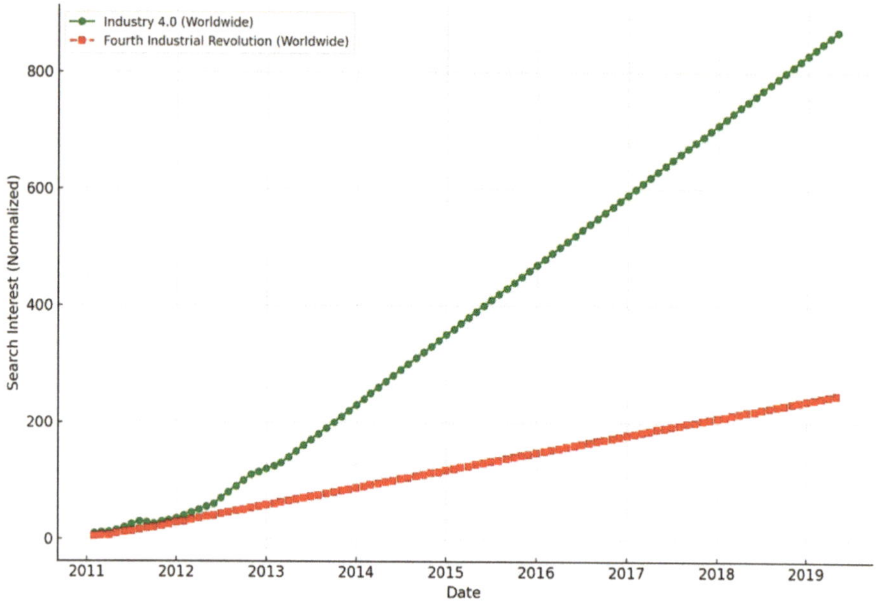

FIGURE 8.1 Number of Google Searches for Industry 4.0 and Fourth Industrial Revolution (2011–2019).

Public interest in Industry 4.0 has surged in recent years. Data from Google Trends indicate that global searches for "Industry 4.0" and "Fourth Industrial Revolution" began in 2012 and 2015, respectively, with a notable increase in searches since 2016 (see Figure 8.1). Common queries include "What is Industry 4.0," "IoT," and "What is the Fourth Industrial Revolution," highlighting the need for a deeper understanding of these concepts. Consulting firms like IBM, McKinsey, and Deloitte have published numerous reports explaining Industry 4.0 and its implications [4–7].

A search through the Scopus citation database for academic articles featuring "Industry 4.0" or "Fourth Industrial Revolution" in their titles, abstracts, or keywords between 2012 and 2018 revealed 2,320 publications containing "Industry 4.0" and 329 articles mentioning the "Fourth Industrial Revolution" (see Figure 8.1) Most of these articles appeared in engineering (25%), computer science (11%), and business (8%) journals. Similar trends were observed for articles containing the "Fourth Industrial Revolution."

In conclusion, Figures 8.1 and 8.2 illustrate that Industry 4.0 is an emerging topic with nascent research. This presents a golden opportunity for the operations management (OM) research community to delve into the implications of this new industrial revolution, identify novel research questions, and explore the conditions under which these emerging technologies can generate economic, environmental, and social value.

8.1.2 Overview of Technologies Associated with Industry 4.0

This section overviews the various technologies associated with Industry 4.0. For an in-depth discussion on three-dimensional (3D) printing, refer to [8, 9]. For a

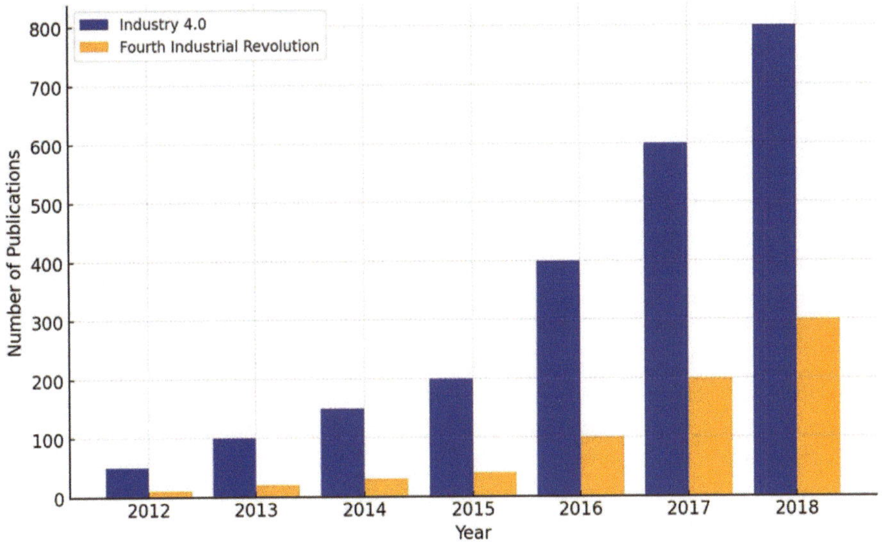

FIGURE 8.2 Number of publications on Industry 4.0 and Fourth Industrial Revolution (2012–2018).

comprehensive analysis of Blockchain's strengths and weaknesses, see [10, 11]. For further insights on other technologies, consult [10, 11].

8.1.2.1 Additive Manufacturing (3D Printing)

This technology involves creating a physical object from a digital 3D model by depositing successive layers of plastic, resin, stainless steel, and ceramics. Additive manufacturing enables faster, cheaper, and more efficient development of prototypes and personalized products, including hearing aids, knee replacements, and toys [8, 12, 13]. Research by Song J and Zhang Y [5] indicates that additive manufacturing adds more value as the number of parts increases. Westerweel and his colleagues also found that reduced logistical costs and shorter lead times can offset higher design costs [14]. Strategically, Dong demonstrated that adopting additive manufacturing allows firms to offer greater product variety to stay competitive [15]. Hu and Sun studied the balance between producing and selling self-replicating 3D printers inspired by RepRap [16]. For more detailed discussions on 3D printing services, refer to [17]. Recent studies by Hedenstierna explore the role of 3D printing in outsourcing contexts [18].

8.1.2.2 Advanced Robotics

Advances in communication technologies, sensor capabilities, and AI are making robots more intelligent and safer to work alongside humans. For instance, BMW integrates robots and human workers on its assembly line in South Carolina. Wearable robotics, such as exoskeletons, help reduce repetitive motion injuries in warehouses and agricultural fields. Companies like Caterpillar and GE are exploring wearable robotics to enhance worker safety. JD, China's largest online retailer, has implemented advanced robots in 500 warehouses for stacking products and packing merchandise [19]. By 2019, JD launched the world's first automated warehouse in Shanghai, featuring Mujin robots [20].

8.1.2.3 Drones

Uncrewed aerial vehicles, or drones, can be remotely controlled and equipped with various sensors to record visual and audio data for monitoring and surveying operations. They can also carry robotic arms for pick-and-drop operations or automated sprays for agriculture. Drones are an integral part of the IoT. Drones are being used in search and rescue missions.

8.1.2.4 Internet of Things (IoT)

IoT refers to a network of devices (e.g., sensors) communicating and interacting over the internet, allowing remote monitoring and control. IoT adoption has surged with significant reductions in sensor costs, increased processing speeds, and advances in measurement and communication technologies. According to Columbus's study, industries such as discrete manufacturing, transportation, logistics, and utilities will spend $40 billion each on IoT platforms, systems, and services by 2020 [21].

8.1.2.5 Blockchain

Blockchain is a secure and distributed ledger system. As described by Olsen and Tomlin, it is distributed because it can be accessed and written by multiple authorized entities, with data stored on a peer-to-peer network [10, 11]. It is secure because once a "block" is added to the chain, it cannot be altered unilaterally. Despite its potential, Babich and Hilary highlight several weaknesses, including privacy concerns, garbage-in-garbage-out issues, and inefficiencies [22, 23]. Wang conducted interviews with 14 supply chain experts to explore how blockchain technology could transform future supply chain operations [24].

8.1.2.6 Artificial Intelligence (AI)

Unlike natural intelligence, AI uses computers to interpret external data, learn from it, and perform descriptive, predictive, or prescriptive analyses. IBM Watson, for example, can answer questions posed in natural language. GE uses sensors to collect data from gas turbines and windmills via its Predix Cloud platform and employs machine learning and deep learning algorithms for preventive maintenance.

As these technologies continue to develop, many companies are exploring ways to exploit the exciting possibilities of Industry 4.0. Simultaneously, OM researchers are defining new research agendas to investigate these emerging technologies further. Figures 8.1 and 8.2 illustrate the growing interest and research activity in Industry 4.0, underscoring the transformative potential of these advancements.

8.1.3 Logistics: A Diamond in the Rough

Within the framework of Industry 4.0, OM research has predominantly focused on the manufacturing applications of these transformative technologies [10, 11]. This trend is understandable, given that the term "Industry 4.0" was initially introduced by the German government in 2011 to promote the digitalization of manufacturing processes. The emphasis on manufacturing is evident from a recent survey conducted by Deloitte, which involved 1,600 C-level executives across 19 countries. According to this survey, 73% of respondents are developing Industry 4.0 technology initiatives to enhance operations, primarily in manufacturing. However, only 6% of these initiatives focus on logistics [25]. This disparity suggests that many firms undervalue the strategic role of logistics as a competitive lever or a business model.

Company executives' underestimation of logistics' strategic potential has motivated this examination of the logistics function in the Industry 4.0 era. Specifically, this chapter argues that companies can leverage Industry 4.0 technologies to create economic, environmental, and social value by transforming logistics into:

1. A competitive lever;
2. A creator of social value; and
3. An enabler of sustainability.

Logistics originated from military operations and is defined more broadly here to include the transportation of humans and goods. Logistics encompasses the detailed coordination of complex operations involving humans, materials, equipment, information, and finance. This coordination often involves moving materials, humans, and equipment, exchanging information among humans and devices, and financial transactions among entities.

This chapter presents several real-world examples to illustrate how various Industry 4.0 technologies can simultaneously enable firms, governments, or NGOs to achieve economic success and social good. In addition, we propose research questions for OM researchers to explore further.

By harnessing the potential of advanced technologies such as the IoT, AI, blockchain, and advanced robotics, the logistics function can be transformed to deliver unprecedented value. For instance, IoT can enhance supply chain visibility, AI can optimize route planning, blockchain can secure supply chain transactions, and advanced robotics can automate warehousing operations. These technologies improve efficiency, reduce costs, and contribute to sustainability by minimizing waste and emissions.

In conclusion, logistics' strategic role in the Industry 4.0 era is profound. As companies recognize and exploit this potential, logistics will evolve from a supporting function to a pivotal component of competitive strategy and sustainability initiatives. The examples and research questions presented in this chapter aim to inspire further exploration and innovation in this critical area of OM.

Figures 8.1 and 8.2 provide visual evidence of the growing interest and research activity in Industry 4.0, underscoring the importance of continued investigation into the logistics function's role within this transformative industrial landscape.

8.2 LOGISTICS SERVICE AS A COMPETITIVE LEVER

Logistics is an essential function that ensures the right product reaches the right customer at the right time. However, many executives still perceive logistics merely as a cost to be managed, often overlooking its potential to make or break a company. Consider the downfall of Blockbuster in 2010. Once the world's largest video rental company in 2004, with over 9,000 stores worldwide, Blockbuster allowed customers to rent videos for a fixed fee. Still, they must return the tapes or DVDs to the same store within two days to avoid penalties. Customers initially tolerated this model as a near-monopoly despite the inconvenience and penalties. However, when Netflix emerged in the late 1990s, offering customers the convenience of returning DVDs via prepaid envelopes and receiving the next DVD in their queue by mail, Blockbuster quickly lost its customer base and filed for bankruptcy in 2010. This example underscores the critical importance of logistics from a customer perspective.

Another illustrative case involves the failures of online grocery store Webvan and furniture store Furniture.com, partly due to poor logistics performance. Customers frequently complained about late deliveries and missing items, while the companies struggled with high "last mile" delivery costs. These examples suggest that successful competition in the retail sector hinges on robust logistics.

The strategic importance of logistics is evident in the significant investments by major online retailers such as Amazon and Alibaba's Tmall. To differentiate themselves in the e-tailing sector, these companies emphasize fast, reliable, and often free delivery services. By leveraging its Whole Foods stores, Amazon offers two-hour home grocery delivery services (Amazon Prime Now) to its Prime members in select U.S. locations. In addition, Amazon launched "Amazon Logistics" in 2018 to improve control over last-mile delivery performance, encompassing speed, reliability, and cost. In 2018, Amazon's fulfillment and shipping expenses reached $34.0 billion and $27.7 billion, respectively, a substantial increase from just over $1 billion each in 2007. The total logistics cost of $61.7 billion accounted for 27.5% of its net revenue [26]. Buchman reported that Amazon operates 258 distribution and fulfillment centers in the United States and an additional 486 centers globally, using thousands of trucks and 32 Boeing 767 aircraft [27].

In China, Alibaba has similarly prioritized logistics to enhance its competitive edge. In 2017, Alibaba acquired a majority stake in its logistics subsidiary Cainiao and committed to investing $15 billion over five years to establish a global logistics network. This focus on logistics demonstrates that it should not be viewed merely as a cost center. Instead, logistics services are a strategic weapon that enables firms to compete on speed, reliability, and cost.

The advent of new technologies, such as the IoT, AI, blockchain, and advanced robotics, is fundamentally transforming the logistics function. These technologies can enhance supply chain visibility, optimize route planning, secure transactions, automate warehousing operations, improve efficiency, and reduce costs. For instance, IoT devices can track shipments in real time, AI algorithms can predict demand and optimize inventory, blockchain can ensure the authenticity and security of transactions, and robotics can streamline sorting and packing processes.

Logistics is a critical competitive lever that can significantly influence a company's success. As firms continue to invest in advanced technologies to enhance their logistics capabilities, they will be better positioned to meet customer expectations and sustain a competitive advantage in the market. Figures 8.1 and 8.2 highlight the growing importance of logistics and the transformative impact of Industry 4.0 technologies on this vital function.

8.3 TRANSFORMING LOGISTICS WITH INDUSTRY 4.0 TECHNOLOGIES

The advent of Industry 4.0 technologies is revolutionizing the logistics sector, transforming it from a mere cost center into a strategic competitive lever. Here are some ways these technologies are enhancing logistics:

1. **Faster Speed:** This involves delivery services utilizing drones and robots. Amazon is exploring the usage of drones to deliver small packages and enhance speed and efficiency. In China, Alibaba's food delivery unit, Ele.me began using drones in 2018 to deliver food across 17 routes from over 100 restaurants in Shanghai's Jinshan Industrial Park, covering 58 square kilometers [28]. In 2019, Google approved using its Wing drones for home deliveries in Australia. Researchers are developing operational models to route these drones effectively to different customers [29, 30]. Similar to those tested by Domino's Pizza, ground-based delivery robots also contribute to faster deliveries. In 2014, Amazon filed a patent for "anticipatory shipping," which uses predictive analytics to analyze a customer's shopping history to predict future needs. This system could enable Amazon to ship products (potentially by drones) even before the customer places an order [31].
2. **Higher Reliability:** The storage and retrieval systems using robots is one such example. In 2012, Amazon acquired the Kiva robotic system for $775 million to automate storage and retrieval operations. The Kiva system enhances productivity by tracking items and bringing products directly to employees for picking, packing, and shipping. This system's design and operational strategies have been extensively analyzed [32]. In addition, developing robotic exoskeletons can improve the speed and reliability of employees' pick-and-pack operations and reduce repetitive motion injuries [33].
3. **Lower Operating Costs:** Intelligent sensors are used for Inventory monitoring and replenishment. Traditional physical store inventory management is often inaccurate and costly [34]. Intelligent sensors and "smart shelves" can monitor inventory levels in real time, notify staff when restocking is needed, and alert warehouses or vendors for replenishment. For instance, a start-up company Wasteless Co uses electronic tags to implement dynamic pricing, reducing food waste by offering discounts based on expiration dates. Pilot runs in Italy showed an 89% reduction in food waste [35]. AWM Smart Shelf uses cameras to gather data on shopper behavior and demographics, which AI can analyze to develop personalized videos displayed on shelves. Smart fridges with cameras, like Samsung's, also provide owners with detailed inventory and expiration date information. Real-time inventory data from various warehouses enables online retailers to make dynamic pricing decisions and optimize order fulfillment.

4. **Improved Efficiency:** One such area of increased efficiency is container shipping enabled by blockchain. Ocean freight operations involve numerous entities and extensive paperwork, leading to long delays [36]. In early 2019, when container ship MSC Zoe lost containers off the Dutch and German coasts, it took weeks to determine the exact number of lost containers. To improve efficiency in the $200-billion ocean freight industry, IBM and Maersk developed a blockchain platform in 2017 to automate and digitize shipping documents, allowing for real-time tracking and reducing duplication errors and delays [37]. In April 2019, China Shipbuilding Industry Company Limited signed an agreement with Shanghai Bank to explore blockchain technology for financing upstream suppliers. Beyond ocean freight, Choi examined blockchain's potential benefits in air logistics for risk analysis using the mean–variance theory [38].

As these Industry 4.0 technologies transform logistics into competitive assets, several research questions arise:

1. How should a firm redesign its supply chain structure to align with transformed logistics services in the Industry 4.0 era?
2. How will these emerging technologies affect supply chain communication, coordination, and collaboration? For example, should retailers share smart shelf data with vendors for anticipatory replenishment services?
3. How will 3D printing impact the logistics industry? Could customers download digital files and print products at home, bypassing the physical supply chain?
4. Will the economic value created by blockchain outweigh its implementation costs? Will the efficiency gains in container shipping justify the investment?
5. How does advanced robotics impact job design for human workers? Will robotics complement or substitute human labor?
6. How will drones and robots take over home deliveries? How can unmanned delivery services be organized and regulated to avoid chaos?

Figures 8.1 and 8.2 illustrate the growing impact and potential of these technologies in transforming logistics and enhancing competitive advantage.

8.4 LOGISTICS AS A CREATOR OF SOCIAL VALUE

Beyond transforming logistics into a competitive asset, emerging technologies can significantly enhance logistics operations by generating social value. Consider the challenges faced by the World Food Programme (WFP) when distributing food and cash to refugees in war-torn countries or disaster zones. Aside from the difficulty of securing supplies, the distribution process is fraught with obstacles due to often non-existent infrastructure and incomplete or inaccurate records of legitimate recipients. This leads to widespread fraud and many deserving individuals not receiving aid. In 2017, WFP collaborated with Parity Technologies, led by Ethereum co-founder Gavin Wood, to develop the World Food Programme Building Blocks blockchain platform. This platform uses a distributed ledger to record iris scans of refugees, streamlining and securing the authentication process. As a result, refugees in Jordan can efficiently and accurately receive their rations and cash, improving distribution accuracy and reducing fraud [39].

8.4.1 Faster and Safer Response and Recovery Operations Using Drones

Following natural disasters like hurricanes or forest fires, traditional helicopter response methods can be costly and dangerous, particularly in challenging terrains or at night. For example, after Hurricane Harvey in 2017, the FAA authorized 43 small drones to assess damage to homes, roads, bridges, and power lines. Allstate Insurance deployed drones to capture visual images for claims assessment [40]. Drones with heat-sensitive infrared cameras can also enhance search and rescue missions by quickly identifying human outlines and directing rescue teams to precise locations. In smaller-scale emergencies, drones equipped with automated external defibrillators (AEDs) are used in the US and the Netherlands to provide rapid response to heart attack victims.

8.4.2 Improved Accessibility to Diagnostic Care and Drug Administration via Wearable Medical Devices

In rural areas of developing countries, access to diagnostic care and pharmacy services is limited due to the scarcity or perceived low quality of rural clinics. Wearable medical devices, which collect real-time physiological data and sync with smart devices, enable patients to share medical information with online doctors via telemedicine platforms. Devices like the Apple Watch Series 2 and the Dexcom G6 continuous glucose monitor allow Chinese patients to access affordable telemedicine services [24]. In addition, Covestro launched Makroblend M 525 in 2015, a wearable device that monitors vital signs and administers drugs. Enable Injections, backed by Sanofi, is developing a wearable IV drug delivery system that connects to smartphones via Bluetooth [21].

8.4.3 Enhanced Farmer Productivity Through Drones and Smart Sensors

Improving farmer productivity becomes crucial as the global population grows and arable land decreases. This need has led to innovations in Agriculture 4.0, including drones with sensors and cameras that monitor crops for diseases, assess yields, and identify fertilizer needs. French startup Delair-Tech offers a subscription-based service charging €15 per hectare for precision farming, using drones to enhance agricultural efficiency.

8.4.4 Improved Provenance Using Blockchain

Companies face pressure to disclose supplier information and material provenance in response to food and drug adulteration incidents. Blockchain technology enables firms to track product origins, enhancing transparency and safety. Walmart and other firms, including Nestle, Dole, Tyson Foods, and Unilever, partnered with IBM to use blockchain for food traceability. In 2019, luxury brand conglomerate LVMH launched a blockchain platform to authenticate its products [41].

8.4.5 Improved Mobility via Smart Transportation

Improving social inclusion for the poor, sick, and elderly requires accessible and affordable mobility. Mobility as a service (MaaS) initiatives, such as Helsinki's Whim app, integrate public and private transportation options into a single service, offering travel suggestions based on user preferences and enabling mobile payments [42]. Autonomous buses and smart railways, like those in the Netherlands, use sensors to monitor railcar capacity and notify passengers via mobile apps, further enhancing mobility and efficiency.

8.4.6 Autonomous Vehicles and Smart Public Transportation

While fully autonomous cars face safety challenges, autonomous buses on dedicated lanes and intelligent transportation systems like those tested by NEXT Future Transportation Inc. in Dubai are already in use. The Netherlands Railways uses intelligent sensors to predict railcar capacity and notify passengers, reducing delays and improving service quality. Industry 4.0 technologies facilitate the development of smart cities, integrating intelligent buildings, grids, mobility, and retail to create new social values [43].

8.4.7 Research Questions

The examples above illustrate how Industry 4.0 technologies create social value. However, several research questions remain:

1. How should governments or insurance companies leverage data from wearable medical devices to develop incentives for health improvement?
2. How can service providers monetize precision farming using drones, considering farmers' uncertainty about its value? Should pricing be subscription-based or involve risk-sharing?
3. How can privacy concerns be addressed when firms use blockchain to track supply chain operations? Who should audit the records, and how often?
4. What pricing mechanisms should be used to coordinate supply and demand in MaaS? How should profits be shared among transportation operators?

Figures 8.1 and 8.2 highlight the transformative impact of these technologies on logistics and their potential to create significant social value.

8.5 LOGISTICS AS A SOCIAL VALUE CREATOR

In addition to transforming logistics into a competitive advantage, emerging technologies can significantly enhance logistics operations by creating social value.

For instance, the World Food Programme (WFP) faces immense challenges when distributing food and cash to refugees in conflict zones or disaster-stricken areas. Beyond the logistical hurdles of procuring supplies, distribution is further complicated by the lack of infrastructure and incomplete or inaccurate records of legitimate recipients. This often leads to fraudulent claims, leaving many rightful recipients without aid. To address these issues, the WFP partnered with Parity Technologies in 2017 to develop the "World Food Programme Building Blocks" blockchain platform, which uses a distributed ledger to record the iris images of refugees. This allows for efficient and accurate authentication through a simple iris scan, ensuring legitimate refugees in Jordan receive their rations and cash, particularly benefiting women participating in the UN Women's Cash for Work Programme [39].

8.5.1 Faster and Safer Response and Recovery Operations Using Drones

Traditional helicopter response methods are costly and dangerous after natural disasters like hurricanes or forest fires, especially in challenging terrains or at night. For example, following Hurricane Harvey in 2017, the FAA authorized the use of 43 small drones to assess damage to infrastructure. Allstate Insurance also deployed drones to capture visual images for claims assessment [40]. Drones equipped with heat-sensitive infrared cameras can efficiently conduct search and rescue missions along rivers and coastlines by quickly identifying human outlines, enabling rescue teams to be dispatched directly to precise locations. In more minor emergencies, drones equipped with AEDs have been deployed in cities in the US and the Netherlands to respond to sudden heart attacks, potentially saving lives.

8.5.2 Improved Accessibility to Diagnostic Care and Drug Administration via Wearable Medical Devices

In rural areas of developing countries, access to diagnostic care and pharmacy services is often limited due to a lack of clinics or perceived low-quality care. Wearable medical devices, which collect real-time physiological data and sync with smart devices, enable patients to share medical information with online doctors via telemedicine platforms. Devices like the Apple Watch Series 2 and the Dexcom G6 continuous glucose monitor facilitate affordable telemedicine services for many Chinese patients [44]. In addition, Covestro launched the Makroblend M 525 in 2015, a wearable device that monitors vital signs and administers drugs. Enable Injections, backed by Sanofi, is developing a wearable IV drug delivery system that connects to smartphones via Bluetooth [21].

8.5.3 Enhanced Farmer Productivity Through Drones and Smart Sensors

Improving farmer productivity is crucial as the global population grows and arable land decreases. This need has driven innovations in Agriculture 4.0, including drones with sensors and cameras that monitor crops for diseases, assess yields, and identify fertilizer needs. French startup Delair-Tech offers a subscription-based service charging €15 per hectare for precision farming, using drones to enhance agricultural efficiency.

8.5.4 Improved Provenance Using Blockchain

Companies face pressure to disclose supplier information and material provenance in response to food and drug adulteration incidents. Blockchain technology enables firms to track product origins, enhancing transparency and safety. Walmart and other firms, including Nestle, Dole, Tyson Foods, and Unilever, partnered with IBM to use blockchain for food traceability. In 2019, luxury brand conglomerate LVMH launched a blockchain platform to authenticate its products [41].

8.5.5 Improved Mobility via Smart Transportation

Improving social inclusion for the poor, sick, and older adults requires accessible and affordable mobility. MaaS initiatives, such as Helsinki's Whim app, integrate public and private transportation options into a single service, offering travel suggestions based on user preferences and enabling mobile payments [42]. Autonomous buses and smart railways, like those in the Netherlands, use sensors to monitor railcar capacity and notify passengers via mobile apps, further enhancing mobility and efficiency.

8.5.6 Autonomous Vehicles and Smart Public Transportation

While fully autonomous cars face safety challenges, autonomous buses on dedicated lanes and intelligent transportation systems like those tested by NEXT Future Transportation Inc. in Dubai are already in use. The Netherlands Railways uses intelligent sensors to predict railcar capacity and notify passengers, reducing delays and improving service quality. Industry 4.0 technologies facilitate the development of intelligent cities, integrating intelligent buildings, grids, mobility, and retail to create new social values [43].

8.5.7 Research Questions

The examples above illustrate how Industry 4.0 technologies create social value. However, several research questions remain:

1. How should governments or insurance companies leverage data from wearable medical devices to develop incentives for health improvement?
2. How can service providers monetize precision farming using drones, considering farmers' uncertainty about its value? Should pricing be subscription-based or involve risk-sharing?
3. How can privacy concerns be addressed when firms use blockchain to track supply chain operations? Who should audit the records, and how often?
4. What pricing mechanisms should be used to coordinate supply and demand in MaaS? How should profits be shared among transportation operators?

Figures 8.1 and 8.2 highlight the transformative impact of these technologies on logistics and their potential to create significant social value.

8.6 LOGISTICS AS A SUSTAINABILITY ENABLER

In addition to economic and social benefits, there is an increasing emphasis on sustainability, with calls for renewable energy, reducing carbon footprints, protecting endangered species, and maintaining ecosystem balance. This concern is justified given that the growing and aging global population (expected to rise from 7 billion in 2017 to over 9 billion by 2050) far exceeds the shrinking supply of natural resources (arable land, water, oil, gas, etc.). This imbalance has heightened public awareness of environmental issues (climate change, deforestation, pollution) and social issues (poverty, hunger, inequality, gender equity, population growth) over the last two decades. Such concerns are also reflected in the Millennium Development Goals (United Nations, 2000). For example, researchers from Rutgers University analyzed global fisheries data and ocean temperature maps from 1930 to 2010, finding that global warming has reduced fish stocks by 4% [21]. Moreover, destructive fishing practices like trawling in West Africa further harm the ecosystem. Without immediate intervention and stringent law enforcement, the planet's ecosystem risks collapsing alarmingly. Consequently, firms and governments increasingly explore leveraging emerging technologies to address sustainability challenges.

8.6.1 Protecting Endangered Species with Drones and AI

Illegal trade in endangered species, valued at $7 to $10 billion annually, poses a significant threat to species such as elephants, rhinoceroses, and tigers. From 2007 to 2017, over 7,245 African rhinos were poached. Organizations such as the World Wildlife Fund (WWF) and Save the Rhino have raised funds to recruit more rangers, but this approach is costly and inefficient. Scientists are now employing innovative methods to protect these animals. For instance, IBM scientists discovered that zebras, which often move alongside rhinos, behave differently when encountering poachers than predators. By fitting zebras with radio collars, scientists can monitor their movements and alert rangers to potential poaching activities. Machine learning and deep learning technologies also analyze poaching patterns, allowing rangers and drones to patrol high-risk areas more effectively. In 2012, Google contributed $5 million to support WWF's conservation drone program in Africa and Asia [21].

8.6.2 Reducing Water Waste in Farming with Drones and Data Analytics

In Chile, researchers are developing drones equipped with spectral sensors and cameras to help farmers monitor crop conditions. Farmers can assess moisture content, groundwater levels, plant health, pest infestations, and growth rates by analyzing sensor data. This enables targeted water and pest control, conserving resources and improving yields [21]. Similarly, Spanish startup BrioAgro uses underground sensors to monitor moisture, light, and soil nutrients. The data is analyzed to provide farmers with precise water and fertilizer recommendations via mobile phones, promoting efficient water use through precision farming.

8.6.3 Reducing Emissions with Smart Transportation

Ride-sharing platforms like Uber, Didi, and BlaBlaCar use real-time tracking to facilitate pooled rides, reducing emissions. Autonomous vehicles are expected to increase shared rides further in the future. A Swiss study found that 61% of respondents preferred pooled shared autonomous vehicles over private ones [45]. In addition, crowd-shipping platforms assign parcels to registered drivers for delivery, minimizing traditional delivery van usage and emissions. Platforms facilitating goods transfer between crowd shippers can reduce travel distances [46–48]. Trunk delivery services, where logistics providers like DHL access customers' car trunks for delivery, also reduce emissions by eliminating the need for multiple delivery attempts [24].

8.6.4 Sustaining High-Quality Agricultural Products with Blockchain

Many desirable agricultural products, such as coffee and cocoa beans, are produced by smallholder farmers in developing countries. The complex and opaque supply chain operations and fluctuating market prices often result in low and unstable wages for these farmers. Programs like Starbucks' C.A.F.E. initiative help farmers adopt sustainable practices and combat threats like coffee leaf rust. Denver's Coda Coffee partnered with startup bext 360 to develop blockchain-traced coffee, integrating machine vision, blockchain, cloud computing, and AI to trace every step of the coffee supply chain [49].

These examples demonstrate how emerging technologies can enhance sustainability. As firms and governments continue to apply these technologies, several research questions arise:

1. Various technologies can help reduce poaching of endangered species, but as supply decreases, some buyers may offer higher prices, attracting more poachers. How should governments develop programs to curb demand instead of supply?
2. Since 80% of freshwater withdrawals are used for agriculture, what incentive programs should governments implement to encourage farmers to adopt water conservation practices?
3. Trunk deliveries and crowd shipping offer a promising trend, but trust and safety are concerns. What mechanisms should be developed to ensure trust and safety for all involved parties?
4. While blockchain technology offers promise, potential weaknesses could undermine its benefits. What mechanisms can facilitate commitments and self-enforcement for adopting blockchain technology across different entities?

Figures 8.1 and 8.2 highlight the potential of these technologies to enhance sustainability and create significant social value.

8.7 CONCLUSION

In this chapter, we have explored the transformative potential of emerging technologies such as drones, smart sensors, robotics, blockchain, and AI to revolutionize the logistics function. These technologies can elevate logistics from a mundane operational necessity to a dynamic competitive lever, a creator of social value, and a sustainability enabler.

While the benefits of these technologies are evident, they also pose significant challenges and concerns. Many of these technologies are still unproven and may introduce new problems, including:

- Social unrest due to job losses caused by automation (e.g., autonomous vehicles, robots, and drones);
- Social inequality as these technologies often require access to smartphones and smart devices, which remain inaccessible to low-income people; and
- Income inequality, with the educated class reaping the most benefits.

Researchers must examine the impact of these technologies on the welfare of various socio-economic classes. Despite the advancements supporting Industry 4.0, companies must address several underlying risks:

8.7.1 Cyberattacks

Cybersecurity concerns grow as supply chains become more digitized, relying on real-time communication and coordination among numerous devices (sensors, robots, drones). Industry 4.0 involves many devices communicating through different operating and information systems, making digitalized supply chains vulnerable to cyberattacks, including industrial espionage, internet protocol (IP) leakage, or production sabotage. In severe cases, cyberattacks can cripple entire logistics and transportation systems.

8.7.2 Faulty Data

Smart devices facilitate smooth supply chain operations by sensing, collecting, sharing, and analyzing data. However, if hacked or malfunctioning, these systems can cause disasters. The 2019 Boeing 737 Max crashes, caused by faulty sensor readings, highlight the risks of over-reliance on smart devices. Companies must develop foolproof protocols with human interventions to mitigate such risks.

8.7.3 Safety Regulations

The use of advanced robotics, automated guided vehicles, robotic systems in warehouses, and autonomous trucks and drones necessitates the development of standard safety guidelines and regulations to ensure worker and public safety. For example, unauthorized drones disrupted operations at Heathrow and Gatwick airports in 2018, underscoring the need for stringent regulations.

8.7.4 Privacy Issues

Many sensors record visual and audio data of individuals in various settings, raising legitimate privacy concerns. For instance, intelligent shelves recording shopper demographics and habits necessitate regulatory measures to ensure proper data collection and storage.

In summary, the emerging technologies driving the Industry 4.0 movement are disruptive and can lead to new business models and significant value creation for companies and society. However, they also raise concerns regarding employment, safety, and privacy. Collaboration between the private sector, public sector, and government is crucial to establish mutually beneficial plans. Such collaboration can ensure these technologies enhance company performance while contributing positively to society.

By addressing these challenges and fostering collaboration, we can harness the full potential of Industry 4.0 technologies to create a more efficient, equitable, and sustainable future.

REFERENCES

[1] Rifkin, J., *The third industrial revolution: how lateral power transforms energy, the economy, and the world*. 2011: Macmillan.

[2] Manyika, J., et al., *Unlocking the potential of the internet of things*, Vol. 1. 2015: McKinsey Global Institute.

[3] Frank, A.G., L.S. Dalenogare, and N.F. Ayala, Industry 4.0 technologies: implementation patterns in manufacturing companies. *International Journal of Production Economics*, 2019. 210: p. 15–26.

[4] Qiu, F., et al., Demi-flexible operating policies to promote the performance of public transit in low-demand areas. *Transportation Research Part A: Policy and Practice*, 2015. 80: p. 215–230.

[5] Song, J.-S. and Y. Zhang, Stock or print? Impact of 3-D printing on spare parts logistics. *Management Science*, 2020. 66(9): p. 3860–3878.

[6] Baur, C. and D. Wee, *Manufacturing's next act*, Vol. 6. 2015: McKinsey & Company. p. 1–5.

[7] Insights, D., *The fourth industrial revolution is here—are you ready*. 2018: Deloitte Development LLC.

[8] Berman, B., 3-D printing: the new industrial revolution. *Business Horizons*, 2012. 55(2): p. 155–162.

[9] Berman, B., 3-D printing: the new industrial revolution. *IEEE Engineering Management Review*, 2013. 41(4): p. 72–80.

[10] Olsen, T.L. and B. Tomlin, Industry 4.0: opportunities and challenges for operations management. *Manufacturing & Service Operations Management*, 2020. 22(1): p. 113–122.

[11] Olsen, T. and R. Parker, *Autonomous vehicles: a research agenda for operations management*. 2019, Working Paper: University of Auckland.

[12] Sodhi, M.S. and C.S. Tang, Supply-chain research opportunities with people with low incomes as suppliers or distributors in developing countries. *Production and Operations Management*, 2014. 23(9): p. 1483–1494.

[13] Sodhi, M.S. and C.S. Tang, Supply chains built for speed and customization. *MIT Sloan Management Review*, 2017. 58.

[14] Westerweel, B., R.J. Basten, and G.-J. van Houtum, Traditional or additive manufacturing? Assessing component design options through lifecycle cost analysis. *European Journal of Operational Research*, 2018. 270(2): p. 570–585.
[15] Dong, L., D. Shi, and F. Zhang, *3D printing vs. traditional flexible technology: implications for manufacturing strategy*. Available at SSRN, 2016. 2847731.
[16] Hu, B. and Z. Sun, Managing self-replicating innovative goods. *Management Science*, 2022. 68(1): p. 399–419.
[17] Rogers, H., N. Baricz, and K.S. Pawar, 3-D printing services: classification, supply chain implications and research agenda. *International Journal of Physical Distribution & Logistics Management*, 2016. 46(10): p. 886–907.
[18] Hedenstierna, C.P.T., et al., Economies of collaboration in build-to-model operations. *Journal of Operations Management*, 2019. 65(8): p. 753–773.
[19] Marr, B., How Chinese retailer JD. com uses AI, Big data & robotics to take on Amazon. *Forbes*. August, 2018. 3.
[20] Hornyak, T., The world's first humanless warehouse is run only by robots and is a model for the future. *CNBC*, 2018.
[21] Tang, C.S. and L.P. Veelenturf, The strategic role of logistics in the Industry 4.0 era. *Transportation Research Part E: Logistics and Transportation Review*, 2019. 129: p. 1–11.
[22] Babich, V. and G. Hilary, OM Forum—distributed ledgers and operations: what operations management researchers should know about blockchain technology. *Manufacturing & Service Operations Management*, 2020. 22(2): p. 223–240.
[23] Babich, V. and C.S. Tang, Managing opportunistic supplier product adulteration: deferred payments, inspection, and combined mechanisms. *Manufacturing & Service Operations Management*, 2012. 14(2): p. 301–314.
[24] Liu, C., Q. Wang, and Y.O. Susilo, Assessing the impacts of collection-delivery points to individual's activity-travel patterns: a greener last mile alternative ? *Transportation Research Part E: Logistics and Transportation Review*, 2019. 121: p. 84–99.
[25] Insights, D., *The fourth industrial revolution is here: are you ready?* Deloitte Insights Rep. 2018: Deloitte.
[26] Richter, F., The growing weight of Amazon's logistics costs. *Statista*. March, 2019. 1.
[27] Buchman, E., The rise of Amazon logistics. *Transport Topics*, 2018.
[28] Tao, L. and Y. Yang, Your Chinese takeaway is ready to be delivered by drone. *South China Morning Post*, 2018. 30.
[29] Carlsson, J.G. and S. Song, Coordinated logistics with a truck and a drone. *Management Science*, 2018. 64(9): p. 4052–4069.
[30] Agatz, N., P. Bouman, and M. Schmidt, Optimization approaches for the traveling salesperson problem with a drone. *Transportation Science*, 2018. 52(4): p. 965–981.
[31] Kopalle, P., Why Amazon's anticipatory shipping is pure genius. *Forbes*. January, 2014. 28.
[32] Roy, D., et al., Robot-storage zone assignment strategies in mobile fulfillment systems. *Transportation Research Part E: Logistics and Transportation Review*, 2019. 122: p. 119–142.
[33] Ashley, S., Robotic exoskeletons are changing lives in surprising ways. *NBCnews.com*. February, 2017. 21.
[34] DeHoratius, N., A.J. Mersereau, and L. Schrage, Retail inventory management when records are inaccurate. *Manufacturing & Service Operations Management*, 2008. 10(2): p. 257–277.
[35] Pieters, J., Albert Heijn to combat food waste with "dynamic discounts". *NL Times*. May, 2019. 21.
[36] Lun, Y.V. and J. Hoffmann, Connectivity and trade relativity: the case of ASEAN. *Journal of Shipping and Trade*, 2016. 1: p. 1–13.

[37] Groenfeldt, T., *IBM and Maersk apply blockchain to container shipping*, 2017. Available from: https://www. forbes.com/sites/tomgroenfeldt/2017/03/05/ibm-and-maersk-apply-blockchain-to-container-shipping.
[38] Choi, T.-M., et al., The mean-variance approach for global supply chain risk analysis with air logistics in the blockchain technology era. *Transportation Research Part E: Logistics and Transportation Review*, 2019. 127: p. 178–191.
[39] Juskalian, R., *Inside the Jordan refugee camp that runs on blockchain*. 2018: Massachussets: MIT Technology Review.
[40] Marshall, A., Above devastated Houston, armies of drones prove their worth. Por encima de un Houston devastado, los ejércitos de drones demuestran su valor). *Wired. com*, 2017. 4.
[41] Allison, I., Louis Vuitton owner LVMH is launching a Blockchain to track luxury goods. *Coindesk*. March, 2019. 26.
[42] Goodall, W., et al., The rise of mobility as a service. *Deloitte Review*, 2017. 20(1): p. 112–129.
[43] Qi, W. and Z.J.M. Shen, A smart-city scope of operations management. *Production and Operations Management*, 2019. 28(2): p. 393–406.
[44] Liu, C., China embraces telemedicine. *HSBC Insights*. July, 2018. 26.
[45] Stoiber, T., et al., Will consumers prefer shared and pooled-use autonomous vehicles? A stated choice experiment with Swiss households. *Transportation Research Part D: Transport and Environment*, 2019. 71: p. 265–282.
[46] Sampaio, A., et al., The benefits of transfers in crowdsourced pickup-and-delivery systems. *Optimization Online*, 2018.
[47] Oliveira, A.S., et al., A scenario-based approach for the vehicle routing problem with roaming delivery locations under stochastic travel times. *Optimization Online*, 2019.
[48] Sampaio, A., et al., Crowd-based city logistics. In *Sustainable transportation and smart logistics*. 2019: Elsevier. p. 381–400.
[49] Philips, E., Bringing blockchain to the coffee cup. *Wall Street Journal*, 2018.

9 Revolutionizing Chemical Engineering 4.0

Artificial Intelligence Innovations and Machine Learning

Ali Soofastaei

9.1 INTRODUCTION

The current buzz surrounding artificial intelligence (AI), especially in machine learning (ML), is palpable and infectious. The notion that AI is on the brink of a revolutionary transformation of humanity has sparked both visionary optimism and cautious concern among experts. The commercial potential of AI is drawing massive investments from venture capitalists and state-sponsored initiatives worldwide, with China leading the charge. McKinsey, for example, projects that AI could generate market impacts worth trillions of dollars across various domains. This surge of interest is fueled by the rapid and remarkable advancements in AI over the past decade, exemplified by breakthroughs such as AlphaGo, autonomous vehicles, Alexa, and Watson. These advancements in game playing, robotics, computer vision, speech recognition, and natural language processing are impressive. However, much like the previous waves of AI enthusiasm—such as expert systems in the 1980s and neural networks in the 1990s—there is also a tendency to overestimate the immediate potential of these technologies, as noted by market research firm Gartner and others [1].

The excitement about AI has naturally extended to chemical engineers exploring its potential applications in areas such as catalyst design. This enthusiasm is driven by the promise that AI offers novel solutions to long-standing challenges in chemical engineering. However, it is essential to recognize that the application of AI in chemical engineering is not a new phenomenon. It has been an ongoing endeavor for over 35 years, marked by notable successes.

This chapter is intended for chemical engineers interested in the prospects of AI in their field and researchers new to this area. The objectives are threefold: first, to review the progress made so far, highlighting past efforts that offer valuable lessons for the future; second, to identify current and future opportunities for AI in chemical engineering, drawing on these lessons to provide a realistic assessment of its prospects; third, to document and preserve the early milestones of AI in

DOI: 10.1201/9781003186823-9

chemical engineering for historical purposes, given its increasing role in research and education [2].

Chemical engineering stands at a significant crossroads, undergoing an unprecedented transition that presents challenges and opportunities in modeling and automated decision-making. This transition is driven by the convergence of powerful computing and communications platforms, advancements in molecular engineering, the increasing automation of globally integrated operations, stringent environmental constraints, and business demands for faster delivery of goods and services. One critical outcome of this convergence is generating, using, and managing vast amounts of diverse data, where AI, particularly ML, plays a crucial role.

So, what is AI? The term was coined in 1956 at a mathematics conference at Dartmouth College. Over the years, AI has been defined in various ways, but a simple and visionary definition is: "AI is the study of how to make computers do things at which, at the moment, people are better." This definition implies that AI could eventually excel at all human tasks, achieving super-human performance as AlphaGo and AlphaGo Zero demonstrated. Historically, AI encompasses several branches, including [3]:

- Game playing (e.g., Chess, Go)
- Symbolic reasoning and theorem-proving (e.g., Logic Theorist, MACSYMA)
- Robotics (e.g., self-driving cars)
- Vision (e.g., facial recognition)
- Speech recognition and natural language processing (e.g., Siri)
- Distributed and evolutionary AI (e.g., drone swarms, agent-based models)
- Hardware for AI (e.g., Lisp machines)
- Expert systems or knowledge-based systems (e.g., MYCIN, CONPHYDE)
- ML (e.g., clustering, deep neural nets, Bayesian belief nets)

Some branches focus on specific applications, like game playing and vision, while others are methodological, such as expert systems and ML. These methodological branches are most directly applicable to chemical engineering and have been extensively researched over the past 35 years. While the current buzz is primarily around ML, the expert system framework offers valuable symbolic knowledge representation concepts and inference techniques that could be crucial as we develop more comprehensive solutions beyond the purely data-centric emphasis of ML [4].

Many tasks across these AI branches share standard features, such as pattern recognition, reasoning, and decision-making under complex conditions. They often deal with ill-defined problems, noisy data, model uncertainties, large search spaces, nonlinearities, and the need for quick solutions—features also found in many process systems engineering (PSE) problems in synthesis, design, control, scheduling, optimization, and risk management. Recognizing these similarities, some of us in the early 1980s began exploring these problems from an AI perspective. The excitement about AI at that time, centered on expert systems, was intense and contagious, with high expectations for AI's near-term impact. Significant investments were made in AI startups and within large companies, leading to the development of specialized hardware like Lisp machines and promising proof-of-concept systems in various domains, including chemical engineering. Despite these initial high hopes, AI did not quite meet its early promise, as optimization and model predictive control did.

So, what went wrong? Why did not AI have the anticipated impact? To address this, we need to examine the different phases of AI in chemical engineering.

9.2 PHASES OF AI IN CHEMICAL ENGINEERING

9.2.1 Phase 0: Early Attempts

While significant efforts to develop AI methods for chemical engineering problems began in the early 1980s, it is noteworthy that some pioneering researchers, such as Gary Powers, Dale Rudd, and Jeff Siirola, explored AI in PSE as early as the late 1960s and early 1970s. A landmark development from this period was the Adaptive Initial DEsign Synthesizer (AIDES) system, created by Siirola and Rudd for process synthesis. This system was arguably the first to employ AI methods such as means-ends analysis, symbolic manipulation, and linked data structures in chemical engineering (see Figure 9.1).

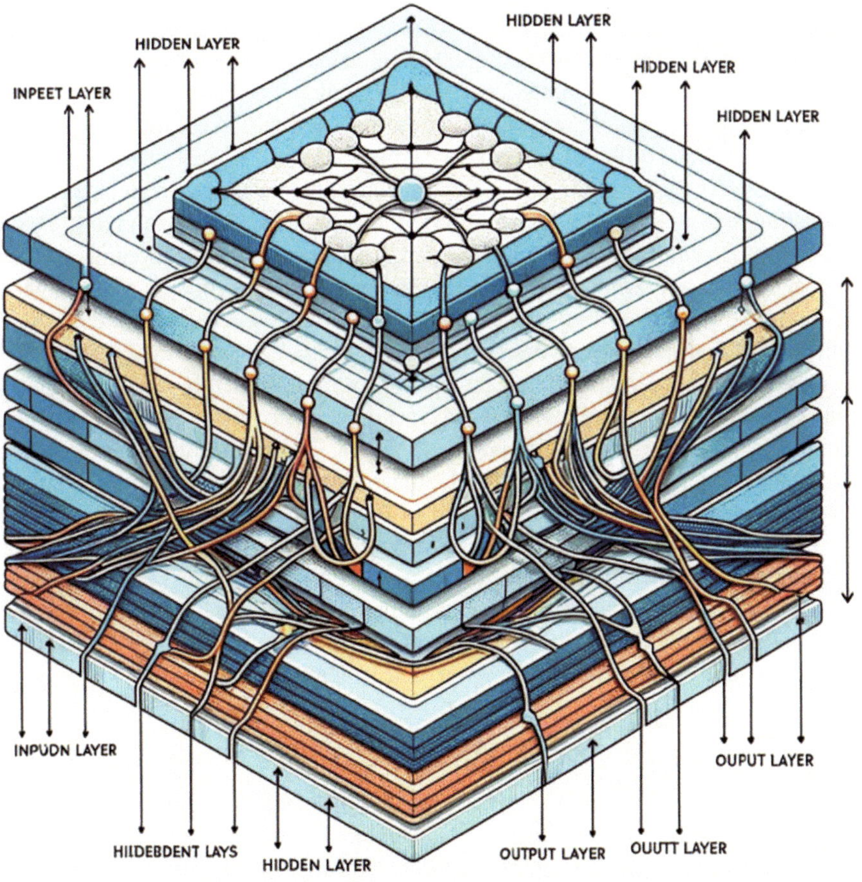

FIGURE 9.1 Architecture of a feedforward neural network.

9.2.2 Phase I: The Expert Systems Era (~1983 to ~1995)

The first significant wave of AI applications in chemical engineering emerged during the Expert Systems Era, which lasted from the early 1980s to the mid-1990s. Expert systems, also known as knowledge-based systems, are computer programs that replicate the problem-solving abilities of human experts within a specific domain. These systems leverage vast amounts of specialized knowledge, often in the form of heuristics, to efficiently narrow the search space and identify solutions by recognizing patterns and applying the appropriate rules of thumb.

The architecture of expert systems was inspired by cognitive psychology's stimulus-response model and symbolic computation's pattern-matching-and-search model. These concepts were rooted in Emil Post's work in symbolic logic and further advanced by Simon and Newell in the 1960s and 1970s through the development of the production system framework. This framework was pivotal in separating domain knowledge from search or inference processes, providing the computational flexibility to tackle ill-structured problems [5].

One of the earliest and most influential expert systems was MYCIN, which was developed at Stanford University to diagnose infectious diseases. This success inspired other systems like PROSPECTOR for mineral prospecting and R1 for configuring Vax computers. The first notable application in chemical engineering was CONPHYDE, developed in 1983 at Carnegie Mellon University to predict the thermophysical properties of complex fluid mixtures, and DECADE, also developed by CMU researchers, followed in 1985 for catalyst design.

The Expert Systems Era saw numerous advances in process synthesis, design, modeling, and diagnosis. Notable contributions included Stephanopoulos' Design-Kit for conceptual design and MODELL.LA is a language for developing process models. Davis and Kramer carried out important work in process fault diagnosis, while my group focused on causal model-based diagnostic expert systems and the potential of learning expert systems.

The Abnormal Situation Management consortium, funded by the National Institute of Standards and Technology's Advanced Technology Program and leading oil companies, was a significant initiative of this era. This consortium laid the groundwork for the Clean Energy Smart Manufacturing Innovation Institute, funded in 2016. In addition, the first course on AI in PSE was taught at Columbia University in 1986 and later at Purdue University, evolving from an expert systems emphasis to include advanced ML topics [6].

Despite the impressive achievements, expert systems faced challenges. Developing and maintaining these systems was costly and time-consuming, limiting their scalability and adaptability to changing industrial applications.

9.2.3 Phase II: The Neural Networks Era (~1990 to ~2008)

As the limitations of expert systems became apparent, the focus shifted to neural networks, marking the beginning of the Neural Networks Era around 1990. This phase represented a crucial shift from the top-down design paradigm of expert systems to the bottom-up approach of neural networks, which automatically acquired knowledge from large datasets, easing the development and maintenance of models.

The reinvention of the backpropagation algorithm by Rumelhart, Hinton, and Williams in 1986 was a key breakthrough. This algorithm allowed feedforward neural networks to learn hidden patterns in input–output data by propagating errors back through the network and adjusting connection weights. While the concept of neural networks dates back to the 1940s with McCulloch and Pitts, the ability to handle nonlinear problems using backpropagation was a significant advancement.

This era saw substantial progress in various domains, including modeling, fault diagnosis, control, and product design. Significant contributions included Kramer's connection between autoencoder architectures and nonlinear principal component analysis and Bakshi and Stephanopoulos's WaveNet architecture. Research also continued on expert systems, genetic algorithms, and multiscale modeling.

Despite neural networks' success in practical applications, some challenging problems in vision, natural language processing, and speech understanding remained unsolved. Researchers realized that deeper neural networks with more hidden layers were needed, but training these networks was nearly impossible with existing techniques. This bottleneck persisted until breakthroughs in training deep neural nets launched the current phase of AI in chemical engineering [7].

9.2.4 Current Phase: The Rise of Deep Learning and Advanced AI

The advent of deep learning and advanced AI techniques characterizes the current phase of AI in chemical engineering. These methods have overcome many of the limitations of earlier approaches, enabling the development of sophisticated models capable of handling complex, nonlinear problems. Integrating deep learning with traditional AI methods drives innovations and applications in the field, setting the stage for a transformative future in chemical engineering [8].

As we continue to explore the potential of AI in chemical engineering, it is essential to learn from past experiences, embrace current advancements, and look forward to the future with a balanced perspective. By doing so, we can harness the full potential of AI to address the pressing challenges and opportunities facing the chemical engineering industry today.

9.2.5 Lack of Impact of AI During Phases I and II

Despite significant efforts spanning two decades, AI did not revolutionize chemical engineering as anticipated. Upon reflection, several factors contribute to this outcome. First, the challenges we addressed remain formidable even today. Second, the era lacked the necessary powerful computing, storage, communication, and programming environments to tackle these problems effectively. Third, data availability was severely limited. Finally, whatever resources were available were prohibitively expensive.

Three primary challenges characterized Phases I and II: conceptual, implementational, and organizational. While substantial progress was made on conceptual issues such as knowledge representation and inference strategies for synthesis, design, diagnosis, and safety, the implementation and organizational hurdles prevented practical application. Essentially, there was no "technology push."

Moreover, there was no significant "market pull" either. During this period, the more accessible benefits in process engineering were achieved through optimization and model-predictive control (MPC) technologies. As algorithms and hardware improved, traditional approaches scaled effectively for problems solvable through first-principles-based models. By contrast, AI-based approaches were essential for problems where such models were difficult or impossible to build (e.g., diagnosis, safety analysis, materials design, and speech recognition). These AI approaches required massive computational power and vast amounts of data, both unavailable at the time. This practical shortfall led to two "AI winters," marked by a significant reduction in AI research funding, slowing progress further [9].

Typically, it takes about 50 years for a technology to mature, penetrate, and have a widespread impact from its discovery to adoption. For example, simulation technology like Aspen Plus took around 50 years to achieve 90% market penetration from its inception in the 1950s. A similar discovery, growth, and penetration cycle occurred for optimization technologies such as mixed-integer linear programming (MILP), mixed-integer nonlinear programming, and MPC. In hindsight, AI as a tool was only about 10–15 years old during Phases I and II, making it premature to expect widespread impact.

This analysis suggests a significant AI impact could be expected around 2030–2035. While predicting technology penetration and impact is not an exact science, this estimate appears reasonable, given the current state of AI. The anticipated impact was premature for those of us who began working on AI in the early 1980s, but the intellectual challenges were both stimulating. Many intellectual challenges, such as developing hybrid AI methods and causal model-based AI systems, persist and continue to be areas of active exploration.

9.2.6 Are Things Different Now for AI to Have an Impact?

The progress of AI over the last decade has been exhilarating, and the resource constraints previously mentioned have been mainly overcome. Implementation difficulties have significantly diminished. Organizational and psychological barriers have also been reduced as society increasingly trusts and accepts recommendations from AI-assisted systems such as Google, Alexa, and Yelp for various tasks. Companies are starting to embrace organizational and workflow changes to integrate AI-assisted processes.

To illustrate the magnitude of this progress, consider this comparison: In 1985, the CRAY-2 supercomputer was arguably the most powerful computer, with a computational speed of 1.9 gigaflops, consuming 150 kW of power. This $16 million machine (equivalent to about $32 million today) was enormous, requiring a large, custom, air-conditioned environment [10].

Fast forward to today, and the CRAY-2's capabilities are dwarfed by devices as small as the Apple Watch (Series 1). The Apple Watch outperforms the CRAY-2, achieving three gigaflops while consuming just 1 watt of power, costing only $300. This represents a 150,000-fold improvement in performance-to-cost ratio, solely on the hardware front. Similarly, dramatic advances have occurred in software, with significant improvements in algorithm performance and high-level programming environments such as MATLAB, Mathematica, Python, Hadoop, Julia, and TensorFlow. Tasks that once required weeks of programming in Lisp can now be accomplished in minutes with just a few lines of code.

Advancements in wireless communication technologies have also been profound. Furthermore, the availability of vast amounts of data, or "big data," in many domains has fuelled stunning advancements in machine learning. These developments are truly game changing.

What has driven this progress? Essentially, Moore's Law has continued to deliver exponential improvements in computing power for over 30 years, far surpassing its anticipated lifespan. This relentless progress has enabled the "technology push" we see today. Concurrently, the "market pull" is also evident. Much of the efficiency gains achievable through optimization and MPC technologies have been realized. We must address more complex decision-making problems requiring AI-assisted solutions to achieve further gains and automation. This convergence of "technology push" and "market pull" sets the stage for transformative change [11].

Looking back over the past 30 years, three milestones stand out in the history of AI: Deep Blue's victory over Garry Kasparov in chess in 1997, Watson's triumph as the Jeopardy champion in 2011, and AlphaGo's surprising win in 2016. The AI advancements that made these remarkable achievements possible are now poised to extend their impact beyond game playing, influencing various fields and applications.

9.3 PHASES OF AI IN CHEMICAL ENGINEERING: CURRENT PHASE III—DEEP LEARNING AND THE DATA SCIENCE ERA (CIRCA 2005 TO PRESENT)

In my view, Phase III of AI in chemical engineering commenced around 2005, marking the dawn of the data science and predictive analytics era. This new phase has been driven by three pivotal technologies: deep or convolutional neural networks (CNNs), reinforcement learning, and statistical ML. These advancements are the backbone of recent AI success stories in game playing, natural language processing, robotics, and computer vision.

Unlike the neural networks of the 1990s, which typically featured only a single hidden layer of neurons, modern deep neural networks boast multiple hidden layers, as illustrated in Figure 9.2. This multilayer architecture enables hierarchical feature extraction for complex pattern recognition tasks. However, training these deep networks using traditional backpropagation or gradient descent algorithms was nearly impossible. The breakthrough came in 2006 with a layer-by-layer training strategy and the advent of powerful graphics processing units. In addition, the convolutional process in neural network training, a filtering technique well known in signal processing, made hierarchical feature extraction feasible. After defining the network architecture and filter parameters (such as size and number), a CNN learns the optimal filters from a vast dataset—a crucial requirement for successful network performance [12].

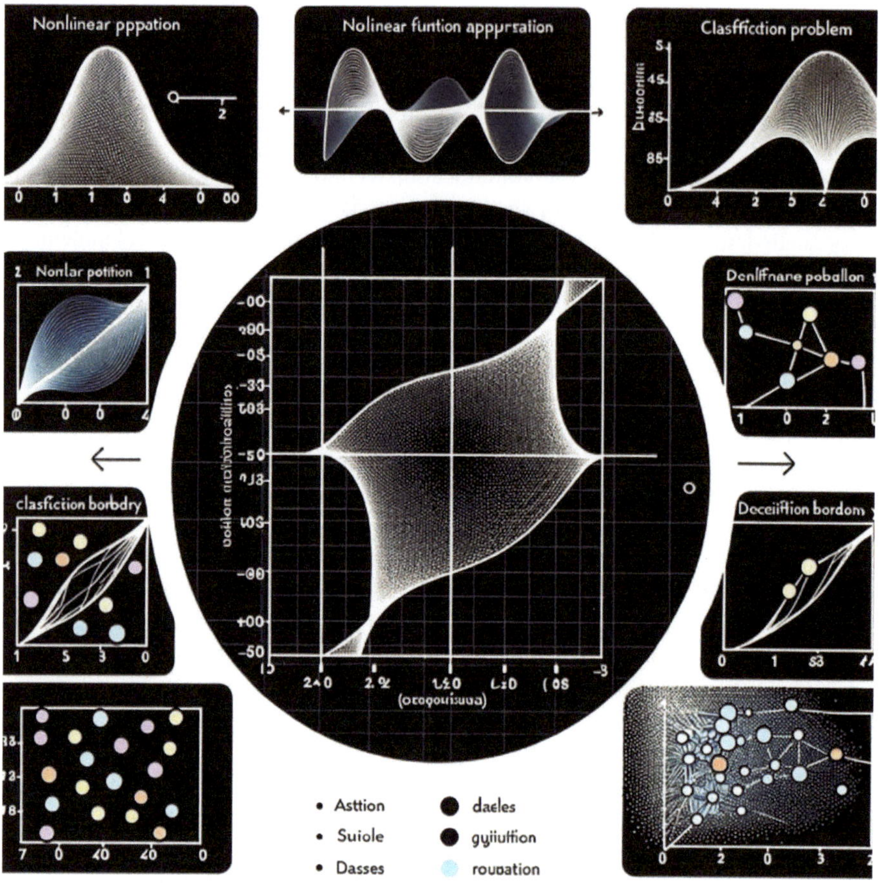

FIGURE 9.2 Examples of nonlinear function approximation and classification problems.

Another significant architectural innovation was the recurrent neural network (RNN). Unlike feedforward neural networks, which lack temporal awareness and only consider the current input, RNNs are designed for sequential data, such as time series, where future predictions depend on past inputs. For instance, predicting the next word in a sentence requires knowledge of the preceding words. RNNs address this by incorporating previous inputs into their current processing, effectively giving the network a form of "memory." This memory capability was further enhanced by the long short-term memory (LSTM) unit, which includes a cell, an input gate, an output gate, and a forget gate. The cell retains values over arbitrary time intervals, while the gates control the flow of information in and out of the cell. LSTM networks are particularly effective for making predictions based on time series data, where unknown durations may separate significant events.

Beyond architectural advancements, reinforcement learning has emerged as a critical concept. It involves learning a sequence of actions to achieve a desired outcome, such as maximizing an objective function. This goal-oriented learning process allows an agent to adapt behavior based on reward-punishment signals received through interaction with its environment. An everyday example is training a dog, where treats reward desired behavior and lack of treats indicates undesired behavior. Repeated reinforcement of these patterns helps the dog adopt the desired behavior. This feedback control-based learning mechanism is essentially Bellman's dynamic programming reimagined for modern ML applications.

Effective reinforcement learning requires millions of training sessions for complex problems like the game of Go. AlphaGo, for instance, played millions of games against itself to master Go, accumulating centuries of human expertise within days. While this is an impressive feat, it is worth noting that game environments can provide unlimited, accurate training data, a luxury often unavailable in scientific and engineering domains. However, this limitation can sometimes be mitigated when data originates from computer simulations, as in certain materials science applications.

Reinforcement learning differs from the dominant learning paradigms—supervised and unsupervised. In supervised learning, the system learns the relationship between input (X) and output (Y) from labeled input–output (X–Y) pairs. In contrast, unsupervised learning involves discovering patterns in a dataset without labels (i.e., no Y). Unsupervised learning identifies similarities among entities to cluster them into groups with similar features, while supervised learning focuses on differences, aiding in classification tasks. For instance, given unlabeled animal features, unsupervised learning might group horses and zebras, whereas supervised learning would distinguish between them based on labels like "horse" and "zebra."

The third essential advancement, statistical ML, integrates mathematical methods from probability and statistics with ML techniques. This combination has produced numerous valuable methods, including least absolute shrinkage and selection operator, support vector machines, random forests, clustering, and Bayesian belief networks.

These three technological advances—deep neural networks, reinforcement learning, and statistical ML—have been propelled by dramatic progress in hardware, communication capabilities, and the availability of extensive datasets. Together, they drive the current AI revolution in game playing, image processing, and autonomous vehicles [13].

However, what does this progress signify for chemical engineering? Is the long-awaited promise of AI in chemical engineering finally within reach after three decades of efforts and setbacks? What would developing a "Watson-like system" for chemical engineering applications take? Before addressing these questions, we must first explore the challenges and considerations of knowledge modeling in the AI era.

9.4 AI'S ROLE IN MODELING KNOWLEDGE: FROM NUMERIC TO SYMBOLIC STRUCTURES AND RELATIONSHIPS

As Phase III progresses and AI concepts and tools become increasingly pervasive, we enter a transformative knowledge acquisition, modeling, and utilization era. While Phases I and II offered glimpses of this transformation, it is yet to be widely realized.

To truly understand AI's place in chemical engineering, one must consider the evolution of knowledge modeling paradigms. Historically, chemical engineering was primarily an empirical and heuristic discipline, lacking quantitative, first-principles-based modeling approaches. This began to change in the 1950s with the advent of the Amundson era, which introduced applied mathematical methods, particularly linear algebra, ordinary differential equations (ODEs), and partial differential equations (PDEs), to develop first-principles-based models of unit operations. Similarly, decision-making in PSE was also largely empirical and heuristic until the 1960s, when mathematical programming methods such as MILP, pioneered by Roger Sargent, revolutionized the field.

The subsequent significant development in this continuum of modeling paradigms is the introduction of knowledge representation concepts and search techniques from AI. This shift arguably began in the early 1980s under the leadership of researchers like Westerberg and Stephanopoulos. After remaining on the fringes for three decades, pursued by a small group of researchers, this knowledge-modeling paradigm is now becoming mainstream.

The Amundson era focused on formal methods for modeling process units, while the Sargent era and the AI era are about modeling the process engineer—formally representing human information processing and decision-making to solve problems in synthesis, design, control, scheduling, optimization, and risk analysis. While some of these problems can be addressed by mathematical programming, others, such as fault diagnosis and process hazards analysis, require causal models-based reasoning, better suited to AI techniques [14].

The conceptual breakthrough of representing and reasoning with symbolic structures and relationships is a crucial contribution of AI. Despite the current excitement about ML, this essential aspect of AI should not be overlooked. As we progress beyond purely data-driven models toward more comprehensive symbolic knowledge processing systems, the significance of this contribution will become more apparent. This development has far-reaching implications as we create hybrid AI systems that combine first principles with data-driven processing, causal models-based explanatory systems, and domain-specific knowledge engines.

AI methods are not merely tools for extracting patterns from large data sets, although that is a significant benefit. Instead, they represent a new knowledge modeling paradigm, the next evolutionary stage in developing formal methods—following applied mathematics (differential and algebraic equations), operations research (mathematical programming), and AI. Applied mathematics models numerical relationships between variables and parameters, mathematical programming models relationships between constraints, and AI models relationships between symbolic variables and structures. While logic was initially considered the best foundation for AI, recent developments suggest that probability, statistics, and network science might be more suitable, depending on the application.

Chemical engineers have always valued their modeling capabilities. However, in this new era, modeling must go beyond differential-algebraic equations (DAEs), the staple of chemical engineering models—the Amundson legacy. Addressing the complex modeling challenges in symbolic reasoning and high-level decision-making requires a broader approach than used in chemical engineering. There is a wide variety of knowledge representation concepts leading to other classes of models that will play an essential role in this new era. While it is not the purpose of this chapter to extensively discuss modeling concepts, it is nevertheless helpful to outline and summarize the issues involved.

Models can be broadly classified into (1) mechanism-driven models based on first principles and (2) data-driven models. Each class can be further categorized into (1) quantitative and (2) qualitative models. Combinations of these classes lead to hybrid models [15].

DAE models are suitable for problems that can be mathematically described and are common in thermodynamics, transport phenomena, and reaction engineering. However, other kinds of knowledge do not lend themselves to such models. For example, reasoning about cause and effect in a process plant is central to fault diagnosis, risk analysis, alarm management, and supervisory control. Traditional DAE models are often unsuitable for generating mechanistic explanations about causal behavior, especially in complex and nonlinear systems with incomplete or uncertain data. This problem often requires hybrid models, such as combinations of graph-theoretical models (e.g., signed digraphs), production system models (e.g., rule-based representations), and data-driven models (e.g., principal component analysis or neural networks).

While we are familiar with ODE/PDE, statistical regression, and mathematical programming models, we are less familiar with other classes widely used in AI. These include graph-theoretical models (used extensively for causal reasoning in identifying abnormal events, diagnosis, and risk analysis), Petri nets (used for modeling discrete event systems), rule-based production system models (used in expert systems for automating higher-order reasoning), semantic network models such as ontologies (used in materials discovery and design, domain-specific compilers), and object-oriented models such as agent-based models (used in simulating the behavior and decision-making choices of independent, interacting entities with complex attributes and decision-making powers). In addition, data-driven quantitative models such as pattern recognition-based models (e.g., neural networks, fuzzy logic), stochastic models (e.g., genetic algorithms, simulated annealing), and equation-free pattern-recognition models in studying nonlinear dynamical systems are becoming increasingly relevant [16].

These AI-based modeling approaches are becoming essential in our modeling arsenal in this new phase of AI. However, considering the emerging challenges, the number of academic researchers developing AI-based models in chemical engineering is inadequate. As I observed in another perspective article a decade ago, this needs to be addressed in our research and education agendas.

It is encouraging to see that the barriers to implementing AI have significantly decreased due to the emergence of relatively easy-to-use software environments, such as R, Python, Julia, and TensorFlow. However, practicing AI correctly involves more than learning to run code in these environments. It requires a deep understanding of AI principles, akin to understanding the theory behind MILP rather than merely knowing how to execute an MILP program in MATLAB. In the past, the absence of user-friendly environments meant that researchers were compelled to learn Lisp, the primary language of AI, in courses that comprehensively covered AI concepts, tools, and techniques. A well-educated applied AI researcher from that era would typically take several courses on AI, ML, natural language processing, databases, and other relevant topics to gain a deep understanding of AI methods. In contrast, modern, user-friendly AI software makes it easy for newcomers to build ML models quickly, often leading to a false sense of mastery in AI or ML. This is a dangerous trap. Just as statistical tools can be misused if one is not careful, the same fate can befall users of ML tools.

Developing AI methods requires more than just keeping up with new developments in computer science and applying them to chemical engineering. While there are some easy gains to be made, many intellectually challenging problems in our industry are not amenable to simple solutions. Nor can these problems be solved by computer scientists alone, as they lack domain knowledge. This would be akin to thinking that mathematicians because they solve transport phenomena problems using differential equations, could address core chemical engineering problems. They can provide us with generic concepts, tools, and techniques, but we must adapt and extend these with domain knowledge to solve our problems effectively. To do this well, one must be well educated in AI fundamentals beyond merely running ML code [17].

It is becoming clear that our undergraduate and graduate students need to become familiar with applied AI techniques. We should develop a dual-level course for junior/senior undergraduate and first-year graduate students that teaches applied AI using chemical engineering examples. This course would be akin to the applied mathematical methods core course required by chemical engineering graduate programs. However, teaching AI properly goes beyond purely data-centric ML. There is a tendency to create cookbook-style courses where students learn to apply different software packages mechanically without a deeper understanding of the fundamentals. The course needs to be firmly grounded in knowledge modeling philosophies, knowledge representation, search and inference, and knowledge extraction and management issues.

We need to differentiate between training and education when designing such a course. Training focuses on "know-how," that is, how to execute a recipe to solve a problem. In contrast, education emphasizes "know-why," understanding why the problem exists from a first-principles-based mechanistic perspective. A significant difference is training someone to repair an air conditioner and teaching them thermodynamics. While the former is functional and has its place, our courses should be more than merely functional. We should avoid the criticism voiced during the formative years of calculus that "the user-friendly approach of Leibniz made it easy for people who did not know calculus to teach those who will never know calculus!" The easy availability of user-friendly ML tools poses a similar predicament today.

Developing AI methods in chemical engineering is not just about applying the latest computer science techniques. It requires a deep understanding of AI fundamentals, an appreciation of the complexities of our industry, and a commitment to integrating AI into our educational and research agendas. This holistic approach will enable us to leverage AI's full potential to address the challenges and opportunities in chemical engineering.

9.5 AI IN CHEMICAL ENGINEERING: RECENT TRENDS AND FUTURE OUTLOOK

The following summary will be a representative survey, providing a helpful starting point for researchers rather than a comprehensive review.

9.5.1 Phase III (2005–Present): The Era of Data-Driven Strategy

In Phase III, we have witnessed the rise of a predominantly bottom-up, data-driven strategy for knowledge acquisition and modeling, utilizing deep convolutional networks, reinforcement learning, and statistical learning. These techniques have facilitated significant advancements in image recognition and speech understanding. However, whether these sophisticated methodologies are necessary to achieve chemical engineering results remains uncertain.

One primary consideration is the need for vast amounts of data for these techniques to be effective. While such data is readily available in fields such as game playing, vision, and speech, it is often lacking in chemical engineering applications—except in scenarios where computer simulations can generate large, reliable datasets. Although data collection has improved in our field, chemical engineering does not yet constitute a "big data" domain like finance, vision, or speech. Furthermore, our systems are governed by fundamental laws and principles of physics, chemistry, and biology, which we should leverage to compensate for the lack of extensive data.

Many of our needs can still be met using Phase I and II techniques, now enhanced with more powerful and user-friendly software and hardware. Therefore, before resorting to deep neural nets or reinforcement learning, it is prudent to revisit earlier approaches. What is truly needed is a way to integrate first-principles knowledge with data-driven models to develop hybrid models more effectively and reliably. Past work on hybrid model development offers a valuable starting point.

9.5.2 Areas of Opportunity in Chemical Engineering

Several topics in chemical engineering stand out for their potential to benefit from renewed AI interest: materials design, process operations, and fault diagnosis. These areas present numerous low-hanging fruits, with proof-of-concept solutions already demonstrated in Phases I and II. The implementation challenges and organizational barriers have significantly diminished, creating fertile ground for innovation. Similar opportunities exist in biomedical and biochemical engineering.

AI is already being utilized in the industry for process operations and diagnosis. For example, General Electric and British Petroleum use ML software to monitor real-time oil-well performance, maximizing production and minimizing downtime. Uptake, a predictive analytics company, has successfully used ML to predict and prevent failures in wind turbines. Italian dairy producer Granarolo implemented ML to forecast production needs, minimize wastage, and maximize profits. In addition, hybrid approaches combining neural networks with first-principles models are being explored for optimization and control.

Materials design is another promising area, where the challenge lies in discovering and designing new materials and formulations with desired properties. This includes various products such as catalysts, nanostructures, pharmaceuticals, additives, polymeric composites, rubber compounds, and alloys. The ultimate goal is to design materials rationally and systematically rather than through the traditional trial-and-error approach. To achieve this, two related problems must be solved: predicting material properties given the structure or formulation (the "forward" problem) and determining the appropriate structure or formulation given the desired properties (the "inverse" problem).

The materials science community has recently recognized the opportunities for AI, advocating for "inverse design" informatics. Elements of this framework were anticipated and demonstrated in the chemical engineering community about two decades ago, with successful industrial applications. What is promising now is the ability to achieve these goals more quickly and easily for more complex materials, thanks to powerful computing environments and abundant data. Recent workshops and conferences highlight various applications, including the design of crystalline alloys, organic photovoltaics, nanoparticle packing, and shape memory alloys.

There is also considerable excitement about ML for catalyst design and discovery. Creating systematic, curated databases for catalytic reaction data, similar to testbeds in computer vision, is crucial for advancing this field. Community-based efforts, such as the Stanford Catalysis-Hub Database and the Atomistic ML Package, are promising developments. The relatively easy problems in materials design involve analyzing extensive data using ML algorithms to understand process-structure-property or process-composition-property relationships. However, the next significant breakthrough—a comprehensive materials discovery system using active learning—remains intellectually challenging.

We must develop domain-specific representations, languages, compilers, ontologies, and molecular structure search engines—domain-specific "Watson-like" knowledge discovery engines to achieve this. This challenging task requires more than a superficial familiarity with AI methods. However, prior proof-of-concept contributions provide potential starting points. Recent developments in automatic reaction network generation and reaction synthesis planning are promising.

9.5.3 The Importance of Symbolic Relationships and Ontologies

Automating higher levels of decision-making highlights the need to model symbolic relationships, in addition to numeric ones, between concepts or entities. An ontology, which explicitly describes domain concepts and their relationships, is crucial for this purpose. Entities may include materials, objects, properties, or variables, and their relationships are often best captured through graph and network-based models. Most domain knowledge exists as a vast network of interdependent relationships, which ontologies aim to represent.

Recent work on ontologies in process engineering has made significant strides, but much more remains to be done, especially for materials design. Initiatives like the Novel Materials Discovery Laboratory and using text-mining to uncover relationships among materials science concepts are promising.

9.5.4 The Need for Mechanistic Understanding

While data is essential, it is not raw data that we seek. We aim for a first-principles-based understanding of underlying phenomena to inform rational decision-making. In materials design, for example, it is not enough to discover a formulation that works; we also need to understand why it works from a mechanical perspective. This understanding, grounded in the fundamental principles of physics and chemistry, offers many benefits.

The problem of explicability or interpretability in AI is gaining attention as black-box models undermine trust. In process diagnosis and control, hybrid AI approaches combining first-principles understanding with data-driven techniques have addressed similar concerns. At present, we lack satisfactory methods for embedding explanations in deep-learning systems. This may lead to a preference for more transparent systems, such as Bayesian networks, in some applications.

9.5.5 The Cognition Gap

A fundamental issue with purely data-driven models like deep neural networks is their lack of "understanding" of underlying knowledge. For instance, a self-driving car can navigate traffic but does not "understand" concepts like mass, momentum, acceleration, and force as humans do. This cognition gap is a fundamental mystery in AI and cognitive science. Understanding this gap will have significant implications for our domain.

In conclusion, while AI presents many opportunities for chemical engineering, it also poses challenges requiring a deep understanding of AI fundamentals and domain-specific knowledge. By integrating first-principles knowledge with data-driven models, we can develop hybrid approaches that leverage the strengths of both methodologies, paving the way for innovative solutions in chemical engineering.

9.6 BEYOND DATA SCIENCE: PHASE IV—EMERGENCE IN LARGE-SCALE SYSTEMS OF SELF-ORGANIZING INTELLIGENT AGENTS

This challenge has a profound system engineering aspect at its core. The fundamental question is: How do we predict a system's macroscopic properties and behaviors based on the microscopic properties of its components? For instance, a single neuron lacks self-awareness, yet the entire system exhibits spontaneous self-awareness when 100 billion neurons are interconnected in a specific manner. What accounts for this phenomenon?

We often understand how to transition from the parts to the whole, mainly when dealing with nonrational or purpose-free entities like molecules. This is the essence of statistical mechanics. However, the scenario becomes more complex when the entities exhibit rational, purposeful, intelligent behaviors.

The approach of moving from the parts to the whole starkly contrasts with the reductionist paradigm that dominated the 20th-century science. Reductionism seeks to understand and predict macroscopic properties by uncovering deeper, fundamental mechanisms and principles. It is a top-down methodology, beginning at the macro level and delving progressively more deeply to the micro-level, nano-level, and beyond (i.e., from the whole to the parts). This paradigm has driven remarkable achievements in physics and chemistry, leading to breakthroughs like quantum mechanics, the general theory of relativity, quantum field theory, the Standard Model, and string theory. Biology, too, achieved significant milestones within this framework, elucidating heredity through molecular structures and phenomena such as the double helix and the central dogma.

However, many of the grand challenges of the 21st-century science are characterized by bottom-up phenomena—emergent properties that arise from the interaction of simpler components. Examples include predicting phenotypes from genotypes, assessing the impact of human behavior on global climate, understanding the emergence of economic inequality, and quantifying and analytically predicting consciousness and self-awareness. By its nature, reductionism falls short in addressing these issues, as it does not typically concern itself with teleology or purposeful behavior. Instead, modeling bottom-up phenomena necessitates a new paradigm that embraces teleological properties, which often manifest at macroscopic levels, even in ostensibly purpose-free entities.

We require a bottom-up analytical framework, an emergentist or constructionist approach, as the antithesis to the reductionist perspective, to bridge the gap from the parts to the whole. This framework should not merely uncover hidden patterns, such as those identified by deep-learning neural networks through complex statistical correlations in vast datasets. Instead, it should offer a comprehensive mathematical framework that explains and predicts macroscopic behavior and phenomena from fundamental principles and mechanisms. Such a theory should predict significant qualitative and quantitative emergent macroscopic features and also elucidate why and how these features arise (and not other potential outcomes). Present deep-learning AI systems lack this explanatory power.

Developing this theoretical framework should be the ultimate goal of AI, marking the path toward systems that can reason using first-principles-based mechanisms. The journey is long, but pursuing self-organization and emergence represents the next critical phase in AI.

9.7 SUMMARY

Having been involved in the initial phases of the AI "revolution"—expert systems in the 1980s and neural networks in the 1990s—I approach the current hype surrounding AI with some skepticism. However, as discussed earlier, the current phase does seem different. We may finally have the conditions necessary for AI to impact chemical engineering significantly. We are at the cusp of a transformative era in the acquisition, modeling, utilization, and management of diverse types of knowledge.

As we develop AI-based models, it is crucial to acknowledge, as Jordan and Rahimi caution, that the current state of ML resembles alchemy—a collection of ad hoc methods. Just as alchemy evolved into the rigorous and formal chemistry and chemical engineering disciplines, ML must also evolve. This evolution can be facilitated by integrating first-principles knowledge whenever possible, which can impose rigor and discipline on purely data-driven models.

Numerous applications are ready to succeed quickly in this new data science phase of AI. However, the most intellectually stimulating and challenging problems lie in developing conceptual frameworks such as hybrid models, mechanism-based causal explanations, domain-specific knowledge discovery engines, and analytical theories of emergence. Achieving these breakthroughs requires moving beyond purely data-centric ML despite the current excitement. It necessitates leveraging other knowledge representation and reasoning methods from earlier AI phases and integrating symbolic reasoning with data-driven processing.

This journey is long, adventurous, and intellectually stimulating, and we are only at its beginning. Our progress will revolutionize all aspects of chemical engineering, leading to unprecedented advancements and innovations.

REFERENCES

[1] Tseng, M.-L., et al., Sustainable industrial and operation engineering trends and challenges toward Industry 4.0: a data driven analysis. *Journal of Industrial and Production Engineering*, 2021. 38(8): p. 581–598.

[2] López-Guajardo, E.A., et al., Process Intensification 4.0: a new approach for attaining new, sustainable and circular processes enabled by machine learning. *Chemical Engineering and Processing-Process Intensification*, 2022. 180: p. 108671.

[3] Oveissi, F. and A.E. Ghadi. Preparing chemical engineers for Industry 4.0: an interactive education approach. In *REES AAEE 2021 conference: engineering education research capability development*. 2021: Perth, WA: Engineers Australia.

[4] Yan, Y., T. Borhani, and P. Clough, Machine learning applications in chemical engineering. In *Machine learning in chemistry: the impact of artificial intelligence*. 2020: Royal Society of Chemistry.

[5] Sadat Lavasani, M., et al., Big data analytics opportunities for applications in process engineering. *Reviews in Chemical Engineering*, 2023. 39(3): p. 479–511.

[6] Chiang, L.H., et al., Towards artificial intelligence at scale in the chemical industry. *AIChE Journal*, 2022. 68(6): p. e17644.

[7] Qin, S.J. and L.H. Chiang, Advances and opportunities in machine learning for process data analytics. *Computers & Chemical Engineering*, 2019. 126: p. 465–473.

[8] Reddy, P.S. and P.K. Ghodke, Image analysis using artificial intelligence in chemical engineering processes: current trends and future directions. In *Image processing and intelligent computing systems*. 2023: CRC Press. p. 79–100.

[9] Piccione, P.M., Realistic interplays between data science and chemical engineering in the first quarter of the 21st century: facts and a vision. *Chemical Engineering Research and Design*, 2019. 147: p. 668–675.

[10] Ghodke, P.K., et al., Artificial Intelligence in the digital chemical industry, its application and sustainability. In *Recent trends and best practices in Industry 4.0*. 2023: River Publishers. p. 1–29.

[11] Qian, Y., S. Vaddiraju, and F. Khan, Safety education 4.0–a critical review and a response to the process Industry 4.0 need in chemical engineering curriculum. *Safety Science*, 2023. 161: p. 106069.

[12] Patwardhan, R.S., et al., Applications of advanced analytics at Saudi Aramco: a practitioners' perspective. *Industrial & Engineering Chemistry Research*, 2019. 58(26): p. 11338–11351.

[13] Reis, M.S. and P.M. Saraiva, Data-centric process systems engineering: a push towards PSE 4.0. *Computers & Chemical Engineering*, 2021. 155: p. 107529.

[14] Bortz, M., et al., AI in process industries–current status and future prospects. *Chemie Ingenieur Technik*, 2023. 95(7): p. 975–988.

[15] Örs, E., et al., A conceptual framework for AI-based operational digital twin in chemical process engineering. In *2020 IEEE international conference on engineering, technology and innovation (ICE/ITMC)*. 2020: IEEE.

[16] Udugama, I.A., et al., Digitalisation in chemical engineering: industrial needs, academic best practice, and curriculum limitations. *Education for Chemical Engineers*, 2022. 39: p. 94–107.
[17] Mateo, F.W. and A. Redchuk, Artificial intelligence as a process optimization driver under Industry 4.0 framework and the role of IIoT, a bibliometric analysis. *Journal of Industrial Integration and Management*, 2022: p. 1–16.

10 Harvesting Tomorrow
The Future of Agriculture in Industry 4.0

Ali Soofastaei

10.1 INTRODUCTION

In recent decades, technological advancements have been reshaping industries across the globe, revolutionizing the way we work, produce, and consume. At the forefront of this technological revolution lies Industry 4.0, often called the Fourth Industrial Revolution, a paradigm shift characterized by integrating digital technologies into manufacturing and production processes. However, the transformative potential of Industry 4.0 extends far beyond factory floors, reaching into sectors vital for sustaining human life, including agriculture.

10.1.1 Defining Industry 4.0 and Its Impact on Various Sectors

Industry 4.0 represents the convergence of several cutting-edge technologies, including the Internet of Things (IoT), artificial intelligence (AI), big data analytics, and automation. This integration enables the creation of "smart" systems that communicate, analyze, and act upon vast amounts of data in real time. While initially applied in manufacturing, Industry 4.0 principles have transcended traditional boundaries, infiltrating the healthcare, transportation, and agriculture sectors.

The impact of Industry 4.0 is profound in agriculture. It heralds a new era of "smart farming," where data-driven insights and automation revolutionize every aspect of agricultural operations. From precision planting to optimized irrigation and predictive maintenance of machinery, Industry 4.0 technologies promise to increase efficiency, reduce waste, and enhance productivity in the agricultural sector [1].

10.1.2 The Role of Advanced Analytics in Transforming Agriculture

At the heart of Industry 4.0 lies the power of advanced analytics—the ability to harness vast amounts of data to extract meaningful insights and drive informed decision-making. In agriculture, this translates into using sophisticated algorithms and machine learning techniques to analyze data collected from sensors, drones, satellites, and other IoT devices.

Advanced analytics holds immense promise for transforming agriculture in several key areas. First, it enables farmers to gain deeper insights into soil health,

DOI: 10.1201/9781003186823-10

weather patterns, and crop conditions, allowing for precise resource allocation and optimized cultivation practices. Moreover, predictive analytics can forecast crop yields, identify potential pest infestations, and mitigate risks, empowering farmers to make proactive decisions and maximize harvests.

Furthermore, integrating AI-driven technologies, such as computer vision and robotics, enables the automation of labor-intensive tasks, leading to increased operational efficiency and cost savings. From automated harvesting to autonomous weed control, these innovations alleviate labor shortages and promote sustainable farming practices by minimizing the use of pesticides and fertilizers [2].

In summary, the convergence of Industry 4.0 technologies and advanced analytics holds tremendous potential for revolutionizing agriculture as we know it. By embracing these innovations, farmers can usher in a new era of sustainable, data-driven farming practices that ensure food security, environmental stewardship, and economic prosperity for future generations.

10.2 EVOLUTION OF AGRICULTURE: FROM TRADITIONAL TO MODERN METHODS

Agriculture, the bedrock of human civilization, has undergone a remarkable transformation over millennia, evolving from rudimentary practices to highly sophisticated techniques driven by technological innovation. This section delves into the historical trajectory of agricultural practices, juxtaposing traditional methods with modern advancements and highlighting the emergence of precision agriculture as a defining paradigm shift.

10.2.1 Historical Overview of Agricultural Practices

Since the dawn of civilization, humans have relied on agriculture to sustain themselves and build societies. Early agricultural practices, dating back thousands of years, were characterized by manual labor, simple tools, and subsistence farming. Ancient civilizations such as Mesopotamia, Egypt, and China developed rudimentary irrigation systems and crop cultivation techniques to cultivate staple crops like wheat, barley, and rice.

The Agricultural Revolution, from approximately 10,000 BCE to 3,000 BCE, marked a pivotal turning point in human history. With the domestication of plants and animals, humans transitioned from nomadic hunter-gatherer lifestyles to settled agricultural societies. During this period, they witnessed the advent of more efficient farming methods, including plows, crop rotation, and animal traction, leading to increased food production and population growth [3].

Throughout the medieval and early modern periods, agriculture remained labor-intensive, largely dependent on manual labor and traditional knowledge passed down through generations. However, the 18th and 19th centuries Industrial Revolution catalyzed significant changes in agricultural practices. Fueled by steam power and later by internal combustion engines, mechanization revolutionized farming, increasing productivity and scale.

10.2.2 Introduction to Modern Agricultural Techniques

In the 20th century, we witnessed unprecedented advancements in agricultural science and technology, ushering in modern agriculture. Innovations such as synthetic fertilizers, pesticides, and hybrid seeds dramatically boosted crop yields, enabling farmers to feed growing populations. Moreover, the mechanization of farming operations, including the widespread adoption of tractors, combines, and other machinery, further enhanced efficiency and productivity.

Biotechnological breakthroughs, such as developing genetically modified organisms and precision breeding techniques, offered new avenues for crop improvement and pest resistance. These advancements, improved irrigation systems, and agronomic practices propelled agriculture into a new era of abundance and innovation [4].

10.2.3 The Emergence of Precision Agriculture

In recent decades, digital technologies have revolutionized agriculture again, giving rise to precision agriculture. Precision agriculture leverages data analytics, global positioning system (GPS) technology, remote sensing, and automation to optimize farming practices and maximize resource efficiency. By precisely tailoring inputs such as water, fertilizers, and pesticides to specific crop needs, farmers can minimize waste and environmental impact while maximizing yields.

Integrating satellite imagery, drones, and sensor networks enables real-time monitoring of crops, soil conditions, and environmental parameters, allowing for proactive decision-making and targeted interventions. Furthermore, advanced analytics and predictive modeling empower farmers to anticipate and mitigate risks, from weather-related disasters to pest outbreaks, ensuring the resilience and sustainability of agricultural systems [5].

In conclusion, agriculture's evolution from traditional to modern methods reflects humanity's relentless quest for innovation and efficiency. As we stand on the cusp of the Industry 4.0, precision agriculture emerges as a beacon of hope, promising a more sustainable, productive, and resilient food system for future generations.

10.3 FUNDAMENTALS OF INDUSTRY 4.0 IN AGRICULTURE

The convergence of Industry 4.0 technologies with agriculture has opened up unprecedented opportunities for farmers to optimize operations, improve efficiency, and enhance sustainability. This section delves into the fundamentals of Industry 4.0 in agriculture, exploring its underlying principles, the integration of critical technologies, and the associated benefits and challenges.

10.3.1 Understanding Industry 4.0 Principles

Industry 4.0 represents a paradigm shift in manufacturing and production processes driven by digitalization, connectivity, and automation. It is characterized by several fundamental principles at its core [6]:

- **Interconnectivity:** The ability of machines, devices, and systems to communicate and exchange data in real time forms the foundation of Industry 4.0. In agriculture, this translates into seamlessly integrating sensors, drones, machinery, and other IoT devices across the farm ecosystem.
- **Data Transparency:** Industry 4.0 emphasizes the importance of transparent, accessible data to enable informed decision-making. Farmers can gain valuable insights into crop health, soil conditions, weather patterns, and more by collecting and analyzing data from various sources.
- **Decentralized Decision-Making:** Industry 4.0 empowers decentralized decision-making by leveraging AI algorithms and machine learning models to process data and autonomously execute tasks. This enables farmers to respond quickly to changing conditions and optimize resource allocation.
- **Smart Automation:** Automation lies at the heart of Industry 4.0, enabling the automation of repetitive tasks and the deployment of autonomous systems. In agriculture, intelligent automation technologies such as robotic arms, automated irrigation systems, and uncrewed aerial vehicles (UAVs) streamline operations and reduce labor costs.

10.3.2 Integration of IoT, AI, Big Data, and Automation in Agriculture

The integration of IoT, AI, big data, and automation holds immense potential for revolutionizing agriculture [7]:

- **IoT in Agriculture:** IoT devices such as soil moisture sensors, weather stations, and GPS trackers collect real-time data on crop conditions, environmental parameters, and equipment status. This data is transmitted to centralized platforms for analysis and decision-making.
- **AI in Agriculture:** AI algorithms process vast amounts of agricultural data to generate insights and recommendations for farmers. From predictive analytics for crop yields to image recognition for pest detection, AI enables precision farming practices that optimize resource utilization and maximize productivity.
- **Big Data in Agriculture:** Big data analytics harnesses large datasets to identify patterns, trends, and correlations that inform strategic decision-making. Farmers can analyze historical and real-time data to optimize planting schedules, predict market demand, and mitigate risks.
- **Automation in Agriculture:** Automation technologies streamline farming operations, reducing manual labor and improving efficiency. From automated harvesting and sorting to autonomous tractors and drones, automation enhances productivity while minimizing human error.

10.3.3 Benefits and Challenges of Implementing Industry 4.0 in Agriculture

Implementing Industry 4.0 technologies in agriculture offers a myriad of benefits, including [8]:

- **Increased Efficiency:** Industry 4.0 technologies optimize farming processes, leading to higher yields, lower costs, and improved resource efficiency.
- **Enhanced Productivity:** By leveraging data-driven insights and automation, farmers can boost productivity and output while minimizing waste.
- **Sustainable Practices:** Industry 4.0 enables precision agriculture practices that promote environmental sustainability by reducing chemical usage, conserving water, and minimizing soil erosion.

However, implementing Industry 4.0 in agriculture also presents several challenges, including:

- **Cost and Accessibility:** The upfront costs of implementing Industry 4.0 technologies can be prohibitive for small-scale farmers, while access to reliable internet connectivity and infrastructure may be limited in rural areas.
- **Data Security and Privacy:** Collecting and storing sensitive agricultural data raises concerns about data security, privacy, and ownership, requiring robust cybersecurity measures and transparent data governance frameworks.
- **Skills and Training:** Adopting Industry 4.0 technologies requires farmers to acquire new skills and competencies in data analytics, digital literacy, and technology integration, highlighting the need for ongoing training and support.

In conclusion, Industry 4.0's fundamentals in agriculture herald a new era of innovation and transformation, offering unprecedented opportunities to enhance efficiency, productivity, and sustainability in farming practices. However, realizing the full potential of Industry 4.0 requires addressing challenges related to cost, accessibility, data security, and skills development, ensuring that the benefits of digitalization are accessible to all farmers, regardless of scale or location.

10.4 DATA-DRIVEN FARMING: LEVERAGING ADVANCED ANALYTICS

Data-driven farming, powered by advanced analytics, has emerged as a cornerstone of modern agriculture. It enables farmers to make informed decisions, optimize resources, and maximize yields. This section explores the importance of data in modern agriculture, the processes involved in collecting and analyzing agricultural data, and the application of predictive analytics for crop management and yield optimization.

10.4.1 Importance of Data in Modern Agriculture

In today's interconnected world, data has become a valuable asset in agriculture, providing farmers with insights into crop health, soil conditions, weather patterns, market trends, and more. The importance of data in modern agriculture can be attributed to several key factors [9]:

- **Precision Farming:** Data enables precision farming practices, allowing farmers to tailor inputs such as water, fertilizers, and pesticides to specific crop needs. Farmers can increase efficiency, reduce waste, and minimize environmental impact by optimizing resource allocation.

- **Decision Support:** Agricultural data serves as a tool, helping farmers make informed decisions about planting schedules, irrigation strategies, pest control measures, and crop rotation plans. Farmers can anticipate challenges and proactively manage risks by analyzing historical and real-time data.
- **Yield Optimization:** Data-driven insights enable farmers to optimize crop yields by identifying factors that affect productivity, such as soil fertility, moisture levels, and pest infestations. By monitoring crop performance and implementing targeted interventions, farmers can maximize yields and profitability.

10.4.2 Collection, Processing, and Analysis of Agricultural Data

The collection, processing, and analysis of agricultural data involve several steps [9]:

- **Data Collection:** Agricultural data is collected from various sources, including IoT sensors, drones, satellites, weather stations, and farm machinery. These devices gather data on crop health, soil moisture, temperature, humidity, rainfall, and more.
- **Data Processing:** Raw agricultural data is processed and aggregated using data management systems and software platforms. This may involve cleaning, filtering, and transforming the data to ensure accuracy and consistency.
- **Data Analysis:** Once processed, agricultural data is analyzed using advanced analytics techniques such as statistical analysis, machine learning, and predictive modeling. This analysis generates insights into crop performance, identifies trends and patterns, and informs decision-making.

10.4.3 Predictive Analytics for Crop Management and Yield Optimization

Predictive analytics plays a crucial role in crop management and yield optimization [10]:

- **Crop Health Monitoring:** Predictive analytics models can forecast crop health indicators such as disease outbreaks, pest infestations, and nutrient deficiencies based on historical data and environmental factors. Early detection enables farmers to take preventive measures and minimize crop damage.
- **Yield Prediction:** Predictive analytics models leverage historical yield data, weather forecasts, and agronomic factors to predict crop yields for upcoming seasons. By accurately forecasting yields, farmers can plan harvest schedules, allocate resources, and negotiate contracts with buyers more effectively.
- **Risk Assessment:** Predictive analytics helps farmers assess and mitigate risks associated with climate variability, market fluctuations, and input costs. By simulating different scenarios and analyzing potential outcomes, farmers can develop risk management strategies and safeguard their operations against unforeseen challenges.

In conclusion, data-driven farming, underpinned by advanced analytics, represents a paradigm shift in agriculture. It empowers farmers with actionable insights and decision-making capabilities. By harnessing the power of data, farmers can optimize resources, mitigate risks, and drive sustainable productivity gains, ensuring the resilience and prosperity of agricultural systems in an increasingly complex and dynamic world.

10.5 IoT APPLICATIONS IN SMART FARMING

Integrating IoT devices in agriculture, known as intelligent farming, has revolutionized how farmers monitor, manage, and optimize their operations. This section explores the diverse applications of IoT devices in agriculture, including crop management, livestock monitoring, and environmental sensing.

10.5.1 Introduction to IoT Devices in Agriculture

IoT devices in agriculture encompass various sensors, actuators, and communication technologies that enable real-time monitoring and control of farm operations. These devices collect data on various environmental parameters, crop conditions, and livestock health and transmit it wirelessly to centralized platforms for analysis and decision-making.

Examples of IoT devices commonly used in agriculture include [11]:

- **Soil Moisture Sensors:** These sensors measure soil moisture levels at different depths, helping farmers optimize irrigation schedules and prevent water waste.
- **Weather Stations:** Weather stations collect data on temperature, humidity, rainfall, wind speed, and solar radiation, providing farmers with valuable insights into weather patterns and forecasting.
- **Crop Health Sensors:** These sensors monitor crop health indicators such as leaf temperature, chlorophyll levels, and nutrient concentrations, enabling early detection of pests, diseases, and nutrient deficiencies.
- **Livestock Monitoring Tags:** IoT-enabled tags or collars worn by livestock collect data on animal behavior, activity levels, and health parameters, allowing farmers to monitor individual animals and detect signs of illness or distress.

10.5.2 Monitoring and Control Systems for Crop Management

IoT devices are critical in crop management, providing farmers real-time insights into crop conditions and environmental factors. Monitoring and control systems powered by IoT technology enable farmers to [11]:

- **Optimize Irrigation:** Soil moisture sensors and weather data help farmers precisely manage irrigation, ensuring that crops receive the right amount of water at the right time to maximize yield and minimize water waste.
- **Monitor Crop Health:** Crop health sensors and imaging technologies enable farmers to monitor plant stress, disease outbreaks, and nutrient

deficiencies, allowing for targeted interventions such as pest control measures and fertilizer application.

- **Automate Pest Detection:** IoT devices equipped with cameras and image recognition algorithms can detect signs of pest infestation in crops, enabling early intervention and preventing crop damage.

10.5.3 IoT Applications in Livestock Management and Environmental Monitoring

In addition to crop management, IoT devices are increasingly used in livestock management and environmental monitoring on farms [12]:

- **Livestock Monitoring:** IoT-enabled tags and collars track the location, activity levels, and health status of individual animals, allowing farmers to monitor livestock behavior, detect signs of illness, and optimize feeding and breeding practices.
- **Environmental Monitoring:** IoT devices such as water quality sensors, air quality monitors, and GPS trackers enable farmers to monitor environmental parameters and track changes over time. This information helps farmers maintain environmental sustainability and compliance with regulations.
- **Precision Livestock Farming:** IoT technologies enable precision livestock farming practices, such as automated feeding systems, climate-controlled environments, and remote health monitoring, improving animal welfare and productivity.

In conclusion, IoT applications in intelligent farming offer transformative opportunities for farmers to increase efficiency, reduce costs, and enhance sustainability across all aspects of agricultural production. By harnessing the power of IoT devices, farmers can make data-driven decisions, optimize resource utilization, and adapt to changing environmental conditions, ensuring the long-term viability of agricultural systems in an increasingly interconnected world.

10.6 AI IN AGRICULTURE

AI is revolutionizing agriculture by empowering farmers with advanced decision-making capabilities, enabling proactive pest management, and driving the development of autonomous farming systems. This section explores AI's multifaceted role in agriculture, from supporting farmers in making decisions to implementing machine learning algorithms for crop health monitoring and the emergence of autonomous farming systems.

10.6.1 Role of AI in Decision-Making for Farmers

AI plays a pivotal role in enhancing decision-making for farmers by analyzing vast amounts of data and generating actionable insights. Critical aspects of AI-driven decision support in agriculture include [13]:

- **Predictive Analytics:** AI algorithms analyze historical and real-time data on weather patterns, soil conditions, crop health, and market trends to predict future outcomes and inform decision-making. Farmers can optimize planting schedules, resource allocation, and marketing strategies by forecasting crop yields, market demand, and input requirements.

- **Prescriptive Recommendations:** AI systems provide prescriptive recommendations tailored to specific farm conditions and goals. They guide farmers on optimal crop varieties, planting densities, irrigation schedules, and pest management strategies, helping them maximize productivity, minimize risks, and achieve sustainable outcomes.
- **Risk Management:** AI tools assess and mitigate risks associated with climate variability, pest outbreaks, market fluctuations, and input costs by simulating different scenarios and analyzing potential outcomes. By identifying risks and developing contingency plans, farmers can protect their investments and ensure the resilience of their operations.

10.6.2 Machine Learning Algorithms for Crop Disease Detection and Pest Control

Machine learning algorithms are increasingly used in agriculture to detect crop diseases and pests early, enabling timely intervention and minimizing crop losses. Critical applications of machine learning in crop health monitoring include [14]:

- **Image Recognition:** Machine learning models analyze images of crops captured by drones, satellites, or smartphones to identify signs of disease, nutrient deficiencies, and pest damage. By automating the detection process, farmers can quickly identify and treat affected areas, preventing the spread of diseases and minimizing yield losses.
- **Sensor Data Analysis:** Machine learning algorithms process data from IoT sensors, such as temperature, humidity, and leaf wetness sensors, to detect anomalies indicative of disease or pest infestations. Machine learning models can accurately predict disease outbreaks and recommend appropriate control measures by correlating sensor data with historical patterns and environmental conditions.
- **Disease Prediction Models:** Machine learning models leverage historical data on crop diseases, weather conditions, and agronomic practices to develop predictive models that forecast disease risk and severity. Farmers can integrate these models into decision support systems to proactively manage disease outbreaks and minimize crop damage.

10.6.3 Autonomous Farming Systems Powered by AI

Autonomous farming systems powered by AI and robotics are revolutionizing agricultural operations by automating labor-intensive tasks and improving efficiency. Critical components of autonomous farming systems include [15]:

- **Robotic Harvesting:** AI-enabled robots with computer vision systems and robotic arms harvest fruits and vegetables with precision and efficiency, reducing labor costs and minimizing crop damage.

- **Autonomous Tractors:** AI-powered tractors equipped with GPS, sensors, and actuators can autonomously navigate fields, plow, plant, and apply inputs, optimizing field operations and reducing fuel consumption.
- **Drone Technology:** AI-enabled drones equipped with cameras, sensors, and AI algorithms monitor crops from the air, identifying areas of stress, disease, or nutrient deficiencies. By providing aerial imagery and real-time insights, drones enable farmers to make data-driven decisions and optimize crop management practices.

In conclusion, AI is transforming agriculture by empowering farmers with advanced decision-making capabilities, enabling proactive pest management, and driving the development of autonomous farming systems. By harnessing the power of AI-driven technologies, farmers can optimize productivity, minimize risks, and achieve sustainable outcomes in an increasingly complex and dynamic agricultural landscape.

10.7 ROBOTICS AND AUTOMATION IN AGRICULTURAL PRACTICES

Integrating robotics and automation technologies in agriculture revolutionizes traditional farming practices, increases efficiency, and reduces labor costs. This section explores the latest automation trends in agriculture, robotics for planting, harvesting, and sorting, and drone applications in precision agriculture.

10.7.1 Automation Trends in Agriculture

Automation has become a prominent agricultural trend, driven by robotics, AI, and sensor technology advancements. Critical trends in agricultural automation include:

- **Autonomous Vehicles:** Self-driving tractors and machinery equipped with GPS, sensors, and AI algorithms can navigate fields, perform tasks such as plowing, planting, and spraying autonomously, and optimize field operations.
- **Robotic Systems:** Robots equipped with robotic arms, cameras, and sensors are increasingly used for planting, harvesting, sorting, and packaging crops, reducing labor costs and increasing efficiency.
- **IoT Integration:** Integrating IoT devices and sensor networks enables real-time monitoring and control of agricultural operations, allowing farmers to optimize resource utilization and minimize waste.

10.7.2 Robotics for Planting, Harvesting, and Sorting

Robotics plays a crucial role in automating labor-intensive tasks such as planting, harvesting, and sorting crops.

Critical applications of robotics in agriculture include [15]:

- **Planting:** Robotic planters equipped with precision seeding mechanisms can plant crops accurately and consistently, optimizing planting densities and minimizing seed wastage.

- **Harvesting:** Robotic harvesters equipped with computer vision systems and robotic arms can harvest fruits, vegetables, and grains precisely and efficiently, reducing labor costs and minimizing crop damage.
- **Sorting:** Robotic sorting systems equipped with cameras and sensors can sort fruits, vegetables, and grains based on size, color, ripeness, and quality, ensuring uniformity and consistency in the final product.

10.7.3 Drones and Their Applications in Precision Agriculture

Drones, also known as UAVs, are revolutionizing precision agriculture by providing farmers with aerial imagery, real-time data, and insights into crop health and field conditions. Critical applications of drones in precision agriculture include:

- **Aerial Imaging:** Drones equipped with cameras, multispectral sensors, and thermal imaging cameras can capture high-resolution aerial imagery of crops, soil, and terrain, providing farmers with valuable insights into crop health, nutrient deficiencies, and water stress.
- **Field Monitoring:** Drones can monitor fields and crops from the air, identifying areas of stress, disease, or pest infestations and enabling farmers to take timely corrective actions to mitigate crop losses.
- **Crop Spraying:** Drones equipped with precision spraying systems can accurately and efficiently apply fertilizers, pesticides, and herbicides to crops, reducing chemical usage, minimizing environmental impact, and optimizing crop yields.

In conclusion, robotics and automation technologies are transforming agricultural practices by increasing efficiency, reducing labor costs, and improving productivity. From autonomous vehicles and robotic systems to drones and IoT integration, the future of agriculture is increasingly automated, enabling farmers to achieve sustainable outcomes and meet the growing demand for food in an ever-changing world.

10.8 SUSTAINABILITY AND ENVIRONMENTAL IMPACT

Sustainability and environmental stewardship are becoming increasingly important considerations in modern agriculture, especially in Industry 4.0. This section explores how Industry 4.0 enables sustainable agriculture practices, reduces the environmental footprint through advanced analytics, and promotes biodiversity and soil health in modern farming.

10.8.1 Sustainable Agriculture Practices in Industry 4.0

Industry 4.0 technologies offer innovative solutions for promoting sustainability in agriculture:

- **Precision Farming:** Precision agriculture practices, enabled by IoT, AI, and automation, optimize resource use by targeting inputs such as water, fertilizers, and pesticides to crops' needs. This reduces waste, minimizes environmental impact, and improves resource efficiency.

- **Conservation Tillage:** Advanced analytics and machine learning algorithms can analyze soil data to optimize tillage practices, such as reduced or no-till farming. Conservation tillage helps preserve soil structure, reduce erosion, and enhance carbon sequestration, contributing to soil health and sustainability.
- **Crop Rotation and Diversity:** AI-driven decision support systems can analyze crop performance data and recommend optimal crop rotation schedules to enhance soil fertility, reduce pest pressure, and minimize disease outbreaks. Crop rotation and diversity promote ecosystem resilience and long-term sustainability.

10.8.2 Reducing Environmental Footprint Through Advanced Analytics

Advanced analytics play a crucial role in reducing the environmental footprint of agriculture:

- **Efficient Resource Management:** Data-driven insights enable farmers to optimize resource use, such as water, energy, and fertilizers, by identifying inefficiencies and implementing targeted interventions. This reduces resource waste and environmental pollution, contributing to sustainable agricultural practices.
- **Pollution Prevention:** Predictive analytics models can forecast environmental risks, such as nutrient runoff, pesticide drift, and soil erosion, based on weather patterns, soil conditions, and farming practices. By identifying potential sources of pollution, farmers can implement preventive measures to protect water quality and ecosystem health.
- **Lifecycle Assessment:** Advanced analytics tools can conduct lifecycle assessments of agricultural products, analyzing the environmental impact of production, transportation, and consumption. This enables farmers to identify opportunities for reducing greenhouse gas emissions, energy consumption, and waste generation throughout the supply chain.

10.8.3 Promoting Biodiversity and Soil Health in Modern Farming

Biodiversity and soil health are essential components of sustainable farming practices:

- **Habitat Preservation:** IoT devices and drones can monitor wildlife habitats and biodiversity hotspots on farms, identifying conservation-value areas and implementing habitat enhancement measures. Preserving biodiversity promotes ecosystem services such as pollination, pest control, and soil fertility, essential for agricultural productivity and resilience.
- **Soil Health Management:** AI algorithms analyze soil data to assess soil health indicators, such as organic matter content, nutrient levels, and microbial activity. By optimizing soil management practices, such as cover

cropping, crop rotation, and composting, farmers can improve soil structure, fertility, and resilience to drought and erosion.

- **Carbon Sequestration:** Sustainable farming practices, such as agroforestry, conservation tillage, and perennial cropping systems, enhance carbon sequestration in soils and vegetation. Agriculture can mitigate climate change and contribute to the global carbon balance by sequestering atmospheric carbon dioxide.

In conclusion, sustainability and environmental impact are critical considerations in modern agriculture, and Industry 4.0 technologies offer innovative solutions for addressing these challenges. Farmers can ensure agricultural systems' long-term viability and resilience in an increasingly complex and interconnected world by promoting sustainable agriculture practices, reducing environmental footprint through advanced analytics, and promoting biodiversity and soil health.

10.9 CHALLENGES AND FUTURE DIRECTIONS

As agriculture continues to embrace technological advancements and transform in the era of Industry 4.0, several challenges and ethical considerations must be addressed. In addition, exploring future trends and opportunities that will shape the agricultural sector in the coming years is essential.

10.9.1 Addressing Technological and Infrastructural Challenges

Despite the promise of Industry 4.0 technologies in agriculture, several challenges hinder widespread adoption and implementation [14]:

- **Access to Technology:** Ensuring equitable access to technology and digital infrastructure remains challenging, particularly for smallholder farmers in remote or underserved regions. Addressing barriers to access, such as affordability, connectivity, and digital literacy, is essential for democratizing the benefits of advanced analytics and automation in agriculture.
- **Data Integration and Interoperability:** Integrating data from disparate sources, such as IoT devices, drones, and farm management software, poses technical challenges related to data standardization, interoperability, and compatibility. Developing open-source platforms and data exchange protocols can facilitate seamless data integration and interoperability, enabling holistic decision-making and optimization of agricultural operations.
- **Cybersecurity and Data Privacy:** As agriculture becomes increasingly data-driven, ensuring the security and privacy of agricultural data is paramount. Farmers must safeguard sensitive information, such as crop yields, soil data, and farm management practices, from cyber threats and unauthorized access. Implementing robust cybersecurity measures, encryption protocols, and data governance frameworks is crucial for protecting agricultural data integrity and confidentiality.

10.9.2 Ethical Considerations in Data-Driven Agriculture

The adoption of data-driven technologies in agriculture raises ethical considerations related to data ownership, privacy, and algorithmic bias [15]:

- **Data Ownership and Control:** Farmers must retain ownership and control over their agricultural data, including data generated by IoT devices, drones, and automated machinery. Clear data ownership agreements and consent mechanisms should ensure farmers have autonomy over how their data is collected, used, and shared.
- **Privacy and Consent:** Collecting and analyzing agricultural data may inadvertently infringe on individual privacy rights, particularly when data is collected from neighboring farms or shared with third-party service providers. Farmers should obtain informed consent from stakeholders and implement data anonymization and aggregation techniques to protect individual privacy while enabling data-driven decision-making.
- **Algorithmic Bias and Fairness:** AI algorithms used in agriculture may perpetuate biases or inequalities, leading to inequitable outcomes for marginalized communities or vulnerable populations. Farmers and developers must prioritize fairness, transparency, and accountability in algorithm design, implementation, and validation to mitigate bias and ensure equitable access to agricultural technologies and opportunities.

10.9.3 Future Trends and Opportunities in the Agricultural Sector

Looking ahead, several trends and opportunities are poised to shape the future of agriculture [7]:

- **Adoption of Emerging Technologies:** Emerging technologies such as blockchain, quantum computing, and synthetic biology hold promise for revolutionizing agriculture by enhancing traceability, optimizing genetic breeding, and developing sustainable alternatives to conventional farming practices.
- **Climate Resilience and Adaptation:** Climate change poses significant challenges to agricultural productivity and food security, requiring innovative solutions for climate resilience and adaptation. Sustainable farming practices, precision irrigation systems, and climate-smart crops can help mitigate the impact of climate variability and extreme weather events on agricultural systems.
- **Circular Economy and Sustainable Supply Chains:** The transition to a circular economy in agriculture emphasizes resource efficiency, waste reduction, and closed-loop systems. Circular agriculture practices such as nutrient recycling, bioenergy production, and regenerative agriculture promote sustainability throughout the agricultural value chain, from production to consumption.

In conclusion, navigating data-driven agriculture's challenges and ethical considerations requires collaborative efforts from farmers, policymakers, researchers, and technology developers. By addressing technological and infrastructural challenges, upholding ethical principles, and embracing future trends and opportunities, the agricultural sector can harness Industry 4.0's transformative potential to build resilient, sustainable, and equitable food systems for future generations.

10.10 CASE STUDIES AND SUCCESS STORIES

This section delves into real-world examples of Industry 4.0 implementation in agriculture, showing case studies demonstrating increased efficiency, profitability, and sustainability. These success stories offer valuable lessons learned and best practices for aspiring innovative farmers leveraging advanced technologies to optimize their operations.

10.10.1 Real-World Examples of Industry 4.0 Implementation in Agriculture

10.10.1.1 Case Study 1: FarmBot

FarmBot is an automated precision farming system that leverages robotics, IoT, and AI to enable small-scale farmers to grow crops with precision and efficiency. With FarmBot, farmers can remotely plan, monitor, and control planting, watering, and weeding operations using a user-friendly web interface. By automating repetitive tasks and optimizing resource use, FarmBot helps farmers maximize yields while minimizing labor costs and environmental impact (see Figures 10.1–10.4).

10.10.1.2 Case Study 2: John Deere's Precision Agriculture Solutions

John Deere, a leading agricultural machinery manufacturer, offers precision agriculture solutions that integrate IoT, GPS, and automation technologies to optimize farming operations. John Deere's precision planting systems enable farmers to achieve precise seed placement, spacing, and depth, improving crop emergence and yield uniformity. In addition, John Deere's autonomous tractors and machinery

FIGURE 10.1 FarmBot Express and Express XL automate the veggie garden [16].

FIGURE 10.2 Precision Ag Technology | Data Management | John Deere Australia [17].

FIGURE 10.3 AeroFarms—a sample of vertical farm [18].

can perform field operations such as plowing, seeding, and spraying autonomously, reducing operator fatigue and increasing productivity.

10.10.2 Case Studies Showing Increased Efficiency and Profitability

10.10.2.1 Case Study 3: AeroFarms

AeroFarms is a vertical farming company that utilizes aeroponic technology, IoT, and data analytics to grow leafy greens and herbs indoors with minimal water and

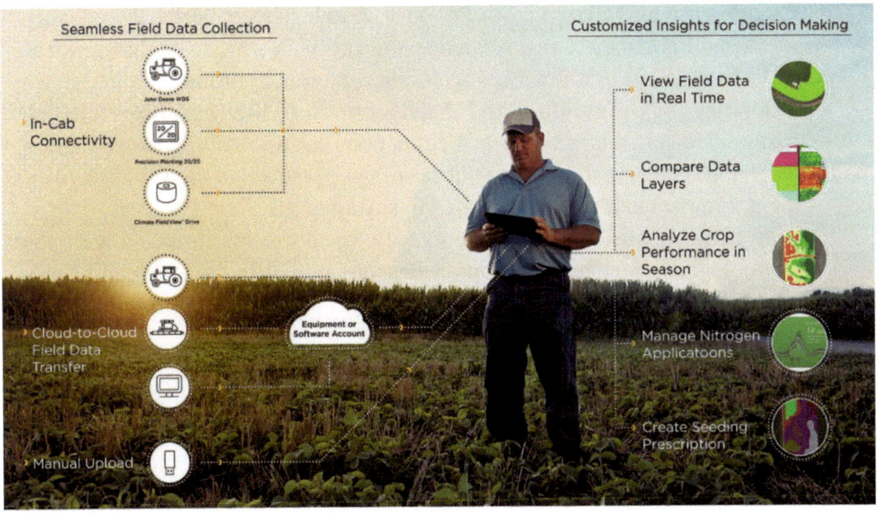

FIGURE 10.4 The Climate Corporation announces multiple data connectivity agreements, making the Climate FieldView™ platform the most broadly connected in the industry [19].

energy usage. By vertically stacking crops in controlled environments, AeroFarms achieves higher yields per square foot than traditional farming methods, using 95% less water. In addition, AeroFarms' data-driven approach enables precise monitoring and optimization of environmental conditions, leading to consistent crop quality and increased profitability.

10.10.2.2 Case Study 4: The Climate Corporation's Climate FieldView

The Climate Corporation, a subsidiary of Bayer, offers Climate FieldView. This digital agriculture platform integrates satellite imagery, weather data, and machine learning algorithms to provide farmers with actionable insights and decision-support tools. By analyzing field-level data, Climate FieldView helps farmers optimize planting decisions, manage inputs, and monitor crop health in real-time. As a result, farmers can increase efficiency, reduce costs, and maximize profitability while minimizing environmental impact.

10.10.3 Lessons Learned and Best Practices for Aspiring Smart Farmers

- **Start Small, Scale Gradually:** Implement simple, cost-effective technologies and gradually increase their effectiveness as you gain experience and confidence.
- **Embrace Data-Driven Decision-Making:** Invest in data collection and analysis tools to gather insights into crop performance, soil health, and environmental conditions. This will enable informed decision-making and the optimization of farming practices.

- **Collaborate and Learn from Peers:** Join industry networks, attend workshops, and participate in knowledge-sharing platforms to exchange ideas, experiences, and best practices with other intelligent farmers.
- **Prioritize Sustainability and Environmental Stewardship:** Integrate sustainable farming practices, such as conservation tillage, crop rotation, and habitat preservation, into your farming operations to minimize environmental impact and promote long-term sustainability.
- **Stay Flexible and Adapt to Change:** Be open to experimenting with new technologies, practices, and approaches, and be willing to adapt and iterate based on feedback and lessons learned.

In conclusion, case studies and success stories of Industry 4.0 implementation in agriculture highlight the transformative potential of advanced technologies in increasing efficiency, profitability, and sustainability. By learning from these examples, aspiring innovative farmers can glean valuable insights and best practices to optimize their operations and thrive in the future's ever-evolving agricultural landscape.

10.11 CONCLUSION: THE FUTURE LANDSCAPE OF AGRICULTURE

In this final section, we reflect on the transformative potential of Industry 4.0 in agriculture, summarize key insights and findings from the preceding chapters, and issue a call to action for embracing innovation and sustainability in farming practices.

10.11.1 Summary of Key Insights and Findings

Throughout this chapter, we have explored the intersection of advanced analytics, robotics, and automation with agriculture, ushering in the era of Industry 4.0. Key insights and findings include:

- **The Evolution of Agriculture:** From traditional farming methods to modern techniques driven by data, AI, and IoT, agriculture has significantly transformed, increasing productivity, efficiency, and sustainability.
- **Role of Advanced Technologies:** Industry 4.0 technologies such as AI, robotics, and IoT revolutionize farming practices, enabling precision agriculture, data-driven decision-making, and autonomous farming systems.
- **Sustainability and Environmental Impact**: Industry 4.0 offers opportunities to promote sustainability in agriculture by optimizing resource use, reducing environmental footprint, and enhancing biodiversity and soil health.
- **Challenges and Ethical Considerations:** Addressing technological barriers, ensuring data privacy and security, and addressing ethical considerations such as algorithmic bias are critical for realizing Industry 4.0's full potential in agriculture.

10.11.2 Reflections on the Transformative Potential of Industry 4.0 in Agriculture

The convergence of Industry 4.0 technologies with agriculture holds immense promise for addressing global challenges such as food security, climate change, and environmental degradation. By harnessing the power of data, AI, and automation, farmers can optimize resource use, increase productivity, and reduce environmental impact, ensuring the resilience and sustainability of agricultural systems in an increasingly complex and interconnected world.

10.11.3 Call to Action for Embracing Innovation and Sustainability in Farming Practices

As we look to the future landscape of agriculture, farmers, policymakers, researchers, and industry stakeholders must embrace innovation and sustainability in farming practices. This includes:

- **Investing in Research and Development:** Continued investment in research and development is essential for advancing Industry 4.0 technologies, driving innovation, and overcoming technological barriers in agriculture.
- **Promoting Collaboration and Knowledge Sharing:** Collaboration among farmers, researchers, policymakers, and industry stakeholders is crucial for sharing best practices, fostering innovation, and addressing common challenges in agriculture.
- **Embracing Sustainable Practices:** Prioritizing sustainability in farming practices, such as regenerative agriculture, agroecology, and carbon sequestration, is essential for mitigating climate change, preserving natural resources, and ensuring the long-term viability of agricultural systems.
- **Empowering Farmers with Digital Skills:** Providing farmers with access to training and education in digital literacy, data analytics, and technology integration is essential for enabling them to harness the full potential of Industry 4.0 in agriculture and adapt to evolving technological landscapes.

In conclusion, Industry 4.0's transformative potential is shaping the future landscape of agriculture, offering unprecedented opportunities for innovation, sustainability, and resilience. By embracing advanced technologies, fostering collaboration, and prioritizing sustainability, we can build a more efficient, equitable, and environmentally sustainable agricultural sector for future generations.

REFERENCES

[1] Aditya Shastry, K. and H. Sanjay, Data analysis and prediction using big data analytics in agriculture. In P.K. Pattnaik, R. Kumar, & S. Pal (Eds.), *Internet of things and analytics for agriculture*, Vol. 2. 2020: Springer. p. 201–224.

[2] Akhter, R. and S.A. Sofi, Precision agriculture using IoT data analytics and machine learning. *Journal of King Saud University-Computer and Information Sciences*, 2022. 34(8): p. 5602–5618.

[3] Byrum, J., et al., Advanced analytics for agricultural product development. *Interfaces*, 2016. 46(1): p. 5–17.
[4] Chergui, N., M.-T. Kechadi, and M. McDonnell, The impact of data analytics in digital agriculture: a review. In *2020 International multi-conference on: "organization of knowledge and advanced technologies" (OCTA)*. 2020: IEEE.
[5] Choudhury, A., et al., *Smart agriculture automation using advanced technologies: data analytics and machine learning, cloud architecture, automation and IoT*. 2022: Springer.
[6] Elijah, O., et al., An overview of the Internet of Things (IoT) and data analytics in agriculture: benefits and challenges. *IEEE Internet of Things Journal*, 2018. 5(5): p. 3758–3773.
[7] Gupta, D.S.K. and S. Malik, Application of predictive analytics in agriculture. *TTID-MKD*, 2022. 2(4): p. 1–5.
[8] Krisnawijaya, N.N.K., et al., Data analytics platforms for agricultural systems: a systematic literature review. *Computers and Electronics in Agriculture*, 2022. 195: p. 106813.
[9] Naqvi, R.Z., et al., Big data analytics and advanced technologies for sustainable agriculture. In C.M. Hussain & P. Di Sia (Eds.), *Handbook of smart materials, technologies, and devices: applications of Industry 4.0*, 2020: Springer International Publishing. p. 1–27.
[10] Niu, H. and Y. Chen, Smart big data in digital agriculture applications: acquisition, advanced analytics, and plant physiology-informed artificial intelligence, 2023: Springer Nature.
[11] Pattnaik, P.K., R. Kumar, and S. Pal, *Internet of things and analytics for agriculture*, Vol. 2. 2020: Springer.
[12] Sekhar, C.C., et al., Effective use of big data analytics in crop planning to increase agriculture production in India. *International Journal of Advanced Science and Technology*, 2018. 113: p. 31–40.
[13] Shakoor, N., et al., Big data-driven agriculture: big data analytics in plant breeding, genomics, and remote sensing technologies to advance crop productivity. *The Plant Phenome Journal*, 2019. 2(1): p. 1–8.
[14] Sharma, R., S.S. Kamble, and A. Gunasekaran, Big GIS analytics framework for agriculture supply chains: a literature review identifying the current trends and future perspectives. *Computers and Electronics in Agriculture*, 2018. 155: p. 103–120.
[15] Tarik, H. and O.J. Mohammed, Big data analytics and artificial intelligence serving agriculture. In *Advanced intelligent systems for sustainable development (AI2SD'2019), Vol. 2: Advanced intelligent systems for sustainable development applied to agriculture and health*. 2020: Springer.
[16] TechAU, *FarmBot express and express XL automate your veggie garden*. 2024. Available from: https://techau.com.au/farmbot-express-and-express-xl-automates-your-veggie-garden/ (cited 2024, April 6).
[17] Deere, J., *Precision Ag Technology | Data management | John Deere Australia*. 2024. Available from: https://www.deere.com.au/en/technology-products/precision-ag-technology/data-management/data-sync/ (cited 2024, April 6).
[18] Janiec, C., *AeroFarms goes public via SPAC to "set the narrative" in vertical farming*. 2021. Available from: https://www.agriinvestor.com/aerofarms-goes-public-via-spac-to-set-the-narrative-in-vertical-farming/ (cited 2024, April 6).
[19] FieldView, *The climate corporation*. 2016. Available from: https://www.businesswire.com/news/home/20160304005395/en/The-Climate-Corporation-Announces-Multiple-Data-Connectivity-Agreements-Making-the-Climate-FieldView%E2%84%A2-Platform-the-Most-Broadly-Connected-in-the-Industry (cited 2024, April 6).

11 Artificial Intelligence in Insurance

Transforming Risk Management and Customer Experience

Anand S. Rao and Ali Soofastaei

11.1 INTRODUCTION

Over the past few decades, the insurance industry has been at the forefront of adopting digital technologies, evolving through successive waves that have reshaped its landscape. This digital transformation has gained a significant momentum in recent years with the advent of generative artificial intelligence (AI), representing a leap forward in how insurance companies operate and interact with their customers.

11.1.1 The Journey of Digital Transformation

The journey began with the first wave of **digitization**, transforming analog and physical data into digital formats. This fundamental change laid the groundwork for the **big data** era, characterized by massive increases in data availability and analytical capabilities. Following closely were the waves of **automation and analytics**, enhancing operational efficiencies and decision-making processes. With the advent of **Generative AI**, significant productivity improvement is promised.

11.1.2 The Impact of AI on the Insurance Industry

Recent research in AI within the insurance industry indicates a transformative progression across various facets of operations and services. Initially highlighted by [1], significant advancements in AI have optimized the accuracy and efficiency of claims processing through innovative systems that integrate big data and AI technologies. Following this, [2] provides a bibliometric analysis that underscores the widespread adoption and critical impact of machine learning (ML) and AI, particularly in automating claims and enhancing customer service.

Further exploration into the practical applications is provided by [3], which discusses the integration of decision support systems and AI to automate and enhance insurance agent activities, improving decision-making and operational efficiency.

DOI: 10.1201/9781003186823-11

Similarly, [4] focuses on AI-driven enhancements in claims processing, leveraging advanced technologies to streamline operations and personalize customer interactions.

Risk management also benefits from AI as [5] examines the use of supervised ML in accident risk analysis, demonstrating the technology's capability in predictive risk assessments. Finally, the issue of fraud detection is tackled by [6], where a secure AI architecture is developed to efficiently monitor and detect fraud in real time, significantly reducing the incidence of fraudulent claims and enhancing the security of transactions.

Together, these studies paint a comprehensive picture of AI's role in reshaping the insurance industry through enhanced operational efficiency, risk management, and customer interaction, setting a foundational base for future technological integrations.

11.1.3 Digital Waves and Their Synergies

This chapter explores how each digital wave has uniquely contributed to and seamlessly integrated with the next, collectively revolutionizing the insurance industry. We examine specific use cases across various functional insurance areas—from marketing and product development to claims management and risk assessment. Unlike typical approaches that focus on isolated applications of AI or analytics, this narrative delves into how these technologies interplay and build upon one another, progressively transforming the insurance landscape.

11.1.4 Comprehensive Overview of Digital Transformations

We aim to provide a comprehensive overview of digital transformations in the insurance sector, highlighting the synergies between different technological waves and their cumulative impact on the industry. We begin by exploring the initial stages of digital integration in the insurance sector, focusing on transitioning from manual processes to digitized operations. The subsequent sections analyze the increasing complexity of technological applications in insurance, from leveraging big data for enhanced decision-making to automation and the application of AI for predictive analytics. The fifth wave of AI follows this. The final parts discuss the ethical implications of AI in insurance and outline potential areas for future research, emphasizing the importance of responsible AI practices in sustaining industry growth and trust.

11.2 FIVE WAVES OF DIGITAL TRANSFORMATIONS

In today's fast-paced and complex insurance industry, companies increasingly recognize the profound impact of digital transformation, which unfolds across five successive waves—from essential digitization to the sophisticated realms of AI. This evolution transcends a mere technological upgrade, representing a strategic overhaul that addresses unique industry challenges such as regulatory compliance, risk management, and customer experience enhancements. Digitization boosts operational productivity, streamlines intricate processes, and significantly enhances revenue generation and profitability within the insurance sector. Navigating these five waves

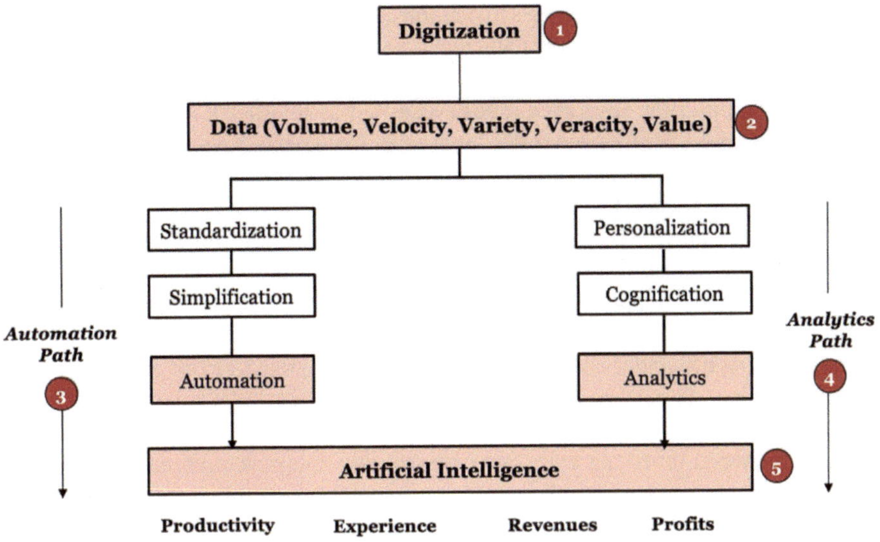

FIGURE 11.1 Five waves of digital transformations.

of digital transformation [7] allows insurance companies to refine risk assessment models, optimize claims processing, and engage with policyholders more effectively, delivering personalized services and fostering higher customer loyalty.

Accompanying this exploration of digital transformation in the insurance industry is Figure 11.1, which outlines the progressive stages from essential digitization to the cutting-edge applications of AI. This diagram serves as a roadmap, illustrating how each wave builds upon the previous, enhancing the industry's ability to tackle challenges and seize opportunities through digital innovation.

11.2.1 The First Wave: Digitization

Digitization, the foundational process of converting analog or physical data into digital formats, is pivotal across industries. It sets the groundwork for advancements in big data, automation, analytics, and AI. This transformation is crucial as it facilitates transitioning from traditional operations to modern, technology-driven processes that significantly enhance efficiency and innovation.

Digitizing data is the first critical step toward harnessing the power of big data analytics. It allows organizations to collect, store, and analyze vast amounts of data, unlocking insights that drive more informed decision-making and strategic business planning. In addition, digitization enables the seamless integration of automation technologies that streamline operations and reduce human error, thereby increasing productivity and operational efficiency.

For example, major insurance companies like State Farm and Allianz have implemented extensive digitization initiatives. State Farm has digitized claim-processing systems, allowing mobile app submissions, dramatically reducing processing times, and improving customer satisfaction [8]. By contrast, Allianz has leveraged digitization to integrate blockchain technology for secure and transparent policy management [9].

Digitization in the insurance industry has led to significant shifts in operational efficiencies and a broader economic landscape. Integrating advanced digital technologies has enhanced decision-making, productivity, and a better customer experience while reducing costs [1].

Advancements in AI—especially ML and deep learning—are set to revolutionize the insurance industry. These technologies enable insurers to offer "active" insurance products that adapt to real-time individual behavior changes and circumstances. Insurers adopting these technologies report significant improvement in claims processing efficiency and increased customer satisfaction due to more personalized and timely service.

Further, digitized data is the backbone for deploying AI and ML technologies. These advanced tools can automate complex decision-making processes, optimize data processing, and personalize customer interactions through intelligent systems such as chatbots and virtual assistants. Such applications refine customer service and enhance the capability to analyze and utilize data effectively.

However, transitioning to a digitized environment involves challenges, including significant investments in technology and training. Organizations must thoughtfully integrate new digital tools with existing systems and manage evolving regulatory requirements to safeguard data privacy and security. Embracing digitization requires a strategic approach to technology adoption and a commitment to fostering a culture of continuous innovation and adaptability within organizations. Insurance companies must navigate the data privacy and security challenges while ensuring compliance with global data protection regulations such as general data protection regulation (GDPR) in Europe and California consumer privacy act (CCPA) in California.

In the insurance sector, digitization is becoming increasingly pivotal, transforming traditional business models and operational processes. This digital shift enables insurers to collect, process, and analyze large volumes of data more efficiently, essential for refining risk assessment, pricing strategies, and customer segmentation. Using advanced analytics powered by digitized data, insurance companies can gain deeper insights into customer behavior and preferences, leading to more personalized service offerings. In addition, digitization facilitates the development of new insurance products, such as usage-based insurance models that leverage real-time data from Internet of Things (IoT) devices such as automotive sensors and health monitors.

11.2.1.1 Key Operational Areas Transformed by Digitization

Digitization has profoundly transformed several critical operational areas of insurance by converting analog data into digital formats, leading to significant efficiency gains and improved accuracy.

11.2.1.2 Property and Casualty Insurance (P&C Insurance)

Customer Onboarding and Engagement: Digitization has replaced time-consuming face-to-face meetings and manual data entry with online applications, where customers directly input their personal and medical histories. This speeds up the process and improves data accuracy. Digital tools like mobile apps and online portals allow policyholders to manage their accounts, fostering greater engagement and retention.

Policy Administration and Management: Transitioning from paper-based systems to electronic records has streamlined policy management. Digital platforms automate updates and recalculations related to premiums and benefits, integrating data across systems to enhance accuracy and access, reduce errors, and boost efficiency.

Claims and Benefits Disbursement: Digitization has simplified claims processing, from manual submissions and physical checks to online forms and electronic payments. This change speeds up disbursements, ensures security, and enhances the customer experience during sensitive periods.

11.2.1.3 Life and Annuity

Customer Onboarding and Engagement: The shift from paper-based applications to online forms has streamlined customer onboarding, reduced manual entry errors, and expedited policy issuance. Digital tools like mobile apps and online portals allow customers to manage their policies, enhancing transparency and engagement.

Policy Administration and Management: Transitioning from physical documents to electronic records has transformed policy management. Automated digital systems now update and manage policyholder information more efficiently, ensuring accuracy and ease of access.

Claims and Benefits Disbursement: Digitization has simplified claims processing. Policyholders can submit claims online, with automated systems quickly assessing and processing these submissions. Electronic payments ensure swift and secure benefits disbursement, improving the experience during sensitive times.

11.2.1.4 Reinsurance

Data Exchange and Collaboration: Digitization has enabled reinsurers to streamline data sharing with insurance companies through cloud-based platforms. These digital tools facilitate real-time data exchange and collaboration, enhancing transparency and speeding up decision-making processes.

Risk Modeling and Analysis: Advanced digital technologies like AI and big data analytics have transformed risk assessment in reinsurance. Digitized historical data creates sophisticated models that predict potential losses more accurately, allowing reinsurers to price risks more effectively and manage their portfolios strategically.

Contract Management and Compliance: Digital contract management systems have replaced paper-based documentation, automating contract creation, execution, and storage. This digitization helps reinsurers manage complex contractual terms more efficiently, ensure compliance with regulations, and reduce administrative costs by streamlining audits and compliance checks.

11.2.2 The Second Wave: Big Data

The digital revolution, marked by the advent of the internet and smartphones in the late 1990s and early 2000*s*, introduced a surge in semi-structured and unstructured data. This era increased data volume and shifted its focus from businesses to consumers, encompassing social and behavioral aspects like preferences, attitudes, and interactions. This transformation catalyzed the "big data" revolution, characterized by the five Vs—Volume, Variety, Velocity, Veracity, and Value—each representing challenges and opportunities for the insurance industry.

Today, big data technologies are crucial in processing vast datasets that inform decisions, enhance operational efficiencies, and tailor customer experiences. For example, reinsurers use big data for catastrophe modeling, combining historical data from past disasters with climate models and geographical information systems to refine risk assessment and pricing strategies.

Moreover, the evolution of technologies such as Hadoop has shown that while specific tools may transform, the underlying utility of big data remains robust, especially with the integration of IoT, Industrial IoT, and 5G technologies. These advancements promise a future where sensor data, streaming in real-time from billions of devices, will significantly outpace human-generated data.

Big data has been a game-changer for the insurance industry, helping companies process large volumes of information, increase workflow efficiency, and reduce operational costs [10]. Here are some critical significant data sources and their applications across the three sectors of P&C, life and annuity, and reinsurance.

11.2.2.1 P&C Insurance

Telematics Data: Real-time driver behavior and usage data collected through telematics allow insurers to provide premium discounts and usage-based insurance.

Sensor Data: Data from devices on the IoT, such as drones, smart homes, and cars, provide valuable insights into customer behavior and risks. For example, car insurance companies can use locational data from the global positioning system to create highly personalized customer profiles.

Online Behavior Data: Social media activity, shopping behavior, and browsing activity can be analyzed to create targeted marketing campaigns and acquire new customers.

Traditional Data Sources: Accident statistics, policyholder's personal information, and third-party sources are still used to group people into risk categories, prevent fraud losses, and optimize expenses.

11.2.2.2 Life and Annuity Insurance

Health and Lifestyle Data: Life insurance companies collect data from various sources, including prescription history, motor vehicle records, electronic health records, and financial records. This data helps underwrite policies, verify information, and assess risk.

Wearable Device Data: Data from fitness trackers and health apps creates interactive policies that reward healthy behavior.

Social Media Data: Although less common, some insurers use social media data to identify potential fraud and reduce risk.

IoT Data: Wearables and insurance products often go together, with apps encouraging customers to participate in fitness programs and offering discounts for meeting exercise goals.

11.2.2.3 Reinsurance

Real-Time Data Tools: Reinsurance leaders have identified real-time data management as a "game-changer" for assessing and managing risk.

Data Analytics: Advanced data analytics capabilities enable reinsurers to improve their risk assessment and mitigation strategies.

Centralized Data Repositories: Consolidating reinsurance data into a centralized repository improves carriers' bottom lines and enables better decision-making.

Significant data sources have transformed the insurance value chain, helping companies improve their operational efficiency and better serve their customers. The following lists some spaces where these sources impact each element of the insurance value chain:

Marketing: Big data sources enable insurance companies to create targeted marketing campaigns. By analyzing customer behavior and preferences through social media, online shopping, and browsing, insurers can identify their target audience and craft tailored messages, which helps in customer acquisition and retention.

Product Development: Data from IoT devices, wearables, and health apps allow insurers to develop innovative products that encourage healthy behavior and safe driving practices.

Sales and Distribution: Big data analytics help insurers streamline sales by automating manual tasks and personalizing the customer experience.

Underwriting and Pricing: This is where big data has had the most impact. By analyzing vast data, insurers can more accurately assess risk profiles and set premiums accordingly. Telematics data, for instance, helps car insurance companies provide usage-based insurance and incentivize safe driving.

Policy Administration and Customer Service: Big data improves customer service by enabling insurers to resolve service issues and provide tailored recommendations quickly. It also helps streamline the claims process, making it faster and more accurate.

Claims or Benefits Management: Big data analytics can identify fraudulent claims, reducing insurers' losses. It also helps automate the claims process, leading to faster resolution and improved customer satisfaction.

Asset Management: Big data helps insurers make informed investment decisions by providing real-time data and advanced analytics capabilities.

Administrative Services: While not directly related to a specific insurance product, big data improves administrative tasks by automating manual processes, reducing costs, and improving efficiency across the organization.

11.2.3 The Third Wave: Automation

With a huge volume of data, organizations must first standardize their business processes. As they start the journey of standardization, they will realize that there are several exceptions to processes that may no longer be required or valid. In addition, some steps in the business process can be eliminated. These changes will inevitably lead to the simplification of the business processes. Removing exceptions to processes and changes to processes, which give more control to business users and power users, could further simplify the business processes. Once simplified, these processes can be automated.

Automation in the insurance industry has evolved significantly since its mid-20th-century origins. Conceptualized initially to enhance manufacturing efficiency, automation has grown to encompass a wide range of applications in the digital era, particularly in sectors such as insurance, where it enhances accuracy and efficiency in customer service and administrative tasks.

Today, automation encompasses everything from simple rule-based tasks to complex processes that require advanced robotic process automation (RPA) and intelligent automation systems. These technologies are pivotal in transforming insurance operations, from underwriting and claims processing to customer service and compliance management.

The business benefits of automation are primarily focused on efficiency. Automation reduces the time required to do repetitive tasks. These repetitive tasks can be manual or cognitive. An example of a repetitive manual task is cutting and pasting information from one application to another. An example of a repetitive cognitive task might be a credit analyst in a bank collecting rating agency data, transaction data, government data, etc., before deciding on a loan application. Automating these tasks results in saving time and thereby a reduction of turnaround time on processes. This, in turn, improves staff or labor productivity and decreases overall labor costs. These direct cost (or bottom-line) benefits can also lead to indirect top-line benefits. Reducing the turnaround time can result in greater customer satisfaction and better customer retention, leading to better profitability. Robotic desktops and RPA can automate many of these repetitive manual and cognitive tasks.

Regarding nonrepetitive or variable tasks, we need techniques such as ML and natural language processing—more the domain of AI—to automate them. Process mining uses ML techniques to take a workflow log or transaction log and find exceptions, bottlenecks, resource contention, turnaround time, etc., that can be used to triage the tasks better and assign them to the suitable skill types within an organization. For example, a customer service trouble ticketing system for trading in an investment bank received multiple trouble tickets or service requests daily. These requests had to be handled within a specific duration (e.g., within 24 hours) due to regulatory requirements. The trouble tickets varied from simple fixes to more complex ones that required the skills of subject matter experts or technical staff. This is a classic example where you have some "routine" fixes and some "non-routine" fixes. The investment bank used NLP and ML to parse the trouble tickets, categorize them by level of complexity, and route them to the right pool of experts to expedite the resolution of problems.

Automation can yield significant benefits in reduced task time, greater staff productivity, and reduced labor costs. However, a significant upfront investment is still involved, which can be recouped from savings. Computing the return on investment (ROI) for such initiatives requires a good baseline of time taken for specific activities, and the productivity benefits may not always translate into reduced headcount.

Automation, particularly RPA, is being leveraged by insurance companies to streamline operations, reduce costs, and improve the overall customer experience. Here are some key automation initiatives across the three sectors.

11.2.3.1 P&C Insurance

Claims Processing: RPA can automate the entire workflow, from intake to assessment to settlement, reducing the time and cost. By leveraging RPA to automate property claims processing, the company has achieved impressive results, including a 15% reduction in processing time and a 10% decrease in the cost of claims [11].

Underwriting: Automation can streamline the underwriting process by collecting and analyzing data from multiple sources, determining risks, and generating quotes. Chubb's adoption of RPA has translated into a remarkable 20% reduction in the cost of underwriting [12].

Policy Servicing: Intelligent automation, including RPA and conversational AI, can automate various steps in policy servicing, such as policy processing, endorsements processing, and addressing customer queries.

First Notice of Loss (FNOL): Intelligent automation can automate the FNOL process, including claim intake, claim review, and setting up the claim in the P&C company's claims management system [13].

Back-Office Tasks: Automation can free employees from administrative duties, allowing them to focus on higher-value work.

11.2.3.2 Life and Annuity Insurance

Underwriting: Automated underwriting uses RPA and AI to generate insurance quotes and assess and price risk, eliminating the need for manual underwriting [14].

Product Development: Automation enables insurers to launch new products faster and adapt to changing market demands.

Customer Service: Automation can provide instant customer support through virtual agents, reducing response times and improving customer satisfaction [14].

Marketing: Automation helps streamline marketing efforts by providing a unified view of the customer across channels and geographies.

Sales and Distribution: Automation enables efficient lead generation and management, allowing faster and more targeted sales.

11.2.3.3 Reinsurance

Underwriting: Automation can streamline the data collection and analysis process, reducing the time and effort required for underwriting.

Claims Management: Automation can help reinsurers process claims more efficiently, reducing the time and resources spent.

Back-Office Processing: Automating bank account reconciliation reduces the number of days from 15 to 3 days [15].

11.2.4 The Fourth Wave: Analytics

With huge volumes of data—standardization, simplification, and automation—organizations can also choose an alternative path of exploiting the data. This path uses the data to personalize user interaction and experience. Users here could be your customers, your suppliers, or your employees. The personalization occurs in three distinct phases. First, one uses the data to understand the behavior of users. The understanding phase is followed by predicting the behavior of users. The third and final stage is to change (or nudge) the behavior of users. For example, consider your favorite streaming service—the company first understands your preferences, watching habits, genres of your liking, etc., then predicts what you are likely to watch and makes appropriate recommendations. As you depend more on a particular streaming service, they can subtly change your viewing habits and produce content that will keep you engaged.

Personalization could lead to what Kevin Kelly calls cognification—making things brighter. Imagine the difference between viewing a detailed catalog (running into the millions) of all movies produced to date in all languages and your favorite streaming service that can recommend movies to you based on your interests and the genre (e.g., the latest action movie or the latest Bollywood movie with your favorite hero!!). The recommendation is tailored to our individual preferences and will change over time as our preferences change and the type of content generated changes. This is what makes them "smart." We all understand that this process is happening not just in the consumer world but also in the enterprise world (or business-to-business or business-to-business-to-consumer value chains).

Unlike the automation path, the critical competitive driver in the analytics path is drawing insights from data and turning those insights into better decisions and actions to produce a better outcome for the customer and the company. Therefore, the emphasis is on decision-making or taking actions that are better than what we would have taken in the absence of the data and the analytics on that data.

The business benefits of analytics are primarily focused on effectiveness. They assist in improving users' experience, making better decisions, or taking better actions. This results in better customer retention, better value for customers, and, therefore, better prices/margins for providers and, ultimately, more revenue and profits for companies that can personalize and unify their products and services. Similar to automation, quantifying the ROI on analytics is a nontrivial task as the effectiveness improvement is based on a human baseline performance. Unfortunately, in several cases, we do not have a baseline for human decision-making in many areas.

Analytics is often categorized into four stages of increasing sophistication—descriptive analytics (asking the question, what hint opened?), diagnostic analytics (why did it happen?), predictive analytics (what will happen?), and prescriptive analytics (what can we do?). In some cases, a fifth phase is added called cognitive analytics (how do we adapt to change?). This broad bucket includes all the AI techniques that can be used with some of the traditional analytics techniques. This leads us to the fourth wave of digital transformation.

Analytics, particularly predictive analytics, is a key tool for insurance companies to improve their operations and enhance their understanding of customers. Here are some key analytics initiatives across the three sectors.

11.2.4.1 P&C Insurance

Claims Management: Predictive analytics can help identify high-cost claims early in the process, allowing for cost containment measures. It can also fast-track low-cost claims for quick settlement, reducing claims administration expenses.

Fraud Detection: Analytics can identify suspicious claims, reduce fraud, and lower annual claims payouts. For example, Auto insurer Infinity Property and Casualty uses predictive analytics to identify fraudulent claims and speed up the settlement of valid claims.

Risk Assessment: Analytics can help insurers tap into new data sources, such as telematics, to improve risk evaluation and pricing.

Customer Retention: Analytics can identify customers unhappy with their coverage, allowing insurers to take proactive measures to prevent churn.

Marketing and Distribution: Analytics can help insurers identify new markets and target prospects more effectively.

11.2.4.2 Life and Annuity Insurance

Underwriting: Analytics can improve underwriting by providing insights from vast data sources, including third-party data.

Product Development: Analytics helps insurers develop innovative products that meet evolving customer demands.

Customer Service: Analytics enables insurers to provide tailored recommendations and improve customer experience.

Marketing: Analytics helps insurers create targeted marketing campaigns and identify new business opportunities.

Sales and Distribution: Analytics can help streamline sales processes and identify cross-selling and upselling opportunities.

11.2.4.3 Reinsurance

Risk Management: Reinsurance firms use analytics to protect risk portfolios against natural disasters and improve risk analysis and business performance.

Data Management: Reinsurance providers invest in centralized data repositories to improve their bottom lines and make better decisions.

Underwriting: Analytics helps reinsurers optimize their underwriting processes and make more informed pricing and reserving decisions.

Claims Management: Analytics can help reinsurers streamline their claims processes and improve recovery rates.

11.2.5 The Fifth Wave: AI

As we reach the top of the maturity curve of both automation and analytics, we invariably morph into AI. In automation, as we move toward intelligent automation, we are increasingly using ML and natural language processing, which is the domain of AI. Similarly, as we move into cognification or more into prescriptive and cognitive analytics, we are in the domain of AI.

The definition we like to use for AI is a classic definition of AI articulated by Stuart Russell and Peter Norvig in their book. We consider AI to be any software system or agent in an environment interacting with other humans and machines that can sense, think, and act to achieve a particular purpose or objective.

AI has existed since 1956—the term was first used by a small group of academics and founding fathers of AI at a conference in Dartmouth. However, AI is still considered an emerging technology in many companies. It has gone through at least two boom-bust cycles in the past and has been on an upsurge in popularity since 2007.

The benefits of AI come from both efficiencies and cost reduction, as well as effectiveness and revenue/margin increases. While data powers the automation, analytics, and AI waves, AI is the glue that binds both automation and analytics.

AI is revolutionizing the insurance industry, improving customer experiences, streamlining operations, and reducing costs. Here are some key AI initiatives across the three sectors.

11.2.5.1 P&C Insurance

Customer Service: NLP-driven chatbots and virtual assistants enhance customer service and automate routine tasks. These chatbots can provide 24/7 support, generate human-like text, and offer tailored policy recommendations.

Agency Performance Management: Enhance agent performance evaluation through predictive analytics and real-time monitoring of key performance indicators (KPIs), enabling tailored training and efficiency improvements [3].

Risk Assessment: Computer vision is used to assess property damage, automate remote inspections, and improve risk assessment. For example, drones equipped with computer vision can safely assess damage after incidents [5].

Fraud Management: AI voice recognition technology helps combat insurance fraud by analyzing tone, speech patterns, and emotions to detect fraudulent intent. It also improves customer experiences by offering faster and more automated responses [6].

11.2.5.2 Life and Annuity Insurance

Marketing: Generative AI creates personalized insurance policies tailored to individual needs, such as health, life, or retirement planning.

Underwriting and Pricing: Natural language processing is used to automate data extraction from various sources, including unstructured text data, to improve underwriting and risk assessment.

Claims Processing: Image recognition, computer vision systems, language recognition, and other AI technologies analyze case information and accelerate the speed of insurance claims settlement [4].

Fraud Management: AI voice recognition can detect fraudulent intent by analyzing tone, speech patterns, and emotions.

11.2.5.3 Reinsurance

Product Development: Computer vision, combined with IoT data, helps reinsurers carefully record the state of assets during underwriting and adjust in near real time. This enables dynamic pricing and improved risk management.

Underwriting and Pricing: Advanced catastrophe risk modeling, powered by ML and trained on actual claims data, improves the authenticity of risk assessments. This leads to more accurate exposure predictions and dynamic pricing for clients.

Claims Management: AI-based claims management systems can process data from various sources, including satellite data, HD video, and IoT datasets. This comprehensive data analysis provides a holistic view of on-site assets, enabling faster and more accurate claims settlements.

11.3 RESPONSIBLE AI IN INSURANCE

The insurance industry is undergoing a digital transformation. However, it is essential to be aware of the risks and ethical challenges posed by AI, generative AI, and other digital technologies. While AI offers many benefits, as discussed earlier, it also introduces significant ethical issues that must be addressed.

The use of AI in insurance has progressed through various stages, from essential digitization to advanced AI applications. Each stage presents unique opportunities and challenges. As AI continues to evolve and reshape the industry, it is crucial to understand the risks and develop proactive strategies for responsible AI implementation. Integrating AI into insurance, including the latest advancements in Generative AI, introduces a range of risks and challenges that require a thoughtful and proactive approach to ensure the responsible use of AI. These risks span various domains, from data handling to algorithmic biases and ethical considerations:

Bias and Fairness: AI systems, including those leveraging Generative AI, can inadvertently perpetuate and amplify existing biases, leading to unfair outcomes and discriminatory practices. Bias can creep into any stage of the AI development process, from biased training data to biased algorithms. This risk is particularly pertinent when using Generative AI for personalized insurance offerings, as seen in life, health, and retirement planning. Proactive measures, such as diverse and inclusive training data, are essential to fostering fairness.

Explainability and Interpretability: The "black-box" nature of some AI systems, especially those utilizing complex Generative AI models, can lead to skepticism and mistrust. Enhancing the interpretability and explainability of AI models is crucial for building trust, ensuring regulatory compliance, and fostering user acceptance. This is especially important when dealing with deepfakes and manipulated media, where interpretability can help detect fraudulent claims.

Safety: AI technologies, including Generative AI, carry the risk of unintended consequences. For example, AI-driven systems interpreting human behavior and emotions or generating text and imagery may be manipulated to create deepfakes or influence people's actions. Ensuring AI's safe and ethical use is critical to mitigating potential harm.

Security: As AI systems handle sensitive data, ensuring data security is paramount. With the advent of Generative AI, the potential for sophisticated data breaches and unauthorized access increases. Robust security measures, such as blockchain technology, are vital to safeguarding customer information and preventing fraud.

Privacy: The extensive data collection and analysis facilitated by AI, including Generative AI, heighten privacy concerns. Insurance companies must navigate complex data regulations such as GDPR and CCPA while protecting customer data. Generative AI's ability to create detailed models and scenarios demands more robust privacy safeguards to prevent potential misuse.

Trust and Transparency: Opaque AI systems, particularly those utilizing Generative AI, can lead to uncertainty and skepticism. Enhancing transparency and clarifying AI decision-making processes are crucial for building trust. Generative AI's complexity underscores the need for interpretability and user understanding to foster trust.

Accountability and Oversight: With AI, including Generative AI, making critical decisions and establishing clear accountability and human oversight are essential. Regular reviews and assessments of AI systems help maintain ethical standards and address potential issues arising from autonomous decision-making.

Risk Management: AI technologies introduce new risks that require proactive management. This includes risks related to data privacy, algorithmic biases, and the potential for Generative AI systems to exhibit unanticipated behaviors or generate misleading content.

Adverse Selection and Moral Hazard: AI-driven personalized insurance offerings, enabled by Generative AI and dynamic pricing models, may lead to adverse selection and moral hazards. Generative AI's ability to create tailored insurance products and adjust pricing based on behavior requires careful monitoring to prevent unintended consequences.

The emergence of several Responsible AI frameworks and toolkits (see [16] with an extensive list of these websites) underscores the growing recognition of AI's societal impact and the need for ethical and technically robust solutions. These frameworks offer a socio-technical system perspective, addressing AI deployment's technical and social aspects.

The technical focus of these frameworks aims to ensure AI systems are fair, interpretable, safe, secure, resilient, and robust. They address some of the issues discussed as follows:

Bias and Fairness: Reducing bias and promoting fairness are critical. However, there is no "silver bullet" answer to addressing all the issues in this area. While some aspects of data bias can be detected and rectified through better sampling and enrichment with synthetic data, fairness is more of a social issue of subjective opinions on what is "fair" or "just" and not necessarily an AI issue. These frameworks help navigate the different metrics and make a more considered decision.

Interpretability and Explainability: Enhancing the interpretability and explainability of AI models is critical to building trust. Once again, organizations have to trade off accuracy with inherently interpretable models. The trade-off depends on the use case under consideration, the stakeholders involved, and the impact of these models.

Privacy and Security: With AI involving sensitive data, privacy and security are paramount. Frameworks assist in navigating evolving data regulations and safeguarding data against breaches, especially in an industry vulnerable to cyberattacks and fraud.

Robustness: AI systems should be reliable and safe over the long term. Frameworks guide model selection and data handling to ensure consistent performance and prevent unintended consequences.

The social focus of these frameworks emphasizes ethical principles, regulatory compliance, and risk management:

Ethics and Regulation: Organizations are guided in navigating the evolving regulatory landscape, translating ethical principles into concrete practices, and aligning with human rights laws. This is essential for gaining societal trust in the insurance industry.

Governance: Holistic governance begins with aligning AI strategies and expectations with the organization's priorities. Some frameworks [17] include planning, model development, and technology sourcing for effective governance.

Risk Management: An enterprise-wide approach to risk management is vital. This includes assessing AI for fairness, safety, and reliability and addressing data privacy and protection, as highlighted by the EU's GDPR, CCPA, and the European Insurance and Occupational Pensions Authority [18].

Accountability: Clear accountability and human oversight are established, with mechanisms for reporting and addressing ethical dilemmas and breaches. The concept of "three lines of defense" practiced in regulated industries such as financial services can be a good model for ensuring accountability with AI.

In addition to the principles and frameworks, technology and consulting companies provide practical tools for assessing and mitigating some risks, including bias tools, explainability, and deep fakes. A detailed list of these tools can be found elsewhere [17]. These socio-technical systems provide a comprehensive guide for organizations to navigate the complex world of responsible AI, ensuring ethical and technically sound solutions in the insurance domain.

11.4 CONCLUSION

This chapter has explored the profound transformation of the insurance industry through the progressive adoption of AI, detailing the evolution from essential digital integration to sophisticated AI implementations. By exploring five digital waves—digitization, big data, automation, analytics, and AI—we have illustrated their cumulative impact on the insurance sector, enhancing operational efficiencies, customer interaction, and risk management.

Advancements in AI have particularly optimized the accuracy and efficiency of claims processing and fraud detection, supported by case studies and current research. These innovations not only streamline operations but also personalize customer interactions, thereby enhancing the overall customer experience and transaction security. Our discussion has also addressed the ethical implications of AI, underscoring the importance of responsible AI practices to sustain industry growth and trust.

Future research should continue to investigate the integration of emerging technologies within the insurance sector, focusing on the ethical deployment of AI and its broader societal impacts. By doing so, the insurance industry can navigate the complexities of digital transformation while ensuring fairness, transparency, and accountability in AI applications. This exploration is crucial for developing strategic frameworks that guide responsible and effective digital adoption in insurance.

REFERENCES

[1] McKinsey & Company. (2020). "Digitization in the Insurance Industry." McKinsey.
[2] Forrester. (2020). "The Role of AI in Insurance." Forrester.
[3] Gartner. (2021). "AI in Agency Performance Management." Gartner.
[4] KPMG. (2020). "AI in Claims Processing." KPMG.
[5] IBM. (2021). "Computer Vision in Risk Assessment." IBM.
[6] Accenture. (2021). "AI Voice Recognition in Fraud Management." Accenture.
[7] Deloitte. (2020). "Navigating the Five Waves of Digital Transformation in Insurance." Deloitte.
[8] State Farm. (2020). "State Farm Mobile App: Efficient Claims Processing." State Farm.
[9] Allianz. (2021). "Blockchain Technology in Policy Management." Allianz.
[10] PwC. (2021). "Big Data Technologies and Their Impact on Insurance." PwC.
[11] Deloitte. (2021). "RPA in Claims Processing." Deloitte.
[12] Chubb. (2020). "Automation in Underwriting." Chubb.
[13] Capgemini. (2021). "Intelligent Automation in FNOL Processes." Capgemini.
[14] Accenture. (2020). "Automated Underwriting and Customer Service in Life Insurance." Accenture.
[15] EY. (2021). "Backoffice Automation in Reinsurance." EY.
[16] World Economic Forum. (2020). "Responsible AI: Frameworks and Tools." WEF.
[17] Microsoft. (2021). "Responsible AI: Best Practices and Tools." Microsoft.
[18] European Insurance and Occupational Pensions Authority. (2020). "AI and Data Protection in Insurance." EIOPA.

Index

Note: Page locators in *italic* indicate a figure, and page locators in **bold** indicate a table on the corresponding page.

J

K

L

M

N

O